Prokaryotic Gene Expression

Frontiers in Molecular Biology

SERIES EDITORS

B. D. Hames

Department of Biochemistry
and Molecular Biology
University of Leeds, Leeds LS2 9JT, UK

D. M. Glover

Cancer Research Campaign Laboratories
Department of Anatomy and Physiology
University of Dundee, Dundee DD1 4HN, UK

TITLES IN THE SERIES

1. Human Retroviruses
Bryan R. Cullen

2. Steroid Hormone Action
Malcolm G. Parker

3. Mechanisms of Protein Folding
Roger H. Pain

4. Molecular Glycobiology
Minoru Fukuda and Ole Hindsgaul

5. Protein Kinases
Jim Woodgett

6. RNA–Protein Interactions
Kyoshi Nagai and Iain W. Mattaj

7. DNA–Protein: Structural Interactions
David M. J. Lilley

8. Mobile Genetic Elements
David J. Sherratt

9. Chromatin Structure and Gene Expression
Sarah C. R. Elgin

10. Cell Cycle Control
Chris Hutchinson and D. M. Glover

11. Molecular Immunology (Second Edition)
B. David Hames and David M. Glover

12. Eukaryotic Gene Transcription
Stephen Goodbourn

13. Molecular Biology of Parasitic Protozoa
Deborah F. Smith and Marilyn Parsons

14. Molecular Genetics of Photosynthesis
Bertil Andersson, A. Hugh Salter, and James Barber

15. Eukaryotic DNA Replication
J. Julian Blow

16. Protein Targeting
Stella M. Hurtley

17. Eukaryotic mRNA Processing
Adrian Krainer

18. Genomic Imprinting
Wolf Reik and Azim Surani

19. Oncogenes and Tumour Suppressors
Gordon Peters and Karen Vousden

20. Dynamics of Cell Division
Sharyn A. Endow and David M. Glover

21. Prokaryotic Gene Expression
Simon Baumberg

Prokaryotic Gene Expression

EDITED BY

Simon Baumberg

School of Biology
University of Leeds
Leeds, UK

OXFORD
UNIVERSITY PRESS

OXFORD
UNIVERSITY PRESS

Great Clarendon Street, Oxford OX2 6DP

Oxford University Press is a department of the University of Oxford and furthers the University's aim of excellence in research, scholarship, and education by publishing worldwide in

Oxford New York

Athens Auckland Bangkok Bogotá Buenos Aires Calcutta Cape Town Chennai Dar es Salaam Delhi Florence Hong Kong Istanbul Karachi Kuala Lumpur Madrid Melbourne Mexico City Mumbai Nairobi Paris São Paulo Singapore Taipei Tokyo Toronto Warsaw

and associated companies in Berlin Ibadan

Oxford is a registered trade mark of Oxford University Press

Published in the United States by Oxford University Press Inc., New York

First published 1999

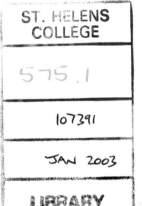

British Library Cataloguing in Publication Data
Data available

Library of Congress Cataloging in Publication Data

Prokaryotic gene expression / edited by Simon Baumberg.
— (Frontiers in molecular biology)
Includes bibliographical references.
1. Microbial genetics. 2. Gene expression. 3. Prokaryotes.
I. Baumberg, S. II. Series.
QH434.P762 1999 579.3′135—dc21 98–52907

ISBN 0 19 963604 4 (Hbk)
ISBN 0 19 963603 6 (Pbk)

Typeset by
Footnote Graphics, Warminster, Wilts

Printed in Great Britain
on acid-free paper by
The Bath Press, Avon

Preface

It was first demonstrated in the bacterium *E. coli*, in the 1960s, that changes in gene expression according to environmental conditions could be mediated by the binding of a specific protein to DNA in the neighbourhood of the gene. Since that time, numerous mechanisms for the control of gene expression in bacteria—in response to a variety of physical, nutritional, and other environmental variables—have been discovered and often exhaustively described. Nevertheless, still more are being investigated, and undoubtedly many yet remain to be revealed. Our knowledge of these mechanisms has been put to use for instance in optimizing expression in bacteria of cloned genes encoding commercially valuable products, such as hormones. Also, the bacterial systems have served as paradigms in the study of mechanisms for control of gene expression in eukaryotes. In metazoa, such mechanisms do not merely enable the organism to respond to environmental fluctuations, but play important roles in embryonic development and morphogenesis.

The three chapters by Peter Stockley, John Helmann, and Steve Busby describe the 'classical' types of regulatory mechanism: they detail the properties of proteins that bind to specific regions of DNA and the mechanisms by which they are able to do so. The following chapter, by Zhiping Gu and Paul Lovett, draws attention to the distinctly non-classical controls exerted at the post-transcriptional level. Next, Karl Drlica and colleagues discuss the subtle effects on gene expression of DNA topology. Regina Hengge-Aronis then shows how the control devices discussed previously can be integrated within a global regulatory network. Marietta Atkinson and Alex Ninfa describe the omnipresent systems involving two protein components, one of which (the sensor) causes the transfer of a phosphate group to the other (the regulator). Jon Saunders features systems in which genes are turned on or off by DNA rearrangements. There then come two examples of complex systems in which a variety of control devices are integrated: bacterial pathogenicity, detailed by Charles Dorman, and sporulation and antibiotic production, described by Michael Yudkin and Keith Chater. My own chapters, an historical overview and a reflection on evolutionary implications, frame the other contributions. One topic that it was hoped to include has had, with regret, to be omitted, namely modulated termination and anti-termination of transcription. This is however a relatively self-contained subject, and has moreover been dealt with in recent years in several excellent reviews (referenced in Chapter 1).

I would like to thank the authors for their patience and for demonstrating, in their contributions, their enthusiasm for one of the most remarkable fields in molecular bacteriology.

Leeds S. B.
December 1998

Contents

The colour plate is between pages 40 and 41.

Contributors

MARIETTE R. ATKINSON
Department of Biological Chemistry, University of Michigan Medical School, M5416 Medical Science 1, 1301 Catherine Road, Ann Arbor, MI 48109–0606, USA.

SIMON BAUMBERG
School of Biology, University of Leeds, Leeds LS2 9JT, UK.

STEPHEN J. W. BUSBY
School of Biochemistry, The University of Birmingham, Edgbaston, Birmingham B15 2TT, UK.

KEITH F. CHATER
John Innes Centre, Norwich Research Park, Colney, Norwich NR4 7UH, UK.

CHANG-RUNG CHEN
The Public Health Research Institute, 455 First Avenue, New York, NY 10016, USA.

CHARLES J. DORMAN
Department of Microbiology, The Moyne Institute of Preventive Medicine, Trinity College, Dublin 2, Ireland.

KARL DRLICA
The Public Health Research Institute, 455 First Avenue, New York, NY 10016, USA.

S. MARVIN FRIEDMAN
Department of Biological Sciences, Hunter College of The City University of New York, 695 Park Avenue, New York, NY 10021, USA.

ZHIPING GU
Department of Biological Sciences, University of Maryland Baltimore County, 5401 Wilkens Avenue, Baltimore, MD 21228–5398, USA.

JOHN D. HELMANN
Section of Microbiology, Wing Hall, Cornell University, Ithaca, NY 14853–8101, USA.

REGINE HENGGE-ARONIS
Department for Microbiology, Free University of Berlin, Koenigin-Luise-Strasse 12–16a, 14915 Berlin, Germany.

SAMUEL KAYMAN
The Public Health Research Institute, 455 First Avenue, New York, NY 10016, USA.

PAUL S. LOVETT
Department of Biological Sciences, University of Maryland Baltimore County, 5401 Wilkens Avenue, Baltimore, MD 21228–5398, USA.

MUHAMMAD MALIK
The Public Health Research Institute, 455 First Avenue, New York, NY 10016, USA.

ALEXANDER J. NINFA
Department of Biological Chemistry, University of Michigan Medical School, M5416 Medical Science 1, 1301 Catherine Road, Ann Arbor, MI 48109–0606, USA.

LIN QIU
The Public Health Research Institute, 455 First Avenue, New York, NY 10016, USA.

J. R. SAUNDERS
Department of Genetics and Microbiology, University of Liverpool, L69 3BX, UK.

PETER G. STOCKLEY
Astbury Centre for Structural Molecular Biology, Faculty of Biological Sciences, University of Leeds, Leeds, LS2 9JT, UK.

JIAN-YING WANG
The Public Health Research Institute, 455 First Avenue, New York, NY 10016, USA.

ERDEN-DALAI WU
The Public Health Research Institute, 455 First Avenue, New York, NY 10016, USA.

CHEN XU
The Public Health Research Institute, 455 First Avenue, New York, NY 10016, USA.

MICHAEL D. YUDKIN
Microbiology Unit, Department of Biochemistry, Oxford University, South Parks Road, Oxford OX1 3QU, UK.

XILIN ZHAO
The Public Health Research Institute, 455 First Avenue, New York, NY 10016, USA.

Abbreviations

ACF	accessory colonization factor
ADP	adenosine 5′-diphosphate
AMP	adenosine monophosphate
cAMP	3′,5′-cyclic-AMP
CAP	catabolite activator protein
CRP	cyclic AMP receptor protein
CSF	competence-stimulating factor
CTD	carboxyl-terminal domain
CTP	cytidine 5′-triphosphate
Dam	deoxyadenosine methylase
DNA	deoxyribose nucleic acid
DSR	downstream sequence region
ECF	extracytoplasmic functions
EIEC	enteroinvasive *E. coli*
EIF	exo-ribonuclease impeding factor
$E\sigma^{54}$	RNAP carrying σ^{54}
FIS	factor for inversion stimulation
FNR	activator essential for expression of fumarate and nitrate reductase (and many other anaerobically induced genes)
G	free energy of the system
gp32	gene 32 protein
GTP	guanosine triphosphate
H	enthalpy of the system
HMG	high mobility group
HPK	histidine protein kinase
HTH	helix–turn–helix
ICM	intracytoplasmic membrane
IHF	integration host factor
LPS	lipopolysaccharide
Lrp	Leucine-responsive regulatory protein
MCP	methyl-accepting chemotaxis proteins
NTD	amino-terminal domain
NTP	nucleoside 5′-triphosphate
ppGpp	guanosine 3′,5′-bispyrophosphate
Pu	purine
Py	pyrimidine
RBD	RNA-binding domain
RC	reaction centre

RF2	release factor 2
RHH	ribbon–helix–helix
RNA	ribonucleic acid
RNAP	RNA polymerase
RR	response regulator
S	entropy of the system
ssDNA	single-stranded DNA
T	absolute temperature
TBP	TATA-binding protein
TCP	toxin co-regulated pilus
ThrS	threonyl-tRNA synthetase
TIR	translational initiation region
TRS	tRNA synthetase
UAS	upstream activating sequence
UP	upstream promoter
UTP	uridine 5′-triphosphate
Vmp	variable major protein
WT	wild type

1 History and overview

SIMON BAUMBERG

1. Introduction

This chapter sets the scene for the rest of the book by describing the early development of the subject and how the main themes were established. The choice of references reflects three not easily reconcilable aims: to enable the reader to follow how the topic developed and to find a way into recent literature (in conjunction with later chapters), and to give credit to at least some of the scientists principally involved. A brief mention, with selected references, is also given of systems which do not introduce novel mechanisms but which apply to prokaryotic properties of fundamental or applied importance.

2. The operon model

Our present view of the control of gene expression—not only in bacteria but also in many eukaryotic systems, microbial and otherwise—as in large part resulting from interactions between proteins and sequences of DNA or RNA to which they bind more or less specifically, can be derived from the 'operon model' put forward in three paradigmatic articles by Jacob and Monod: the first in 1960 (1) and the next two in 1961 (2, 3). (The term 'paradigm' can be legitimately applied here, for a change, in its correct usage—see Reference 4.) Fascinating accounts of the development of the concepts involved in control of bacterial gene expression are given by Lwoff and Ullmann (5), Beckwith (6), and Müller-Hill (7). However, as so often is the case, this crystallization of ideas had been preceded by many years of debate in which many scientists took part. In particular, Karlström (8) in the 1930s had, from observations on carbohydrate catabolism, divided enzymes into *adaptive* and *constitutive*; the former being elicited by presence of substrate, the latter being present at the same level under all conditions.

Monod, at the Institut Pasteur in Paris, turned his attention to enzymic adaptation in the 1940s. Among his earliest subjects of study was a *mutabile* strain of *Escherichia coli*. Such strains are naturally *lac*$^-$, lactose non-utilizing (they are in fact *lacY* mutants deficient in galactoside permease); they owe their name to what we now believe to be a remarkably high frequency of reversion to *lac*$^+$, lactose utilizing, a phenomenon strikingly manifest on indicator lactose plates. The *lac*$^+$ revertants show the normal inducible phenotype. Monod was eventually able to disentangle the effects of mutation and physiological adaptation (9), and introduced the now standard terms *induction*, *inducible*, and *inducer*, relating to the expression of the enzyme β-galactosidase that catabolizes lactose. The K-12 strain of *E. coli* came to replace *mutabile* and other strains following the demonstration in this organism of conjugation and phage P1-mediated transduction. The same characteristics of regulation of expression were recognized later for the galactoside permease protein that transports lactose into the cell (10). An important conceptual advance was the recognition that an inducer need not itself be catabolized by the system induced. This followed from the synthesis of thio-β-D-galactosides such as *iso*propyl-thio-β-D-galactoside (IPTG), a potent inducer unmetabolizable by the lactose-catabolizing enzyme β-galactosidase—a *gratuitous inducer* (11). A further significant finding was the isolation of mutants in which the normal inducible *lac* phenotype gave way to constitutive expression (12). The gene in which the constitutivity mutations mapped, termed *lacI*, was found to lie in a cluster *lacIZYA* with the genes for the proteins whose expression it controlled: *lacZ*, *lacY*, and *lacA* encoding β-galactosidase, galactoside permease, and thiogalactoside transacetylase, respectively (13).

Meanwhile, other bacterial systems were being scrutinized. These were mainly in *E. coli* and *Salmonella typhimurium*, an exception being the β-lactamase of *Bacillus cereus*, which is inducible by its β-lactam substrates, the penicillins and cephalosporins (14, 15). Particularly interesting were systems which showed the reverse phenotype from inducible ones like *lac*, in which gene expression decreased, rather than increased, when a particular compound was present in the medium. This was characteristic of biosynthetic pathways, those for amino acids being the most studied. They included arginine (16), tryptophan (17), histidine (18), branched-chain aliphatic amino acids (19), and methionine (20). The term *repression* was coined by Vogel as the counterpart of induction for these systems (21), and phenotypically similar mutants were isolated in which the normally repressible enzymes became non-repressible by the end-product amino acid; these mutants were again referred to as constitutive (17, 22–27). By the end of the 1950s, therefore, it was clear that in some systems (the inducible ones), gene expression could be 'switched on'; while in others (the repressible ones), it could be 'switched off'. How might this be achieved at the molecular level?

2.1 Partial diploids and the operator

The breakthrough—an appropriate use of an overused word—came about when the *E. coli* geneticist Jacob teamed up with Monod, in particular through the study of partial diploids with two copies of the *lac* region. The first method made use of transient exconjugants from Hfr x F^- crosses (28); they later deployed F-primes to make stable partially diploid strains. First they showed that *lacI*$^+$/*lacI*$^-$ diploids were inducible, i.e. *lacI*$^+$ was dominant to *lacI*$^-$; this was interpreted in terms of the *lacI* gene product being an active *repressor* (2), at first suspected of being RNA rather than a protein (3, 29). Induction was interpreted as representing the conversion of a repressor to an inactive form in the presence—presumably, through the binding—of an inducer; *lacI*$^-$ mutants were held to make an altered repressor that was non-functional in repression (2, 3). This interpretation also readily explained the interesting class of *lacI*^{s} mutations, whose mutant *lacI* products were termed *super-repressors*. The *lacI*^{s} mutations led to non-inducibility of the *lac* proteins—and thus to a Lac$^-$ phenotype—and were dominant to *lacI*$^+$ as well as to *lacI*$^-$; both effects could be simply explained by the super-repressor being unable to interact with inducer (2, 3, 30). Interestingly, it became apparent later that *lacI*$^-$ mutations were not necessarily recessive, and that indeed some of those isolated at the Institut Pasteur were dominant (31). The existence of *trans*-dominant negative mutations was eventually recognized as due to the oligomeric nature of many proteins, including almost all regulatory ones: the presence of such mutant subunits rendered hetero-oligomers non-functional (32, 33). It was proposed that in repressible systems, a similar protein *apo-repressor* became active only via the binding of the end-product amino acid (or related metabolite) which acted as *co-repressor*, the combination being the functional repressor (2, 3). In fact, this interpretation turned out to be only partially correct; as described in Section 1.5.2 below, for several of these amino acid biosynthesis systems, control is exerted through the very different attenuation mechanism, either in addition to or instead of a repressor.

A point on terminology: the word 'repressor' remained ambiguous, as it could refer to a protein that mediates the molecular mechanism of repression, or—as in the case of the amino acid biosynthesis systems referred to above—a low molecular weight compound that elicited the phenotypic repression of enzymes in a particular pathway. In the *E. coli* methionine biosynthetic pathway, for instance, one can say that the physiological repressor is *S*-adenosyl-methionine, which acts by binding to the MetJ apo-repressor (34) to form the mechanistically functional repressor.

The greatest imaginative step, again making use of partial diploids, came in the elucidation of the way such repressors might act. Jacob and Monod reasoned that the *lacI*-encoded repressor must have a target; though this had not been characterized or even shown to exist, they gave it a name, the *operator*. If it existed, it must be capable of mutation; since the one certain property of the operator was that it interacted with repressor, the one certain deficiency of an operator mutant was that it would fail to do so. How to isolate such operator mutants? An operator mutant, being unable to interact with repressor, would be constitutive; but most constitutive mutations were *lacI*$^-$. To avoid these, Jacob and Monod started with a *lacI*^{s}/*lacI*^{s} partial diploid, and selected revertants to a Lac$^+$ phenotype. Such mutants could not carry reversions to either *lacI*$^+$ or *lacI*$^-$, since *lacI*^{s} is dominant to both. They would therefore have to carry mutations in the hypothetical operator. Jacob and Monod isolated such constitutive mutants and showed that the O^c mutant site lay between *lacI* and *lacZ* (2, 3). More importantly, they demonstrated that these mutations showed the hitherto virtually unrecognized property of *cis*-dominance. For example, in a diploid *lacI*$^+$$O^c$*lacZ*$^+$*lacY*$^-$/*lacI*$^+$$O^+$*lacZ*$^-$*lacY*$^+$ strain, β-galactosidase was constitutive (being *cis* to O^c) while galactoside permease was inducible (being *cis* to O^+) (2, 3).

The clear inference from these results was that the operator was a region of nucleic acid, either the DNA next to the first gene in the cluster or the RNA next to the first gene in its transcript. Jacob and Monod argued strongly that the mechanism most economical in cell constituents would have been favoured by evolution. They therefore came down on the side of a DNA region as operator, since this would imply that under conditions of repression, the genes in the cluster would not be transcribed; if the operator was an RNA region, it must be presumed that constitutively produced RNA was prevented by repressor from being translated. They could not at that stage, however, prove the DNA nature of the operator to be correct. The term *operon* was introduced to denote a cluster of genes controlled by a single operator (1–3).

With the development in 1963 of the concept of *protein allostery* by Monod, Changeux, and Jacob (35), it was possible to see the LacI repressor as an allosteric protein, in thermodynamic equilibrium between operator-binding and operator-non-binding conformations. When LacI binds inducer, the equilibrium shifts to strongly favour the operator-non-binding conformation. Conversely with repressible systems, the equivalent equilibrium would shift towards operator binding when the apo-repressor bound the co-repressor.

In their first formulation of the operon model, Jacob and Monod had supposed that the operator was both the site of binding of the repressor and the start-site for transcription—this would later be articulated as the binding site for RNA polymerase. Repression would result from competition between the repressor and RNA polymerase for this site. It was later realized that the sites would be separate, though possibly overlapping; and the RNA polymerase binding site was termed the *promoter* (36, 37).

2.2 Developments of the operon model

The operon model immediately became immensely influential; it was quickly pointed out that it could explain not only microbial adaptation, but also, among numerous other things, metazoan differentiation and development (38). For a time, it became difficult (6) to propose models that differed significantly from the *lac* paradigm involving a protein repressor, a DNA operator, and a cluster forming a single operon. Nevertheless, the first major alternative mechanism, whereby the regulatory protein is an activator rather than a repressor, as proposed for two *E. coli* catabolic systems, L-arabinose (*ara*) by Englesberg and colleagues (39, 40) and maltose (*mal*) by Schwartz and colleagues (41), came to be accepted by the late 1960s.

The initial expectation that the inducer of a catabolic system would be its substrate or an analogous compound was first questioned following studies on the *E. coli* glycerol system (42). It was shown that the inducer here was not glycerol but α-glycerophosphate, the product of the first enzyme in the catabolic pathway. This was seen as an example of *product induction*: the *lac* system was later shown to exemplify the same phenomenon, as the true inducer when cells are grown on lactose is *allo*lactose, produced by the action of β-galactosidase as a galactosyltransferase in transferring a galactose residue to the 6′ position of glucose, as opposed to the 4′ position in lactose itself (43, 44).

The organization of genes in a regulated system also revealed itself to show many variations. Jacob and Monod, in presenting the operon model, had suggested that the *lac* arrangement of three genes in a single transcription unit would be predicted to be the most economical, and thus to be the rule. As noted above, an operon was therefore defined as such a co-transcribed gene cluster under the control of an operator. However, this is far from universal. First, as described below, a cluster of genes forming a transcription unit may not be controlled by an operator in the strict sense of a repressor-binding site. The term 'operon' thus changed its usage to mean simply a co-transcribed gene cluster, whatever the nature of its regulation (if any). Next, a cluster can be made up of two or more operons transcribed independently but in the same direction, as with the *Salmonella/Klebsiella hut* system for histidine catabolism (45). The latter system also exemplifies another common feature, whereby the regulatory protein regulates its own synthesis as well as that of the pathway enzymes and transport proteins, a situation first recognized in the nitrate assimilation system of the fungus *Aspergillus nidulans* and termed *autogenous regulation* or *autoregulation* (46). Alternatively, there may be two contiguous but *divergently* transcribed operons, as with the *E. coli bio* and *arg* clusters (47, 48).

The very large gene clusters in streptomycete secondary metabolite biosynthesis/resistance systems, such as *act* (49) and *str* (50), contain mixtures of convergently and divergently transcribed operons. These are usually under the control of single regulatory proteins. Similarly lengthy pathways for catabolism of unusual or xenobiotic compounds, found particularly in pseudomonads, involve small numbers of gene clusters that show *sequential induction*. An early example (51) was the mandelate pathway of a strain of *Pseudomonas fluorescens*: mandelate induces the first cluster, encoding enzymes that mediate its catabolism down as far as benzoate; the latter both induces the next cluster and represses the first one. References 52 and 53 are later reviews of systems of this type incorporating molecular approaches.

Often, genes that do not constitute a single cluster are regulated together; there may be several clusters, or clusters plus isolated genes. The latter is found for the *E. coli arg* system, where there is a cluster *argECBH* plus isolated genes *argA, D, F, G, I,* and *R*, all regulated by the ArgR apo-repressor in conjunction with arginine: the term *regulon* was introduced for such a set of co-regulated genes introduced by Maas and Clark (54) and has become especially significant in view of the recognition of the importance and universality of *global* regulatory systems (see below) which invariably act on many unlinked genes or clusters.

2.3 The operon model becomes biochemistry

In its initial form, the operon model derived from the formal logic of bacterial genetics. For a few years, sceptics remained who argued that until such entities as the repressor and operator had been shown to be unique molecular species with the properties imputed to them, the model must be regarded as still hypothetical. In 1966, the transition was made from abstract genetics to concrete biochemistry.

Gilbert and Müller-Hill (55) purified the *lac* repressor by one of the two properties it was reasoned to possess, namely ability to bind inducer (the other being ability to bind operator, whatever that might turn out to be). The availability in the *lac* system of gratuitous inducers which were unlikely to bind any other *E. coli* protein provided a means. They used equilibrium dialysis, in which protein fractions in dialysis bags were placed in radiolabelled IPTG; fractions that maintained a higher concentration of IPTG within the dialysis bag than outside it were taken to be enriched in repressor. In this way, repressor protein was purified. It was then shown, in a paper legitimately entitled 'The *lac* operator is DNA' (56), that a λ specialized transducing genome, carrying *lac* operator DNA, bound to purified LacI repressor protein in a density gradient if it was O^+ but not if it was O^c.

At much the same time, Ptashne purified another repressor, this time encoded by the *cI* gene of temperate coliphage λ (57). The action of the cI repressor is essential for repression of λ lytic functions and thus for lysogenization. *In vivo* experiments suggested (58) that cI should bind to two operator regions O_L and O_R which overlap promoters P_L and P_R, leading to repression of transcription from the latter. The λ*vir* mutant has mutations affecting both O_L and O_R. Following purification of cI, its binding to these operators was confirmed *in vitro*: as in the *lac* case, purified cI repressor bound to wild-type λ DNA but not to λ*vir*. Though the purification of the cI repressor was published after that of LacI, binding to operator DNA was published first for cI (59). Ptashne has provided (60) an elegant book on the λ system which describes both early and more recent work.

2.4 The first global regulation system

Monod's early studies had also led to the realization that induction of the *lac* operon was depressed when *E. coli* was growing in the presence of a better carbon source (i.e. allowing faster growth) such as glucose. When glucose was limiting, this could lead to two growth phases; in the first of these glucose was utilized, in the second lactose, a phenomenon termed *diauxie* (9). This was generalized as *catabolite repression* in 1961 by Magasanik (61). In 1968, evidence was obtained (62, 63) that cyclic 3′,5′-AMP (cAMP) was involved: the expression of the *lac* operon was stimulated by this nucleotide, whose level was inversely related to growth rate. It was then shown (64, 65) that the effect was mediated separately from repression, via a DNA site upstream of the operator. At first, this site was equated with the promoter, but it was soon shown that it was a distinct site (66, 67) which represented the binding site for a new regulatory protein, the *catabolite activator protein* (CAP) or *catabolite repression protein* (CRP) (68, 69). This protein, when bound to cAMP, bound to sites upstream of a variety of promoters for catabolic—usually inducible—operons, this binding then resulting in activation of transcription. The mechanism of activation by CAP/CRP is not obvious from this description, and was shown much later to involve protein–protein contact with the RNA polymerase α and σ subunits together with both looping and bending of DNA, as described in Chapters 2 and 4. DNA sites bound by activators have never been given an individual name comparable to 'operator', and one has to refer to the 'CAP-binding site'. (Another terminological peculiarity is that the phenotypic phenomenon of catabolite repression is actually mediated by an activation mechanism.)

3. Regulation via alternative σ factors

Following the elucidation of the subunit structure of *E. coli* RNA polymerase and the role of the σ factor (70, 71), it became apparent that the use of alternative σ factors with different specificities would provide an effective way of switching regulons on and off in systems such as phage development and bacterial sporulation. However, it took some years before the generality of this mechanism could be fully demonstrated.

3.1 Alternative σ factors in phage and spore development

The first indication that *Bacillus* sporulation might involve changes in RNA polymerase promoter selection, later interpretable as σ factor related, came from studies on phage φe, whose genome is incorporated within the developing spore but is only expressed on germination (72). The *B. subtilis* phage SPO1 affords a well-studied example. Transcription of 'early' genes uses the host vegetative σ^A factor, but 'middle' and 'late' genes use alternative σ factors, σ^{28} and σ^{34} (73). The lengthy development of the *Bacillus* heat-resistant endospore also utilizes sequential alternative σ factors (73; see also Chapter 11). In a related strategy found with coliphage T4, the function of RNA polymerase is altered, not only by use of the phage-encoded σ factor gp55, but also by modifications to the core: the α subunits are ADP-ribosylated, and phage-encoded proteins bind to the core (74). Modification of RNA polymerase as a strategy reaches an extreme with coliphage T7, which after the early phase of transcription uses an entirely new and unrelated RNA polymerase consisting of a single polypeptide chain (75).

3.2 Alternative σ factors in metabolic systems

The first indication that alternative σ factors might be involved in metabolic as well as developmental systems came from work on the control of nitrogen metabolism in *E. coli* and *S. typhimurium* and of nitrogen fixation (*nif*) genes in *Klebsiella pneumoniae*. It was found that expression of these genes required a σ factor, σ^{54} (also termed σ^N or RpoN), different from that previously characterized (σ^{70}) as used for other genes in growth phase (76–78). A related σ factor, termed σ^L, was later demonstrated in *B. subtilis* (79). Exploration of the consequences of the realization that most prokaryotes spend most of their existence in environmentally stressed or starved physiological states has led to the recognition of the importance of alternative σ factors in gene expression in these situations, e.g. σ^{32} involved in the heat shock response (80) and the 'stationary phase' σ^S (or RpoS) (81), in both cases of *E. coli* (see also Chapter 7).

4. Separation of signal sensing from control of expression

In the classical picture of a regulatory protein such as LacI or CAP, the same protein (though probably through different domains) would both sense a signal from the environment—in these cases, a low molecular weight compound, inducer, or cAMP, respectively—and carry out the appropriate regulatory function—here, DNA binding leading to repression or activation. The *E. coli* nitrogen assimilation (*gln*) and *Kl. pneumoniae* nitrogen fixation (*nif*) systems also gave the first indications that the two processes of signal sensing and regulatory function could be separately carried out by different proteins. In these systems, the NtrB protein autophosphorylates (determined by another complex regulatory process dependent on the nitrogen status of the cell), and then transfers the phosphate group to the NtrC protein, which in the phosphorylated form activates the promoters *glnAp2* and *nifLp* (NtrB/NtrC are also known as GlnL/GlnG or NR_{II}/NR_{I}) (82, 83). Many other systems were soon found which showed the same kind of separation of sensing and regulatory roles; what is more, there was a common biochemistry of the auto- and trans-phosphorylation processes, and parts of the sequences of the proteins themselves showed homology. Such systems were given the collective name *two-component systems*, and are reviewed here in Chapter 8.

A completely different mode of separation of signal sensing from regulation of gene expression is shown by the λ repressor cI, and also the LexA repressor, as part of the *SOS response* system described first for *E. coli* (84) but of very wide occurrence in prokaryotes. DNA damage, e.g. by UV radiation or chemicals such as mitomycin C, leads to the induction of the SOS regulon of genes, subject to repression by LexA, whose products attempt to neutralize the damage or its consequences. The central player in the SOS response is the RecA protein: a key component in homologous recombination, but here with a separate function. The DNA damage leads to activation of RecA, and the activated RecA protein binds to LexA leading to autoproteolysis and thus to derepression of the SOS regulon. However, phages such as λ make use of this system for their own ends, in that cI too has the capacity for autoproteolysis on binding activated RecA: hence DNA damage leads to derepression, i.e. induction, of the repressed prophage (85–87). Here as with conventional two-component systems, signal sensing (a function of RecA) is separated from control of gene expression (the role of cI).

A final mode of dissociation of signal sensing from gene regulation is found with, among other systems, some streptomycete antibiotic production systems, such as *act* and *str* referred to above (Section 1.2.2). Here, regulatory cascades modulate the expression of the pathway-specific activators ActII-ORF4 and StrR, but the activity of these proteins is not, as far as is known at present, itself *modulated* in any way (Chapter 11).

5. A different kind of regulatory mechanism: termination/anti-termination

A further aspect of the classical operon model was that control of gene expression was seen as being exerted at the level of frequency of *initiation* of transcription. That this might not always be so was first revealed by work on gene expression in coliphage λ-infected cells. This and later studies on other, non-phage, systems showed that there exists a class of regulatory proteins which affect the frequency with which an RNA polymerase molecule, having already initiated transcription at an upstream promoter, proceeds past a *terminator* to transcribe one or more downstream genes.

5.1 Anti-terminator proteins

The analysis of RNA transcripts by DNA–RNA hybridization, developed by Spiegelman and colleagues (88) and applied in depth to control of λ gene expression by Szybalski and his group (89), showed that λ N^- mutants were deficient in all but immediate early transcripts, while Q^- mutants were deficient in late transcripts (90–92). Following infection of a non-lysogen with λ, or induction of a λ lysogen with DNA-damaging agents, expression of two extensive operons is initiated from the P_L and P_R promoters (see Section 1.2.3 above). The first genes in the P_L and P_R operons are *N* and *cro*, respectively, and these are expressed as 'immediate early' genes. However, in the absence of N protein, transcription terminates at the end of *N* and *cro*. As N accumulates, it causes *anti-termination* of these transcripts so that the remainder of both operons is expressed as 'delayed early' genes. *Q* is the final gene in the P_R operon; its accumulation permits the anti-termination of a short transcript produced constitutively from the adjacent $P_{R'}$ promoter, leading to expression of the entire 'late' operon comprising genes for head and tail components, morphogenesis, and eventual cell lysis (93). N and Q are therefore both *anti-terminator* proteins (see discussion in References 93 and 93a). Their function is strictly speaking not modulated, e.g. by binding a small molecule; however, they, like other λ regulatory and replication proteins such as cII, O, and P, are subject to relatively rapid proteolysis, and the effective level of intracellular proteolytic enzymes may therefore exert an additional level of control (94, 95).

Later, anti-termination proteins controlling metabolic systems were found. The first of these was BglG, which regulates the *E. coli bgl* system for β-glucoside catabolism. BglG antagonizes the effect of a terminator situated before the genes encoding β-glucosidase and a β-glucoside transport protein. Its function is modulated by phosphorylation; if a suitable β-glucoside is available, it is phosphorylated as a prerequisite for uptake, but if no such molecule is available, BglG is phosphorylated instead. The phosphorylated BglG is unable to anti-terminate (96). Another metabolic anti-terminator protein is TRAP (encoded by *mtrB*) of *B. subtilis*, which blocks the formation of a transcriptional anti-terminator structure in the leader region of the transcript of the biosynthetic *trp* operon. Intracellular accumulation of tryptophan, indicating repletion for this amino acid, leads to its binding to and activating TRAP, which then binds within the leader region such that the anti-terminator cannot form, leading to diminished expression of the *trp* operon. Remarkably, TRAP also acts translationally, by binding to *trpE* and *trp*G transcripts so as to inhibit initiation of translation (97).

5.2 Attenuation: regulation without specific regulatory components

Many prokaryotic amino acid biosynthesis systems show a remarkable and intricate form of control which uses only the standard components of the protein synthesis machinery. This kind of system, which came to be called *transcriptional attenuation*, was first recognized by Yanofsky and colleagues working on the *E. coli trp* system. The *E. coli trpEDCBA* operon is subject to feedback control of expression by tryptophan by two independent mechanisms: repression, by a conventional apo-repressor TrpR with tryptophan acting as co-repressor (see Section 1.2 above), and attenuation. An initial pointer to the existence of attenuation was the increased expression of *trp* in *rho*$^-$ mutants (see Chapter 3), isolated as suppressors of the *polar* effect of nonsense mutations on the expression of downstream genes within prokaryotic operons (98)—it may be noted here, without further comment, that polarity was for many years often connected with mechanisms for control of gene expression (see e.g. Reference 99). Later work by Yanofsky's group showed that a specific region within the *leader sequence* lying between the promoter and *trpE* is specifically involved in attenuation. The role of this region and the mechanism of attenuation, involving alternative RNA secondary structures, are described in References 93a and 100, which also explain why earlier results had suggested a role for aminoacyl transfer RNA's and/or tRNA synthetases in control of these systems (101). Alternative RNA structures are also involved in tRNA-mediated control of expression of aminoacyl-tRNA synthetase and amino acid biosynthesis genes in Gram-positive bacteria (93a, 102).

6. Post-transcriptional control

Jacob and Monod originally suggested that regulation in prokaryotes would be expected to be exerted at the level of transcription, to avoid the wasteful production of untranslated mRNA. Although this does seem to be generally true, many instances of post-transcriptional control are now known, involving mRNA stability, translational repression, translational attenuation, and anti-sense RNA. These are described in Chapter 5.

7. The effect of DNA topology

The initial concept of the DNA double helix was as an extended rather rigid molecule of unvarying topology. This has come to be replaced by a recognition of the ways in which the canonical structure varies over both length and time, through DNA bending, looping, and supercoiling. The story of the recognition of resulting effects on gene expression differs somewhat from other areas of this subject, in that the theoretical prediction that there ought to be such effects predated their discovery from the study of regulatory phenomena. An account of effects both of DNA topology on gene expression and of gene expression on DNA topology is given in Chapter 6.

8 DNA rearrangements and control of gene expression

These represent the form of control of gene expression furthest removed from the operon model. The first system in which a DNA rearrangement was shown to be involved in the control of gene expression was phase variation in *Salmonella*, incidentally the first research area within bacterial genetics of one of the British pioneers of this subject, the late William Hayes (103). These organisms switch at low frequency between alternative flagellar types, reflecting transcription of genes *H1* and *H2* encoding alternative flagellin subunits. Simon and co-workers provided the explanation: reversible inversion of a promoter-bearing DNA segment upstream of a two-gene operon *H2-rh1*, the latter encoding a repressor of *H1* (104). Chapter 9 gives more details of this system and several others involving DNA rearrangements, in most cases affecting expression of genes for surface components of bacteria and phages. In some cases, also described in Chapter 9, expression is also affected by methylation status of sequences upstream of regulated genes.

Programmed deletions in developmental systems also come into this category of mechanisms. An example is provided by nitrogen fixation (*nif*) genes in the filamentous cyanobacterium *Anabaena*. This organism fixes nitrogen only in non-dividing heterocysts, thus avoiding exposure of the oxygen-sensitive nitrogenase to O_2. In the vegetative genome, one of the *nif* clusters *nifHDK* contains 11 kbp of unrelated DNA within *nifD*, and in another cluster the *nifS* and *nifB* genes are separated by 35 kbp of unrelated DNA. In heterocyst development, these segments are removed by site-specific recombination events that utilize short flanking direct repeats (105, 106).

9. Autoinducers

A theme in prokaryotic gene regulation that has recently come into prominence is *autoinduction*. Autoinducers are low molecular mass compounds that are synthesized and secreted at a constant rate per cell. Where diffusion of secreted substances is partially restricted, the external concentration of such an autoinducer surrounding a growing colony is therefore a measure of how long the colony has been developing and the number of cells present. Attainment of a threshold autoinducer concentration results in activation or repression of expression of one or more sets of genes.

The first autoinducer to be characterized (107) was the butyrolactone A-factor which acts as a signal for the expression of sporulation functions and streptomycin production in *Streptomyces griseus* (Chapter 11). It is now known (108, 109) that A-factor works by binding to and inactivating a conventional repressor, the A-factor binding protein. Sporulation and secondary metabolism appear to be triggered by similar autoinducers in many other streptomycetes, but the generality of the phenomenon is still uncertain.

It has now become apparent that autoinduction is extremely widespread in Gram-negative bacteria. The involvement of a homoserine lactone autoinducer was first shown by Silverman's group (110) for expression of *lux* genes determining the luminescence of marine *Vibrio* species. Later, it was found that autoinducers control expression of important determinants such as those for human pathogenicity determinants in *Pseudomonas aeruginosa*, plant pathogenicity determinants and antibiotic biosynthesis in *Erwinia*, and cell–cell communication in such genera as *Serratia*, *Rhizobium*, and *Rhodobacter* (111–118).

10. Molecular interactions in control of prokaryotic gene expression

Many of the systems referred to have been studied by structural biologists. The first X-ray diffraction structures obtained for regulatory proteins, both with and without bound operator DNA fragments, were those of the λ cI repressor (119, 120) and the Cro protein which binds to the same sites as cI but with different affinities (121). This clarified how a protein could bind specifically to a given DNA sequence, and also provided the first example of a DNA-binding domain—the helix–turn–helix (HTH)—which has since appeared in many such proteins, both prokaryotic and eukaryotic. The next structures to be determined were for *E. coli* CAP, alone, with cAMP bound, and bound to both CAP and a DNA fragment with its specific binding site. These once again showed the HTH domain and also gave examples of both the conformational change attendant on binding the cAMP effector, and the bending of DNA resulting from binding the cAMP–CAP complex (122–124). Many other such structures are now available; in some of these, other kinds of DNA binding domain have been observed, such as the β-ribbon motif found in the *E. coli* MetJ repressor (125). This topic is discussed in Chapter 2.

11. Regulation in specific systems

This book is aimed primarily at describing the varied mechanisms for control of prokaryotic gene expression, rather than the numerous systems in which control of expression is an important feature of a prokaryotic property of interest in itself. Through this primary aim, a high proportion of important prokaryotic systems have been touched on, and the references provided will enable the reader to delve further. Some further significant systems that have not otherwise been alluded to, or only in passing, are mentioned in Table 1.

Table 1.1

System	Noteworthy features	References
Response to oxidative stress, enterobacteria	Involves two regulatory systems mediated by SoxRS and OxyR, respectively. SoxR/SoxS are both DNA-binding proteins (i.e. they do not constitute a two-component system). SoxR may respond to O_2^- and NO. OxyR is also a DNA-binding protein, and may be post-translationally regulated following exposure to H_2O_2	128
Free-living nitrogen fixation, *Klebsiella* and *Azotobacter*	One remaining question is the role of NifL. In *Azotobacter*, there are three distinct regulated nitrogenase systems	129, 130
Symbiotic nitrogen fixation, rhizobia	In the symbiotic nitrogen fixing organisms, there are specific regulation systems separately regulating the *nod* genes for nodulation and the *nif* genes for nitrogen fixation	131, 132
Catabolite repression, *Bacillus*	A very different system from that in the enterobacteria: cAMP is not involved; there is a specific catabolite repressor protein; and there is a connection with post-exponential growth controls of gene expression	133–136
Copper resistance, enterobacteria	Whereas mercury (control of resistance to which is mentioned in Chapter 4) is solely toxic, copper is required in minute quantities but is toxic in larger amounts. Its accumulation has therefore to be finely controlled	137
β-Lactam resistance, *Bacillus*	An update on one of the first control systems to be studied in detail, in the 1950s	138
Virulence gene expression in *Staphylococcus*	The latest 'pheromone' system to be discovered	139–141
Development, *Caulobacter*	Organelle formation in this organism constitutes a developmental system	142
Light-inducible functions: carotenoid biosynthesis, *Myxococcus*	A light-inducible system in a non-photosynthetic organism (as opposed to the *Rhodobacter* systems mentioned in Chapter 6)	143

12. Concluding remarks

The reader will realise from this book that we now have a remarkably detailed knowledge of many prokaryotic regulatory mechanisms. The thrust has, however, so far been almost entirely reductive: understandably so, in view of the power of the molecular genetic and biochemical techniques readily available. However, there have been few attempts to integrate the isolated mechanisms; the preliminary approach of Maaløe (126) does not seem to have often been taken up by others, and there have been very few attempts to integrate this knowledge to produce quantitative models whose properties could be tested against experiment. An interesting recent approach is that of Neidhardt and Savageau (127). It is interesting to reflect that in the physical sciences, no model would be regarded as complete until it had satisfied such a test. Now that many prokaryotic genomes have been completely sequenced, there is talk of attempting, as the next mega-task, the modelling of the complete working of a bacterial cell.

References

1. Jacob, F., Perrin, D., Sanchez, C., and Monod, J. (1960) L'opéron: groupe de gènes à expression coordonée par un opérateur. *Compt. Rend. Acad. Sci.*, **250**, 1727–1729.
2. Jacob, F. and Monod, J. (1961) Genetic regulatory mechanisms in the synthesis of proteins. *J. Mol. Biol.*, **3**, 318–356.
3. Jacob, F. and Monod, J. (1961) On the regulation of gene activity. *Cold Spring Harb. Symp. Quant. Biol.*, **26**, 193–211.
4. Kuhn, T. S. (1996) *The structure of scientific revolutions* (3rd edn). University of Chicago Press, Chicago.
5. Lwoff, A. and Ullmann, A. (eds). (1979) *Origins of molecular biology. A tribute to Jacques Monod.* Academic Press, New York.
6. Beckwith, J. (1996) The operon: an historical account. In Escherichia coli *and* Salmonella typhimurium*: Cellular and molecular biology* (2nd edn) (editor-in-chief F. C. Neidhardt). American Society for Microbiology, Washington, DC, pp. 1227–1231.
7. Müller-Hill, B. (1996) *The* lac *operon. A short history of a genetic paradigm.* Walter de Gruyter, Berlin.
8. Karlström, H. (1938) Enzymatische Adaptation bei Mikroorganismen. *Ergebnisse der Enzymforschung*, **7**, 350–376.
9. Monod, J. and Audureau, A. (1946) Mutation et adaptation enzymatique chez *Escherichia coli-mutabile*. *Ann. Inst. Past.*, **72**, 868–879.
10. Cohen, G. N. and Monod, J. (1957) Bacterial permease. *Bact. Rev.*, **21**, 169–194.
11. Perrin, D., Jacob, F., and Monod, J. (1960) Biosynthèse induite d'une protéine génétiquement modifiée, ne présentant pas d'affinité pour l'inducteur. *Compt. Rend. Acad. Sci.*, **251**, 155–157.
12. Cohen-Bazire, G. and Jolit, M. (1953) Isolement par sélection de mutants d'*E. coli* synthétisant spontanément l'amylomaltase et la β-galactosidase. *Ann. Inst. Past.*, **84**, 1–9.
13. Zabin, I., Képès, A., and Monod, J. (1959) On the enzymatic acetylation of isopropyl-β-D-thiogalactoside and its association with galactoside-permease. *Biochem. Biophys. Res. Comm.*, **1**, 289–292.
14. Kogut, M., Pollock, M. R., and Tridgell, E. J. (1956) Purification of penicillin-induced penicillinase of *Bacillus cereus* NRRL 569. A comparison of its properties with those of a similarly purified penicillinase produced spontaneously by a constitutive mutant strain. *Biochem. J.*, **62**, 391–401.
15. Citri, N., and Pollock, M. R. (1966) The biochemistry and function of β-lactamase (penicillinase). *Adv. Enzymol.*, **28**, 238-
16. Gorini, L. and Maas, W. K. (1958) Feed-back control of the formation of biosynthetic enzymes. In *The chemical basis of development* (ed. W. D. McElroy and B. Glass), pp. 469–478. Johns Hopkins Press, Baltimore, Maryland.
17. Cohen, G. N. and Jacob, F. (1959) Sur la répression de la synthèse des enzymes intervenant dans la formation du tryptophane chez *E. coli*. *Compt. Rend. Acad. Sci.*, **248**, 3490–3492.

18. Ames, B. N., Garry, B., and Herzenberg, L. A. (1960) The genetic control of the enzymes of histidine biosynthesis in *Salmonella typhimurium*. *J. Gen. Microbiol.*, **22**, 369–378.
19. Freundlich, M, Burns, R., and Umbarger, H. E. (1962) Control of isoleucine, valine and leucine biosynthesis. I. Multivalent repression. *Proc. Natl. Acad. Sci. USA*, **48**, 1804–1808.
20. Cohn, M., Cohen, G. N., and Monod, J. (1953) L'effect inhibiteur spécifique de la méthionine dans la formation de la méthionine-synthetase chez *Escherichia coli*. *Compt. Rend. Acad. Sci.*, **236**, 746–748.
21. Vogel, H. J. (1957) Repression and induction as control mechanisms of enzyme biogenesis: the 'adaptive' formation of acetylornithinase. In *The chemical basis of development* (ed. W. D. McElroy and B. Glass), pp. 276–289. Johns Hopkins Press, Baltimore, Maryland.
22. Vogel, H. J. (1961) Aspects of repression in the regulation of enzyme synthesis: pathway-wide control and enzyme-specific response. *Cold Spring Harb. Symp. Quant. Biol.*, **26**, 163–172.
23. Gorini, L., Gundersen, W., and Burger, M. (1961) Genetics of regulation of enzyme synthesis in the arginine biosynthetic pathway of *Escherichia coli*. *Cold Spring Harb. Symp. Quant. Biol.*, **26**, 173–182.
24. Maas, W. K. (1961) Studies on repression of arginine biosynthesis in *Escherichia coli*. *Cold Spring Harb. Symp. Quant. Biol.*, **26**, 183–191.
25. Ames, B. N. and Hartman, P. E. (1963) The histidine operon. *Cold Spring Harb. Symp. Quant. Biol.*, **28**, 349–356.
26. Patte, J. C., LeBras, G., and Cohen, G. N. (1967) Regulation by methionine of the synthesis of a third aspartokinase and a second homoserine dehydrogenase in *Escherichia coli* K12. *Biochim. Biophys. Acta*, **136**, 245–257.
27. Lawrence, D. A., Smith, D. A., and Rowbury, R. J. (1968) Regulation of methionine biosynthesis in *Salmonella typhimurium*: mutants resistant to inhibition by analogues of methionine. *Genetics*, **58**, 473–492.
28. Pardee, A. B., Jacob, F., and Monod, J. (1959) The genetic control of cytoplasmic expression of 'inducibility' in the synthesis of β-galactosidase by *E. coli*. *J. Mol. Biol.*, **1**, 165–178.
29. Pardee, A. B. and Prestidge, L. S. (1959) On the nature of the repressor of β-galactosidase synthesis in *Escherichia coli*. *Biochim. Biophys. Acta*, **36**, 545–547.
30. Willson, C., Perrin, D., Cohn, M., Jacob, F., and Monod, J. (1964) Non-inducible mutants of the regulator gene in the lactose system of *Escherichia coli*. *J. Mol. Biol.*, **8**, 582–592.
31. Davies, J. and Jacob, F. (1968) Genetic mapping of the regulator and operator genes of the *lac* operon. *J. Mol. Biol.*, **36**, 413–417.
32. Weber, K., Platt, T., Ganem, D., and Miller, J. H. (1972) Altered sequences changing the operator-binding properties of the *lac* repressor: Colinearity of the repressor protein with the *I*-gene map. *Proc. Natl. Acad. Sci. USA*, **69**, 3624–3628.
33. Schlotmann, M., and Beyreuther, K. (1979) Degradation of the DNA-binding domain of wild-type and I^{-d} *lac* repressor in *Escherichia coli*. *Eur. J. Biochem.*, **95**, 39–49.
34. Ahmed, A. (1973) Mechanism of repression of methionine biosynthesis in *Escherichia coli*. I. The role of methionine, *S*-adenosylmethionine and methionyl transfer ribonucleic acid in repression. *Molec. Gen. Genet.*, **123**, 299–324.
35. Monod, J., Changeux, J.-P., and Jacob, F. (1963) Allosteric proteins and cellular control systems. *J. Mol. Biol.*, **6**, 306–329.
36. Jacob, F., Ullmann, A., and Monod, J. (1964) Le promoteur, élément génétique nécessaire à l'expression d'un opéron. *Compt. Rend. Acad. Sci.*, **258**, 3125–3128.
37. Scaife, J. and Beckwith, J. R. (1966) Mutational alteration of the maximal level of *lac* operon expression. *Cold Spring Harb. Symp. Quant. Biol.*, **31**, 403–408.

38. Monod, J. and Jacob, F. (1961) General conclusions: teleonomic mechanisms in cellular metabolism, growth, and differentiation. *Cold Spring Harb. Symp. Quant. Biol.*, **26**, 389–401.
39. Englesberg, E., Irr, J., Power, J., and Lee, N. (1965) Positive control of enzyme synthesis by gene C in the L-arabinose system. *J. Bacteriol.*, **90**, 946–957.
40. Englesberg, E. and Wilcox, G. (1974) Regulation: positive control. *Ann. Rev. Genet.*, 219–242.
41. Hatfield, D., Hofnung, M., and Schwartz, M. (1969) Genetic analysis of the maltose A region in *Escherichia coli*. *J. Bacteriol.*, **98**, 559–567.
42. Cozzarelli, N. R., Freedberg, W. B., and Lin, E. C. C. (1968) Genetic control of the L-α-glycerophosphate system in *Escherichia coli*. *J. Mol. Biol.*, **31**, 371–387.
43. Müller-Hill, B., Rickenberg, H. V., and Wallenfels, K. (1964) Specificity of the induction of the enzymes of the Lac operon in *Escherichia coli*. *J. Mol. Biol.*, **10**, 303–318.
44. Burstein, C., Cohn, M., Képès, A., and Monod, J. (1965) Rôle du lactose et de ses produits métaboliques dans l'induction de l'opéron lactose chez *Escherichia coli*. *Biochim. Biophys. Acta.*, **95**, 634–639.
45. Magasanik, B. (1980) Regulation in the *hut* system. In *The operon* (2nd edn) (ed. J. H. Miller and W. S. Reznikoff), pp. 373–387. Cold Spring Harbor Laboratory, Cold Spring Harbor, NY.
46. Cove, D. J. (1974) Evolutionary significance of autogenous regulation. *Nature*, **251**, 256.
47. Guha, A., Saturen, Y., and Szybalski, W. (1971) Divergent orientation of transcription from the biotin locus of *E. coli*. *J. Mol. Biol.*, **56**, 53–62.
48. Elseviers, D., Cunin, R., Glansdorff, N., Baumberg, S., and Ashcroft, E. (1972) Control regions within the *argECBH* gene cluster of *Escherichia coli* K12. *Molec. Gen. Genet.*, **117**, 349–366.
49. Hopwood, D. A. and Sherman, D. H. (1990) Molecular genetics of polyketides and its comparison to fatty acid biosynthesis. *Ann. Rev. Genet.*, **24**, 37–66.
50. Piepersberg, W. (1995) Streptomycin and related antibiotics. In *Biochemistry and genetics of antibiotic production* (ed. L. C. Vining and C. Stuttard), pp. 531–570. Butterworth–Heinemann, Newton, Massachusetts.
51. Mandelstam, J. and Jacoby, G. A. (1965) Induction and multi-sensitive end-product repression in the enzymic pathway degrading mandelate in *Pseudomonas fluorescens*. *Biochem. J.*, **94**, 569–577.
52. Ornston, L. N. and Parke, D. (1977) The evolution of induction mechanisms in bacteria: insights derived from the study of the β-ketoadipate pathway. *Curr. Top. Cell. Reg.*, **12**, 209–262.
53. Ramos, J. L., Marqués, S., and Timmis, K. N. (1997) Transcriptional control of the *Pseudomonas* TOL plasmid catabolic operons is achieved through an interplay of host factors and plasmid-encoded regulators. *Ann. Rev. Microbiol.*, **51**, 341–373.
54. Maas, W. K. and Clark, A. J. (1964) Studies on the mechanism of repression of arginine biosynthesis in *Escherichia coli*. I. Dominance of repressibility in diploids. *J. Mol. Biol.*, **8**, 365–370.
55. Gilbert, W. and Müller-Hill, B. (1966) Isolation of the *lac* repressor. *Proc. Natl. Acad. Sci. USA*, **56**, 1891–1898.
56. Gilbert, W. and Müller-Hill, B. (1967) The *lac* operator is DNA. *Proc. Natl. Acad. Sci. USA*, **58**, 2415–2421.
57. Ptashne, M. (1967) Isolation of the λ phage repressor. *Proc. Natl. Acad. Sci. USA*, **57**, 306–313.

58. Ptashne, M. and Hopkins, N. (1968) The operators controlled by the λ phage repressor. *Proc. Natl. Acad. Sci. USA*, **60**, 1282–1287.
59. Ptashne, M. (1967) Specific binding of the λ phage repressor. *Nature*, **214**, 232–234.
60. Ptashne, M. (1992) *A genetic switch* (2nd edn). Cell Press/Blackwell Scientific Publications, Cambridge, MA.
61. Magasanik, B. (1961) Catabolite repression. *Cold Spring Harb. Symp. Quant. Biol.*, **26**, 249–256.
62. Pastan, I. and Perlman, R. L. (1968) The role of the *lac* promoter locus in the regulation of β-galactosidase synthesis by cyclic AMP. *Proc. Natl. Acad. Sci. USA*, **61**, 1336–1342.
63. Ullmann, A. and Monod, J. (1968) Cyclic AMP as an antagonist of catabolite repression in *Escherichia coli*. *FEBS Lett.*, **2**, 714–717.
64. Silverstone, A. E., Magasanik, B., Reznikoff, W. S., Miller, J. H., and Beckwith, J. R. (1969) Catabolite sensitive site on the *lac* operon. *Nature*, **221**, 1012–1014.
65. Silverstone, A. E. and Magasanik, B. (1972) Polycistronic effects of catabolite repression on the *lac* operon. *J. Bacteriol.*, **112**, 1184–1191.
66. Beckwith, J. R., Grodzicker, T., and Arditti, R. (1972) Evidence for two sites in the *lac* promoter region. *J. Mol. Biol.*, **69**, 155–160.
67. Hopkins, J. D. (1974) A new class of promoter mutations in the lactose operon of *Escherichia coli*. *J. Mol. Biol.*, **87**, 715–724.
68. Schwartz, D. and Beckwith, J. R. (1970) Mutants missing a factor necessary for the expression of catabolite-sensitive operons in *E. coli*. In *The lactose operon* (ed. J. R. Beckwith and D. Zipsers), pp. 417–422. Cold Spring Harbor Laboratory, New York.
69. Emmer, M., de Crombrugghe, B., Pastan, I., and Perlman, R. L. (1970) Cyclic AMP receptor protein of *E. coli*. *Proc. Natl. Acad. Sci. USA*, **66**, 480–487.
70. Burgess, R. P., Travers, A. A., Dunn, J. J., and Bautz, E. K. F. (1969) Factor stimulating transcription by RNA polymerase. *Nature*, **221**, 43–47.
71. Travers, A. A. and Burgess, R. P. (1969) Cyclic reuse of RNA polymerase sigma factor. *Nature*, **222**, 537–540.
72. Losick, R. and Sonenshein, A. L. (1969) Change in the template specificity of RNA polymerase during sporulation of *Bacillus subtilis*. *Nature*, **224**, 35–37.
73. Losick, R. and Pero, J. (1981) Cascades of sigma factors. *Cell*, **25**, 582–584.
74. Geiduschek, E. P. (1991) Regulation of expression of the late genes of bacteriophage T4. *Ann. Rev. Genet.*, **25**, 437–460.
75. Sousa, R., Patra, D., and Lafer, E. M. (1992) Model for the mechanism of bacteriophage T7 RNAP transcription, initiation and termination. *J. Mol. Biol.*, **224**, 319–334.
76. Kustu, S., Santero, E., Keener, J., Popham, D., and Weiss, D. (1989) Expression of σ^{54} (*ntrA*) dependent genes is probably united by a common mechanism. *Microbiol. Rev.*, **53**, 367–376.
77. Merrick, M. J. (1993) In a class of its own—the RNA polymerase sigma factor σ^{54} (σ^{N}). *Molec. Microbiol.*, **10**, 903–909.
78. Magasanik, B. (1996) *Regulation of nitrogen utilization*. In Escherichia coli *and* Salmonella typhimurium*: Cellular and molecular biology* (2nd edn) (editor-in-chief. F. C. Neidhardt), pp. 1344–1356. American Society for Microbiology, Washington, DC.
79. Débarbouillé, M., Martin-Verstraete, I., Kunst, F., and Rapoport, G. (1991) The *Bacillus subtilis sigL* gene encodes an equivalent of σ^{54} from Gram-negative bacteria. *Proc. Natl. Acad. Sci. USA*, **88**, 9092–9096.
80. Kolter, R., Siegele, D. A., and Tormo, A. (1993) The stationary phase of the bacterial life cycle. *Ann. Rev. Microbiol.*, **47**, 855–874.

81. Gross, C. A. (1996) Function and regulation of the heat shock proteins. In Escherichia coli *and* Salmonella typhimurium*: Cellular and molecular biology* (2nd edn), (editor-in-chief F. C. Neidhardt), pp. 1382–1399. American Society for Microbiology, Washington, DC.
82. Gussin, G. N., Ronson, C. W., and Ausubel, F. M. (1986) Regulation of nitrogen fixation genes. *Ann. Rev. Genet.*, **20**, 567–591.
83. Ninfa, A. J. and Magasanik, B. (1986) Covalent modification of the *glnG* product, NR_I, by the *glnL* product, NR_{II}, regulates the transcription of the *glnALG* operon in *Escherichia coli. Proc. Natl. Acad. Sci. USA*, **83**, 5909–5913.
84. Markham, B. E., Harper, J. E., and Mount, D. W. (1985) Physiology of the SOS response—kinetics of *lexA* and *recA* transcriptional activity following induction. *Molec. Gen. Genet.*, **198**, 207–212.
85. Sauer, R. T., Ross, M. J., and Ptashne, M. (1982) Cleavage of the λ and P22 repressors by RecA protein. *J. Biol. Chem.*, **257**, 4458–4462.
86. Little, J. W. (1984) Autodigestion of LexA and phage λ repressors. *Proc. Natl. Acad. Sci. USA*, **81**, 1375–1379.
87. Kim, B. and Little, J. W. (1993) LexA and λ cI repressors as enzymes: specific cleavage in an intermolecular reaction. *Cell*, **73**, 165–1173.
88. Hall, B. D. and Spiegelman, S. (1971) Sequence complementarity of T2-DNA and T2-specific RNA. *Proc. Natl. Acad. Sci. USA*, **47**, 137–146.
89. Szybalski, W., Bøvre, K., Fiandt, M., Hayes, S., Hradecna, Z., Kumar, S., Lozeron, H. A., Nijkamp, H. J. J., and Stevens, W. F. (1971) Transcriptional units and their controls in *Escherichia coli* phage λ: Operons and scriptons. *Cold Spring Harb. Symp. Quant. Biol.*, **35**, 341–355.
90. Roberts, J. W. (1969) Termination factor for RNA synthesis. *Nature*, **224**, 1168–1175.
91. Luzzatti, D. (1970) Regulation of λ exonuclease synthesis: Role of the *N* gene product and repressor. *J. Mol. Biol.*, **49**, 515–519.
92. Roberts, J. W. (1975) Transcription termination and late control in phage lambda. *Proc. Natl. Acad. Sci. USA*, **72**, 3300–3304.
93. Das, A. (1993) Control of transcription termination by RNA-binding proteins. *Ann. Rev. Biochem.*, **62**, 893–930.
93a. Henkin, T. M. (1996) Control of transcription termination in prokaryotes. *Ann. Rev. Genet.*, **30**, 35–57.
94. Gottesman, S., Gottesman, M., Shaw, J. E., and Pearson, M. L. (1981) Protein degradation in *Escherichia coli*: the *lon* mutation and bacteriophage λ N and cII protein stability. *Cell*, **24**, 225–233.
95. Gottesman, S. (1996) Proteases and their targets in *Escherichia coli. Ann. Rev. Genet.*, **30**, 465–506.
96. Schnetz, K. and Rak, B. (1990) β-Glucoside permease represses the *bgl* operon of *Escherichia coli* by phosphorylation of the antiterminator protein and also interacts with glucose-specific enzyme III, the key element in catabolite control. *Proc. Natl. Acad. Sci. USA*, **87**, 5074–5078.
97. Babitzke, P. (1997) Regulation of tryptophan biosynthesis: Trp-ing the TRAP or how *Bacillus subtilis* reinvented the wheel. *Molec. Microbiol.*, **26**, 1–9.
98. Korn, L. J. and Yanofsky, C. (1976) Polarity suppressors defective in transcription termination at the attenuator of the tryptophan operon of *Escherichia coli* have altered Rho factor. *J. Mol. Biol.*, **106**, 231–241.
99. Umbarger, H. E. (ed.) (1963) Synthesis and structure of macromolecules. *Cold Spring Harb. Symp. Quant. Biol.*, Vol 28.

100. Landick, R., Turnbough, C. L., Jr., and Yanofsky, C. (1996) Transcription attenuation. In Escherichia coli *and* Salmonella typhimurium: *Cellular and molecular biology* (2nd edn) (editor-in-chief F. C. Neidhardt), pp. 1263–1286. American Society for Microbiology, Washington, DC.

101. Roth, J. R., Silbert, D. F., Fink, G. R., Voll, M. J., Anton, D., Hartman, P. E., and Ames, B. N. (1966) Transfer RNA and the control of the histidine operon. *Cold Spring Harb. Symp. Quant. Biol.*, **31**, 383–392.

102. Henkin, T. M. (1994) tRNA-directed transcription antitermination. *Mol. Microbiol.* **13**, 381–387.

103. Hayes, W. (1947) The nature of somatic phase variation and its importance in the serological standardisation of *O* suspensions of *Salmonella*. *J. Hyg.*, **45**, 111–119.

104. Zieg, J., Silverman, M., Hilmen, M., and Simon, M. (1980) The mechanism of phase variation. In *The operon* (2nd edn) (ed. J. H. Miller and W. S. Reznikoff), pp. 411–423. Cold Spring Harbor Laboratory, Cold Spring Harbor, NY.

105. Haselkorn, R. (1986) Organization of the genes for nitrogen fixation in photosynthetic bacteria and cyanobacteria. *Ann. Rev. Microbiol.*, **40**, 525–547.

106. Haselkorn, R., Golden, J. W., Lammers, P. J., and Mulligan, M. E. (1986) Identification and sequence of a gene required for a developmentally regulated DNA excision in *Anabaena*. *Cell*, **44**, 905–911.

107. Khokhlov, A. S., Anisova, L. N., Tovarova, I. I., Kleiner, F. M., Kovalenko, I. V., Krasilnikova, O. I., Kornitskaya, E. Y., and Pliner, S. A. (1973) Effect of A-factor on the growth of asporogenous mutants of *Streptomyces griseus*. *Zeitschrift für Allgemeine Mikrobiologie*, **13**, 647–655.

108. Onaka, H., Ando, N., Nihira, T., Yamada, Y., Beppu, T., and Horinouchi, S. (1995) Cloning and characterization of the A-factor receptor gene from *Streptomyces griseus*. *J. Bacteriol.*, **177**, 6083–6092.

109. Onaka, H., and Horinouchi, S. (1997) DNA-binding activity of the A-factor receptor protein and its recognition DNA sequences. *Molec. Microbiol.*, **24**, 991–1000.

110. Silverman, M., Martin, M., and Engebrecht, J. (1989) Regulation of luminescence in marine bacteria. In *Genetics of bacterial diversity* (ed. D. A. Hopwood and K. F. Chater), pp. 71–86. Academic Press, London/New York.

111. Bainton, N. J., Bycroft, B. W., Chhabra, S. R., Stead, P., Gledhill, L., Hill, P. J., Rees, C. E. D., Winson, M. K., Salmond, G. P. C., Stewart, G. S. A. B., and Williams, P. (1992) A general role for the *lux* autoinducer in bacterial cell signaling—control of antibiotic biosynthesis in *Erwinia*. *Gene*, **116**, 87–91.

112. Jones, S., Yu, B., Bainton, N. J., Birdsall, M., Bycroft, B. W., Chhabra, S. R., Cox, A. J. R., Golby, P., Reeves, P. J., Stephens, S., Winson, M. K., Salmond, G. P. C., Stewart, G. S. A. B., and Williams, P. (1993) The *lux* autoinducer regulates the production of exoenzyme virulence determinants in *Erwinia carotovora* and *Pseudomonas aeruginosa*. *EMBO J.*, **12**, 2477–2482.

113. Pearson, J. P., Gray, K. M., Passador, L., Tucker, K. D., Eberhard, A., Iglewski, B. H., and Greenberg, E. P. (1994) Structure of the autoinducer required for expression of *Pseudomonas aeruginosa* virulence genes. *Proc. Natl. Acad. Sci. USA*, **91**, 197–201.

114. Fuqua, C., Winans, S. C., and Greenberg, E. P. (1996) Census and consensus in bacterial ecosystems—the *luxR-luxI* family of quorum-sensing transcriptional regulators. *Ann. Rev. Microbiol.*, **50**, 727–751.

115. Eberl, L., Winson, M. K., Sternberg, C., Stewart, G. S. A. B., Christiansen, G., Chhabra, S. R., Bycroft, B., Williams, P., Molin, S., and Givskov, M. (1996) Involvement of *N*-acyl-L-

homoserine lactone autoinducers in controlling the multicellular behaviour of *Serratia liquefaciens*. *Molec. Microbiol.*, **20**, 127–136.
116. Gray, K. M., Pearson, J. P., Downie, J. A., Boboye, B. E. A., and Greenberg, E. P. (1996) Cell-to-cell signaling in the symbiotic nitrogen-fixing bacterium *Rhizobium leguminosarum*—autoinduction of stationary-phase and rhizosphere-expressed genes. *J. Bacteriol.*, **178**, 372–376.
117. Stevens, A. M. and Greenberg, E. P. (1997) Quorum sensing in *Vibrio fischeri*: essential elements for activation of the luminescence gene. *J. Bacteriol.*, **179**, 557–562.
118. Puskas, A., Greenberg, E. P., Kaplan, S., and Schaeffer, A. L. (1997) A quorum-sensing system in the free-living photosynthetic bacterium *Rhodobacter sphaeroides*. *J. Bacteriol.*, **179**, 7530–7537.
119. Pabo, C. O., Krovatin, W., Jeffrey, A., and Sauer, R. T. (1982) The N-terminal arms of λ repressor wrap around the operator DNA. *Nature*, **298**, 441–443.
120. Pabo. C. O. and Lewis, E. M. (1982) The operator-binding domain of λ repressor: structure and DNA recognition. *Nature*, **298**, 443–447.
121. Anderson, W. F., Ohlendorf, D. H., Takeda, Y., and Matthews, B. W. (1981) Structure of the *cro* repressor from bacteriophage λ and its interaction with DNA. *Nature*, **290**, 754–758.
122. McKay, D. B. and Steitz, T. A. (1981) Structure of catabolite gene activator protein at 2.9 Å resolution suggests binding to left-handed DNA. *Nature*, **290**, 744–749.
123. Weber, I. T. and Steitz, T. A. (1987) Structure of a complex of catabolite activator protein and cyclic AMP refined at 2.5 Å resolution. *J. Mol. Biol.*, **198**, 311–322.
124. Schultz, S. C., Shields, G. C., and Steitz, T. A. (1991) Crystal structure of a CAP-DNA complex: the DNA is bent by 90°. *Science*, **253**, 1001–1007.
125. Phillips, S. E. V. (1994) The β-ribbon DNA recognition motif. *Ann. Rev. Biophys. Molec. Struct.*, **23**, 671–701.
126. Maaløe, O. (1979) Regulation of the protein-synthesizing machinery—ribosomes, tRNA, factors, and so on. In *Biological regulation and development: Volume 1, Gene expression* (ed. R. F. Goldberger), pp. 487–542. Plenum Press, New York
127. Neidhardt, F. C. and Savageau, M. (1996) Regulation beyond the operon. In Escherichia coli *and* Salmonella typhimurium*: Cellular and molecular biology* (2nd edn) (editor-in-chief F. C. Neidhardt), pp. 1310–1324. American Society for Microbiology, Washington, DC.
128. Lynch, A. S. and Lin, E. C. C. (1996) Responses to molecular oxygen. In Escherichia coli *and* Salmonella typhimurium*: Cellular and molecular biology* (2nd edn) (editor-in-chief F. C. Neidhardt), pp. 1526–1538. American Society for Microbiology, Washington, DC.
129. He, L. H., Soupene, E., and Kustu, S. (1997) NtrC is required for control of *Klebsiella pneumoniae* NifL activity. *J. Bacteriol.*, **179**, 7446–7455.
130. Walmsley, J., Toukdarian, A., and Kennedy, C. (1994) The role of regulatory genes *nifA*, *vnfA*, *anfA*, *nfRX*, *ntrC*, and *rpoN* in expression of genes encoding the three nitrogenases of *Azotobacter vinelandii*. *Arch. Microbiol.*, **162**, 422–429.
131. Batut, J. and Boistard, P. (1994) Oxygen control in *Rhizobium*. *Ant. van Leeuw. Int. J. Gen. Molec. Microbiol.*, **66**, 129–150.
132. Fischer, H. M. (1994) Genetic regulation of nitrogen fixation in rhizobia. *Microbiol. Rev.*, **58**, 352–286.
133. Galinier, A., Haiech, J., Kilhoffer, M. C., Jaquinod, M., Stulke, J., Deutscher, J., and Martin-Verstraete, I. (1997) The *Bacillus subtilis crh* gene encodes a HPr-like protein involved in carbon catabolite repression. *Proc. Natl. Acad. Sci. USA*, **94**, 8439–8444.
134. Martin-Verstraete, I., Stulke, J., Klier, A., and Rapoport, G. (1995) Two different

mechanisms mediate catabolite repression of the *bacillus subtilis* levanase operon. *J. Bacteriol.*, **177**, 6919–6927.
135. Stulke, J., Martin-Verstraete, I., Charrier, V., Klier, A., Deutscher, J., and Rapoport, G. (1995) The Hpr protein of the phosphotransferase system links induction and catabolite repression of the *Bacillus subtilis* levanase operon. *J. Bacteriol.*, **177**, 6928–6936.
136. Strauch, M. A. (1995) AbrB modulates expression and catabolite repression of a *Bacillus subtilis* ribose transport operon. *J. Bacteriol.*, **177**, 6727–6731.
137. Rouch, D. A. and Brown, N. L. (1997) Copper-inducible transcriptional regulation at two promoters in the *Escherichia coli* copper resistance determinant *pco*. *Microbiol.*, **143**, 1191–1202.
138. Hardt, K., Joris, B., Lepage, S., Brasseur, R., Lampen, J. O., Frere, J. M., Fink, A. L., and Ghuysen, J. M. (1997) The penicillin sensory transducer, BlaR, involved in the inducibility of β-lactamase synthesis in *Bacillus licheniformis* is embedded in the plasma membrane via a four-α-helix bundle. *Molec. Microbiol.*, **23**, 935–944.
139. Otto, M., Sussmuth, R., Jung, G., and Gotz, F. (1998) Structure of the pheromone peptide of the *Staphylococcus epidermidis agr* system. *FEBS Lett.*, **424**, 89–94.
140. Vandenesch, F. (1997) Regulation of exoprotein expression in *Staphylococcus aureus*. *Médecine et Maladies Infectieuses*, **27**, 150–158.
141. Ji, G. Y., Beavis, R. C., and Novick, R. P. (1995) Cell density control of staphylococcal virulence mediated by an octapeptide pheromone. *Proc. Natl. Acad. Sci. USA*, **92**, 12055–12059.
142. Wang, S. P., Sharma, P. L., Schoenlein, P. V., and Ely, B. (1993) A histidine protein kinase is involved in polar organelle development in *Caulobacter crescentus*. *Proc. Natl. Acad. Sci. USA*, **90**, 630–634.
143. Hodgson, D. A. and Berry, A. E. (1998) Light regulation of carotenoid synthesis in *Myxococcus xanthus*. In *Microbial responses to light and time. Symposium 56 of the Society for General Microbiology* (ed. M. X. Caddick, S. Baumberg, D. A. Hodgson, and M. K. Phillips-Jones), pp. 185–211. Cambridge University Press, Cambridge, UK.

2 | Protein recognition of specific DNA and RNA binding sites

PETER G. STOCKLEY

1. A summary of the field of protein–nucleic acid interactions

1.1 Introduction

Facetiously, it has been said that there are only two substantive problems in biology. These are: one, to understand how protein molecules recognize and interact with nucleic acids, and, two, to understand the consequences thereof! This is obviously a massive over-simplification, although it does contain a grain of truth. This chapter will address the first of these problems whilst the remainder of this book will deal with aspects of the second. Nucleic acid recognition will be illustrated by reference to specific examples drawn from the regulation of prokaryotic gene expression. This will necessarily not be a comprehensive review given the wealth of knowledge now available in this area. Some discussion of eukaryotic examples will be introduced for comparison purposes. In some cases, the examples chosen to illustrate particular features of nucleic acid recognition are also the topics of separate chapters elsewhere in this book, and in those cases the reader should follow the cross-referencing instructions to obtain more detailed discussions of the biology of the systems being discussed.

1.2 The conformations of nucleic acids

Understanding sequence-specific nucleic acid-protein recognition must start with an understanding of the sequence-dependence of nucleic acid conformations. In proteins, everyone is aware that the amino acid sequence dictates the conformation of the folded structure, which can be defined by measuring the values of just two dihedral angles, ϕ and ψ, associated with each amino acid residue. These parameters define the degrees of conformational freedom in a polypeptide, which are restricted due to the partial double bond character of the peptide bond. As Fig. 1 shows, the

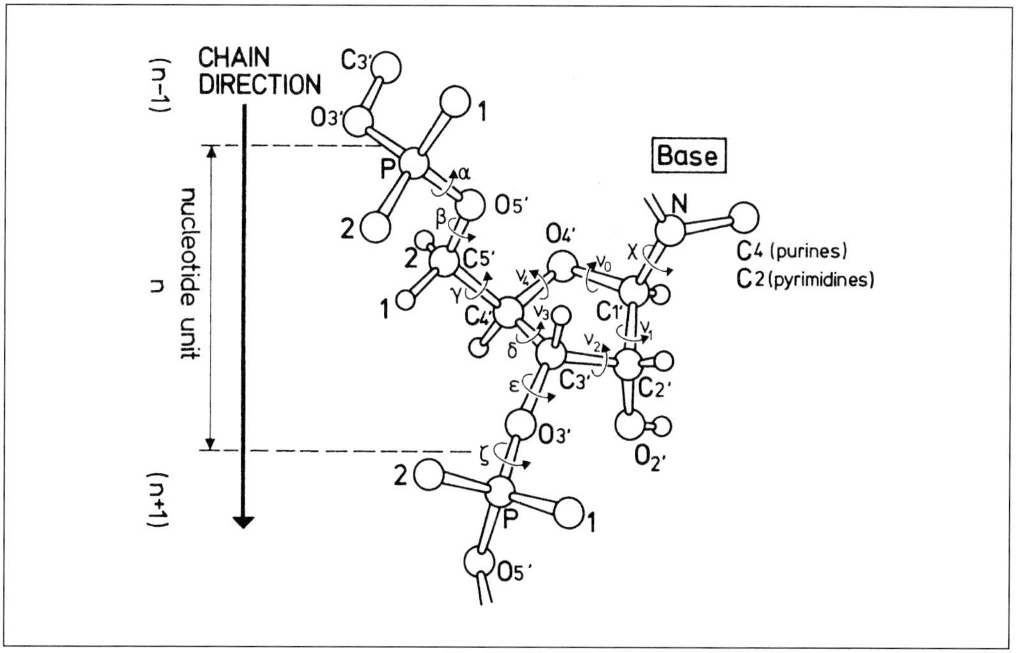

Fig. 1 Conformational parameters of nucleotides. The figure shows the atom numbering and defines the torsional angles for polynucleotides.

situation in nucleotides is much more complicated. There are six torsion angles (α to ξ) associated with the phosphodiester backbone, a further five within the ribose sugar ring (ν_0 to ν_4), and a final degree of freedom defining the angle between the base and the sugar, χ, i.e. the glycosyl bond. Just as with ϕ and ψ, these angles are not freely variable and there are preferred values for each of them. Much of the detailed information available on these parameters has come from X-ray crystal structures of simple nucleotide model compounds, and anyone wishing to understand the fine details should consult an appropriate reference work, such as Saenger's book (1). Here we will concentrate on some of the major consequences of these degrees of freedom for single- and double-stranded polynucleotides.

1.2.1 Sugar pucker

Puckering of the ribose ring is a concept familiar to chemists used to thinking of closed-ring structures. Figure 2.2 shows diagrammatic representations of several possible distortions away from a flat ring structure. The dominant conformations found in polynucleotides are C3′-*endo* and C2′-*endo*, in which either the C2 or the C3 carbon is above the plane of the ring. The sugar pucker is variable in mono-nucleotides or small model compounds but has major consequences for the global structures of double helices. Thus in RNA duplexes, which are essentially all A-type, the C3′-*endo* conformation is generally found, whereas in B-form DNA duplexes, the C2′-*endo* conformation dominates.

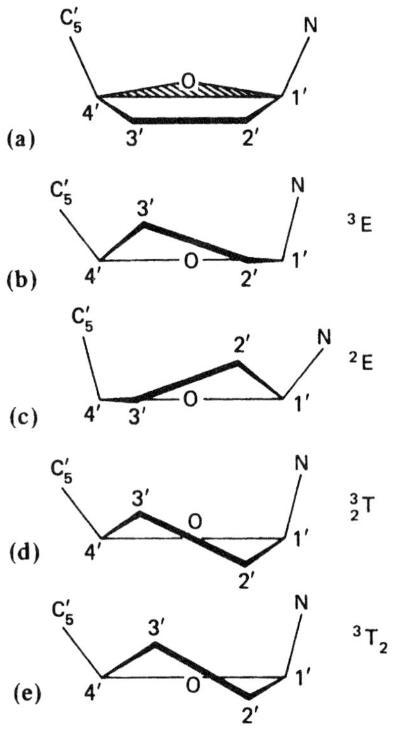

Fig.2 Sugar puckering in ribose rings: (a) shows the ribose as a flat plane; (b) the C3'-*endo* conformer; (c) the C2'-*endo* conformer; (d) the half-chair form, C2'- *exo*–C3'-*endo*; and (e) the C3'-*endo*–C2'-*exo* form.

1.2.2 The *syn*/*anti* orientation about the glycosyl bond

Nucleotide bases can adopt two principal orientations with respect to the sugar moiety. In the *anti* conformation the bases have their small hydrogen atoms above the sugar ring whereas in the *syn* conformation these positions are occupied by the larger oxygen and ring nitrogen atoms. Again the precise detail of each conformer depends on the base concerned.

1.2.3 Flexibility within base pairs

Even when hydrogen bonded within a Watson–Crick base pair, nucleotides retain considerable conformational flexibility. Figure 2.3 shows the degrees of freedom which have been characterized in crystal structures of duplexes. Each base pair unit has three degrees of freedom allowing it to *tilt*, *twist*, and *roll* with respect to the helical axis. Even between the partners of a base pair there is also the potential to undergo *propeller twist*.

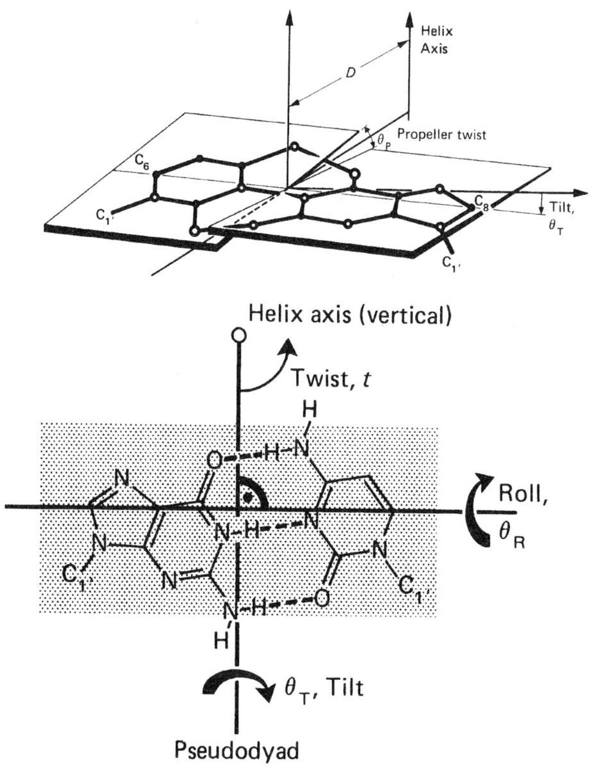

Fig. 3 Conformational flexibility within double helices. Nucleic acid duplexes retain considerable conformational flexibility, as illustrated in this diagram. In solution and in the cell, several conformers are probably in dynamic equilibrium. The precise geometry of the dominant conformer is sequence dependent.

1.2.4 Sequence-dependent DNA conformations

The conformational freedom described above means that the detailed conformation of a nucleic acid duplex is exquisitely dependent on its precise base sequence. In DNA duplexes there is sufficient structural information to allow rules for this dependence to be formulated. In general for B-form duplexes, the minor groove becomes widened in G.C and mixed sequences, but is narrower at hetero- or homopolymer A.T sequences. Propeller twist values are low for G.C sequences, but can be very high for A.T regions, and this effect can be stabilized by additional hydrogen bond contacts to neighbouring base pairs, resulting in significant 'stiffening' of the polynucleotide chain at such sites. One of the major energetic driving forces causing such distortions is the favourable stacking interaction that can occur between neighbouring bases in a duplex. A set of rules for local DNA conformation based upon stacking interactions and the geometrical constraints of nucleotide units has been developed by Drew and Calladine, although it is not foolproof because it neglects effects of electrostatic interactions (2–4).

1.3 Physical principles of molecular recognition

1.3.1 Thermodynamics

It is worth taking a few moments to consider the fundamental forces involved in forming complexes between proteins and nucleic acids, or indeed between any two molecules. In thermodynamic terms, complexes will form and have a defined stability if the overall process leads to a reduction in the free energy (ΔG) of the system. This in turn is related to the changes in enthalpy (ΔH) and entropy (ΔS), which occur during the process, by the equation $\Delta G = \Delta H - T.\Delta S$, where T is the absolute temperature. The enthalpy change is related to the change in bond energies which occur. For our purposes these are almost always non-covalent bonds such as hydrogen bonds and van der Waals interactions. The entropy change is related to the levels of order (disorder) in the system and gives rise to the famous hydrophobic effect, in which large falls in free energy can be achieved by excluding hydrophobic amino acid side chains from contact with water, which is the driving force behind the folding of globular proteins.

Although specialists may disagree, we can simplify both these phenomena for thinking about protein–nucleic acid interactions. Virtually all non-covalent bonding can be thought of as due to electrostatic interactions, such as with positively charged protons (hydrogen bonding) or between polarized electron shells in molecules (van der Waals interactions). Complex formation also leads to a form of hydrophobic effect (5, 6), although in this case each partner in the complex makes a series of contacts with water before complexation and a different set of intermolecular contacts afterwards, leading to effects on enthalpy as well. Because of these two effects it is important to determine the overall area of the surface of the isolated molecules participating in a complex which becomes buried when they interact. For a large number of protein–DNA complexes this value varies between ~1000 and ~3000 $Å^2$ (7, 8). This compares with a value of ~750 $Å^2$ for formation of an antibody-antigen complex (9) and is similar to the burial of surface area in the formation of oligomeric proteins. It is also important to consider the net charges on the interacting surface. This can be done by calculating the electrostatic potential surface across interacting molecules. In general, complexes are only formed when surfaces of opposite potential come together. Attempts are underway to couple the information from thermodynamics to binding target sequence analysis, with the goal of producing predictive algorithms (10).

1.3.2 Sequence-specific recognition

With these ideas in mind, Seeman, Rosenberg, and Rich analysed the recognition potential of Watson–Crick base pairs (11) before structural data became available for protein–nucleic acid complexes. They argued that the properties of the base pairs were such that recognition would occur primarily via hydrogen bonding to base pair edges in the major groove (Fig. 4). Hydrogen bond donor and acceptor groups in this groove are more frequent and variable than those in the minor groove. The methyl group of thymidine also introduces a large steric constraint in this groove. Even so it

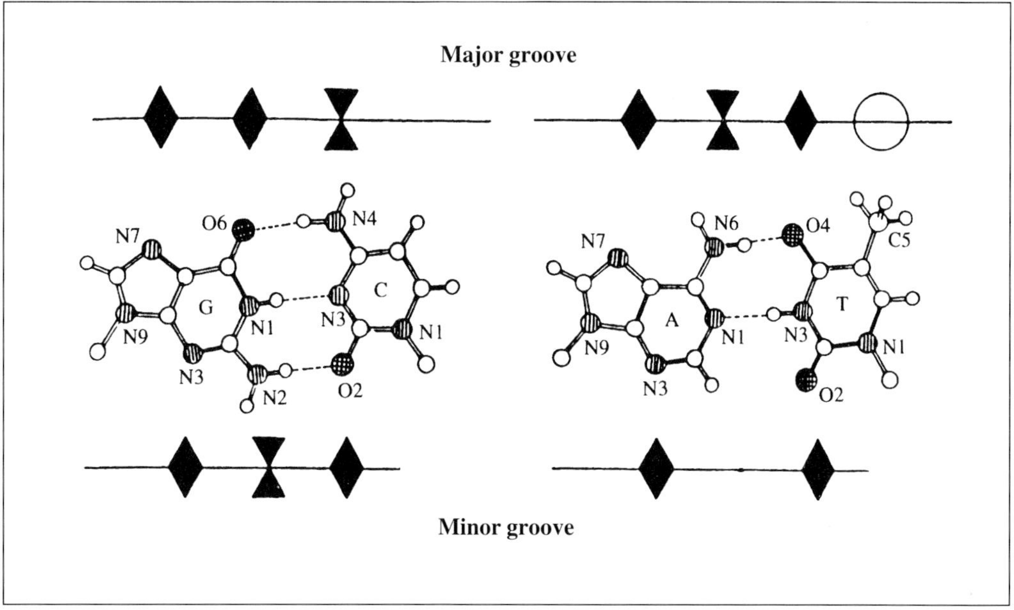

Fig. 4 Hydrogen bonding potentials of the Watson–Crick base pairs. The edges of Watson–Crick base pairs present a series of functional groups capable of making hydrogen bonds with interacting proteins or other ligands. The hydrogen bond acceptors are shown as diamonds, donor sites as inverted diamonds, and the bulky methyl group of thymidine, which can contribute hydrophobic van der Waals interactions or act as a steric blocking group, as a circle. There are more sites facing the major groove of a duplex than the minor groove and recognition is less ambiguous on that face. As a result most DNA sequences are recognized via this groove (11), which is freely accessible in B-form duplexes.

is difficult to imagine interactions with a single base pair being sufficiently discriminatory to provide biological specificity and it was realized that recognition sequences would involve extended sites. This proves to be the case. Recognition sites 12 bp in length are statistically likely to be unique within the *E. coli* genome. In practice, sites are often 10–20 bp in length.

Although these types of interaction have now been shown to exist (12, 13), the true situation is even more complex due to the sequence-dependent flexibility of DNA duplexes described above. This can result in the hydrogen bonding groups along base pair edges being displaced significantly from their positions in standard A- or B-form duplexes. Furthermore, duplexes can be recognized indirectly by their ability to undergo such conformational distortions. The terms 'direct readout' and 'indirect readout' have been coined to describe the situations in which proteins interact directly with base pair edges or indirectly via duplex distortions, respectively (14).

However, DNA sequence recognition can be thought of as occurring in the context of a relatively uniform structure, i.e. the Watson–Crick duplex, usually in a conformation close to a B-form helix. RNA recognition is necessarily distinct because of the preference for ribo-oligonucleotides to adopt A-form rather than B-form helices. The major groove in an A-form helix is almost inaccessible by protein functional groups, although the minor groove is wide and freely accessible. Recognition of specific sequences by contacts in the minor groove is difficult because of the lack of distinctive groups on the different base pairs (Fig. 4). As a result most RNA recognition occurs at non-duplex sites, which occur very frequently in folded RNA molecules. The diversity of RNA folding in three dimensions approaches that seen for proteins and it is therefore not surprising that such molecules can form catalytically active sites (15–17). These conformations are facilitated by RNA stem-loops, single stranded bulges and mismatches, non-Watson–Crick base pairing, and high percentages of post-transcriptionally modified functional groups in the bases and the sugar residues. Recognition thus depends on interaction with specific nucleotide residues in the context of folded structures, which present unique targets for interacting proteins (18, 19).

1.3.3 Sequence-dependent distortions

The conformational flexibility of nucleic acid duplexes also leads to large scale distortions in which the path of the helical axis is changed, i.e. the molecule becomes bent. Bending can be distributed over a number of base pairs each of which adopts a conformation compatible with the larger distortion, or the bending can be restricted to a single base pair step, i.e. a localized kink (20). Such large-scale distortions can be the result of the local nucleotide sequence, as in A.T DNA sequences, or can be induced by interaction with proteins. Alternating A.T sequences have special properties because of the additional hydrogen bonding which can occur along the helical axis between neighbouring base pairs as well as at right angles within a base pair (21). In fact such stretches of the duplex are perfectly straight; however, distortions occur at the junctions with standard B-form duplex, giving the net appearance of bending. Bending in the form of kinks is common in protein complexes, presumably because the protein binds only one of a number of DNA conformers. The free energy of interaction with the protein can compensate for the energy required to distort the duplex. By neutralizing the negatively charged phosphodiester backbone along just one face of a duplex, proteins can promote bending to increase the separation of the phosphates on the other duplex face.

1.3.4 The kinetics of protein–nucleic acid interactions

Many protein–nucleic acid complexes regulate dynamic processes and their formation is therefore reversible (22). This means that the kinetics of target site location by proteins is of functional significance. In a typical prokaryotic DNA genome, such as in *E. coli*, one or a limited number of target binding sites must be identified against a large background of non-specific competitor sites. In some cases, the competitors will have sequences closely related to that of the target. Thus a single, specific binding site must be located within $>10^6$ competing sites at a rate sufficiently rapid to allow meaningful physiological responses to stimuli *in vivo*. It appears that many DNA-binding proteins locate such targets faster than would have been expected for simple bimolecular collisions in solution. Indeed, in many cases the rate is faster than would be allowed by simple three-dimensional diffusion. Such 'facilitated' target site location is believed to be achieved by an initial non-sequence-specific DNA-binding event, which is followed by translocation of the protein along the DNA. The translocation can either be via sliding, a process in which the protein effectively walks along the duplex in a one-dimensional diffusional mechanism, or by hopping, in which the protein exchanges between different domains of the DNA, known as intersegment transfer, or both (23).

1.4 Nucleic acid binding motifs in proteins

In the early 1980s the three-dimensional structures of three prokaryotic transcriptional regulators, or the DNA-binding domains thereof, became available; *E. coli* CRP (24), λ*cro* (25), and λ*cI* (26) (Fig. 5). These were the first examples of structures for sequence-specific DNA-binding proteins, and they were obviously compared to identify common features. All three proteins (domains) are dimers of relatively small polypeptides (≈100 aa in length). Although the topology of the polypeptide fold in each case was distinct, they each contained an unusual element of tertiary structure, which had not been seen in the globular folds of other types of proteins, composed of a surface exposed pair of alpha helices separated by an unusual tight turn of polypeptide. The spatial separation of these helix–turn–helix (HTH) motifs in each dimer was essentially identical and equal to 34 Å, which is also the relative separation of consecutive major grooves of B-form DNA along one face of a duplex. Model building then suggested that it would be straightforward to form a protein–DNA complex by insertion of one of the helices of each HTH 'motif' in adjacent major grooves. Indeed, there is a very good complementarity of fit between the volume of an alpha helix and that of the major groove. Sequence-specificity was then imagined to be due to the direct readout of interactions between amino acid side-chains in the 'recognition helix' and the edges of Watson–Crick base pairs in the DNA.

This proposal was largely substantiated when Ptashne's laboratory carried out the well-known 'helix-swap' experiment, in which it was possible to show that DNA sequence specificity in the HTH proteins could be altered in a defined way by replacing the outward pointing side-chains along the recognition helix (27) (Fig. 6). Thus it appeared that there was a conserved structural feature of sequence-specific DNA-binding proteins, the HTH motif (28). By comparing the sequences of the known HTH proteins it was then possible to generate a consensus amino acid sequence for formation of the motif. Protein sequence database searches then identified hundreds of potential examples of proteins containing such a motif in proteins known to bind DNA and those of unknown function (29, 30). Both prokaryotic and eukaryotic examples were identified, and the concept of DNA-binding motifs (as well as the concept of protein functional domains in general) became firmly established. Shortly thereafter crystal structures of HTH repressor/operator complexes became available confirming much of the previous speculation about the interaction of the HTH motif

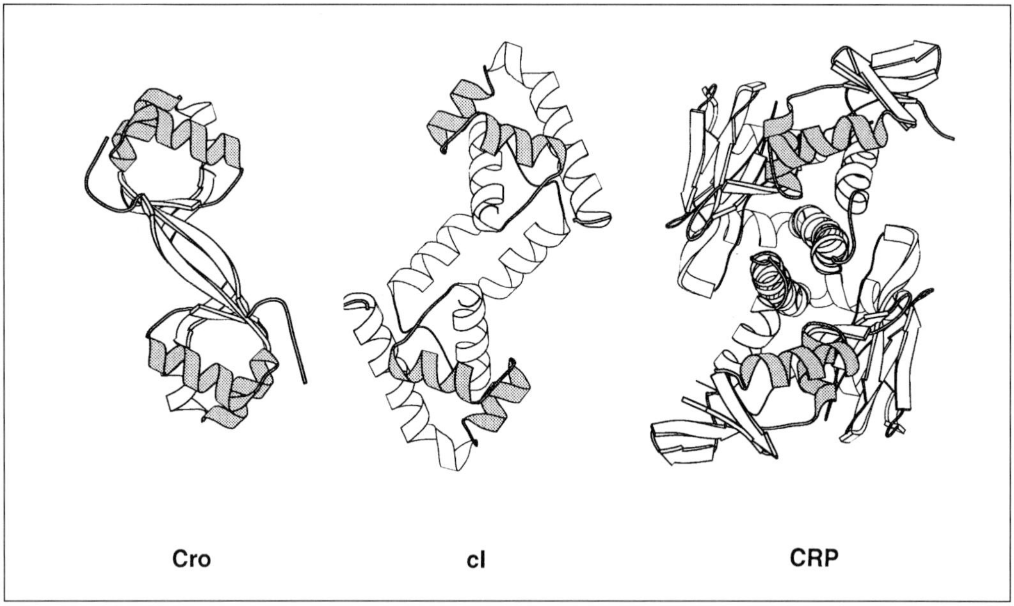

Fig. 5 The helix–turn–helix DNA-binding motif. The figure illustrates the structures of the λ*cro*, *cI*, and the *E. coli* CRP proteins, as ribbon cartoons. Each protein is a dimer of roughly similar size. Although the bulk of the tertiary folds in each case are distinct, they all contain a protruding helix–turn–helix motif (highlighted by shading), two per dimer, separated by 34 Å, allowing interaction with two successive major groove sites along one face of a B-form duplex.

with duplex DNA, although the structures also revealed that the DNA duplex is often distorted by protein binding (31–34) (Fig. 7).

Subsequently, a number of other motifs involved in DNA sequence recognition have been identified by combinations of genetic and structural approaches. Over 150 of these have been studied at the molecular level in complex with their DNA targets, providing a large database for this particular type of intermolecular recognition event (12, 29). Many of these other motifs also contain a recognition alpha helix, as in the helix–loop–helix structures that are common in eukaryotic transcription factors, where the constraints on the region between helices of the HTH are relaxed, and in zinc fingers (35), which again are thus far restricted to eukaryotic transcription factors. A distinct recognition motif based on a pair of polypeptide strands in an anti-parallel β-ribbon has also been identified in the phage P22 repressors Arc and Mnt, the *E. coli* methionine repressor protein, MetJ (see below), and in the bacterial DNA-packaging protein HU (36). The eukaryotic TATA binding protein also uses a β-sheet recognition motif (37). Proteins which are involved in enzymatic transformation of the DNA as well as sequence-specific binding display a wide range of variations on these themes, having recognition clefts, such as in restriction enzymes, and even 'holes' such as in the topoisomerases (38).

To date, our understanding of sequence-specific recognition of RNA molecules is

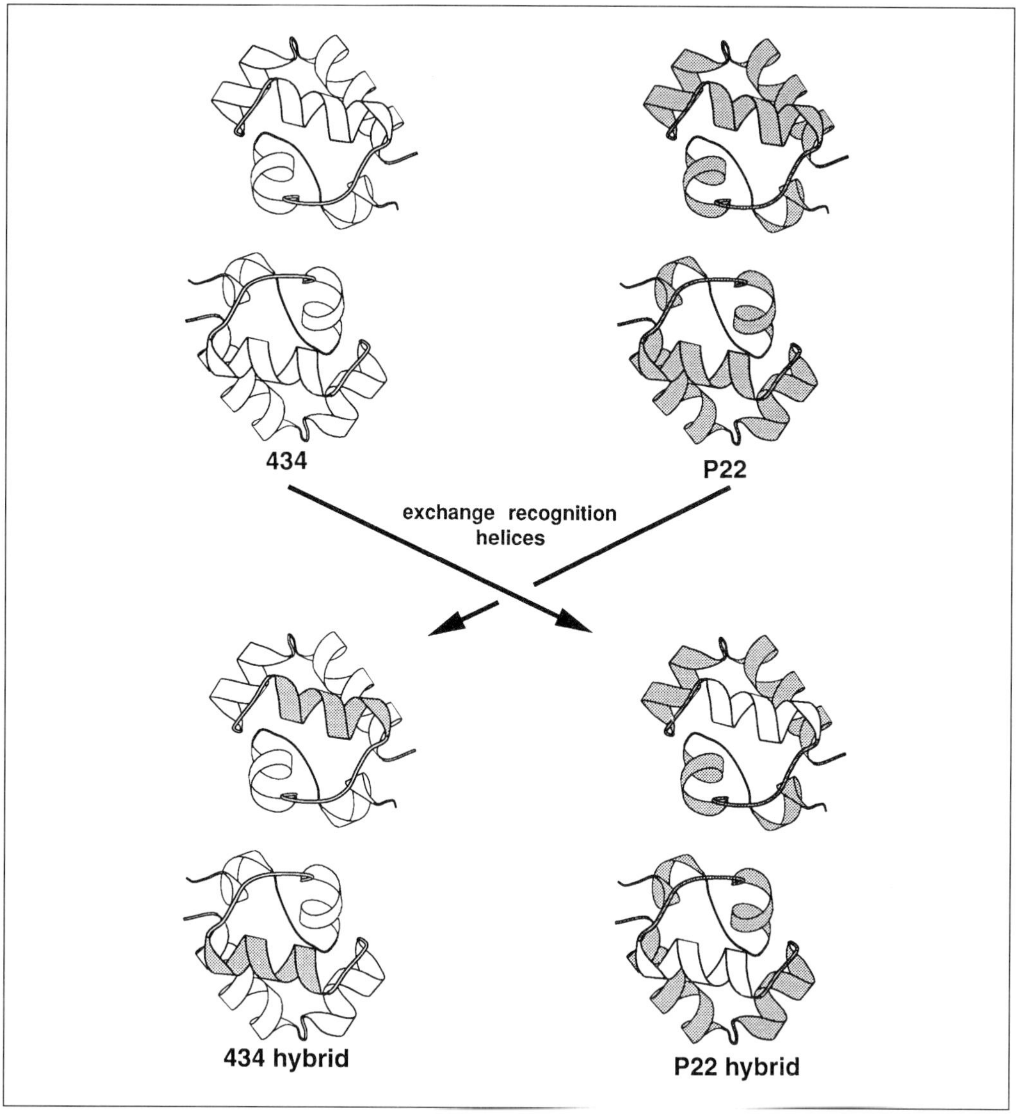

Fig. 6 The helix swap experiment. The figure shows a cartoon representation of the helix swap experiment. The recognition helices from the lambdoid phages P22 and 434 were exchanged genetically, resulting in a pair of chimeric proteins. *In-vivo* genetic assays then confirmed that this protein engineering had also exchanged the DNA operator specificity. (In fact only the outward facing amino acid residues of the second alpha helix in each case were substituted (27).)

much less advanced. Crystal structures are available for only three tRNA synthetase (TRS) complexes (39), the MS2 bacteriophage translational repression complex (40, 41), and the U1A complex from the spliceosome (42, 43) (see below). It is already clear that the recognition is considerably more complex than for most instances with DNA. In particular, it appears that RNA conformations are even more labile than DNA, and

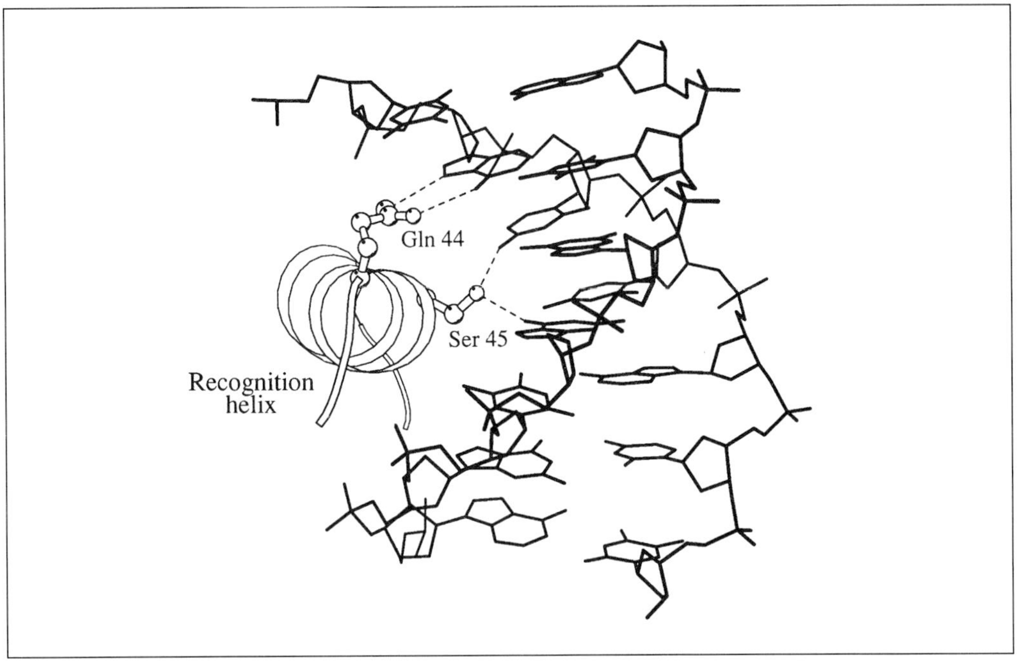

Fig. 7 The helix–turn–helix:DNA complex. Diagram of the recognition helix of the 434 repressor HTH (shown as a ribbon) bound to an operator half-site (shown as a framework model). The shape complementarity of the helix and the major groove are clear. The direct hydrogen bonds between amino acid side-chains and the edges of base pairs are shown as dotted lines.

that the conformation of the bound RNA ligand is often different from the dominant form in solution free of bound protein. This problem is compounded because RNAs exhibit far more complex secondary and tertiary structures than double-stranded DNA. Common target sites for proteins are stem-loops, stem-bubbles, pseudo-knots, and complex sites made up of such smaller sites. Protein motifs which recognize such sites include viral coat protein subunits (44, 45); a conserved RNA-binding domain (RBD) of 90–100 amino acids in length (46–48); proteins containing the RGG box, a 20–25 amino acid region containing several arginine–glycine–glycine sequence repeats interspersed with other, often aromatic, residues; the k homology (KH) domain of ~50 amino acids in length containing the octapeptide, IGX_2GX_2I (where X is any amino acid); and the so-called arginine fork (43). The most common RNA-binding domain is the RNP domain (Fig. 8), which occurs in at least 600 proteins in higher organisms (49–51). Many of these domains are known or are predicted to form β-sheet structures, which are believed to dominate RNA recognition motifs. Often the RNA-binding segments of the protein, such as the arginine-rich peptides retain the ability to bind RNA in isolation; however, their affinity and specificity for the cognate target site is reduced, implying that other regions of the protein contribute significantly to the interactions as well (19).

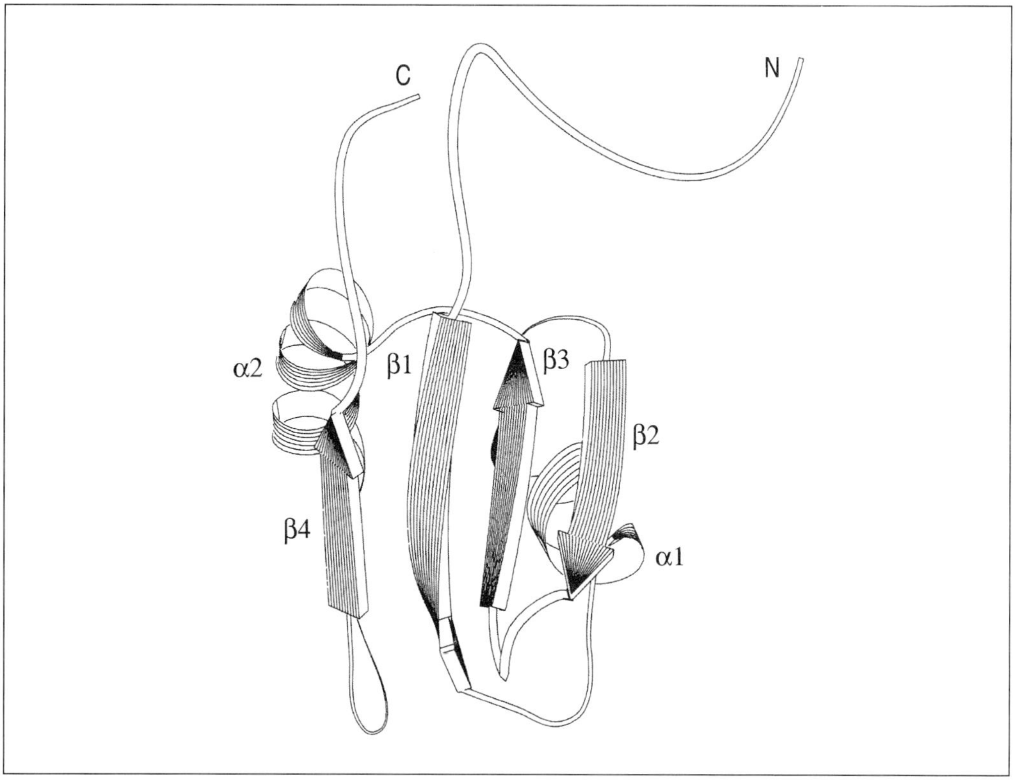

Fig. 8 The structure of the RNP RNA-binding domain. The core RNP domain, consisting of a β-sheet and packing helices, is shown as a ribbon diagram.

2. Examples of protein–nucleic acid complexes

2.1 Sequence-specific DNA–protein interactions

2.1.1 Bacteriophage repressors: λcI, λcro, and 434cI

Figure 2.9 summarizes the interaction between the lambda *cI* repressor N-terminal domain and its operator site, as determined by X-ray crystallography (31). As predicted based on the structure of the unbound protein and the results of genetic experiments, the second alpha helix of the HTH motif is positioned to lie inside the major groove. Amino acid side-chains from this helix make direct contacts to base pairs within the operator. Other regions of the HTH and the rest of the protein fold make contacts with the operator backbone and bases, in particular the N-terminal region is flexible and makes contacts in the minor groove. The repressor forms a non-covalent dimer and interacts with a target DNA fragment having palindromic symmetry, thus increasing the specificity of the interaction. In the cell, *cI* binds to adjacent operators cooperatively due to favourable protein–protein contacts between

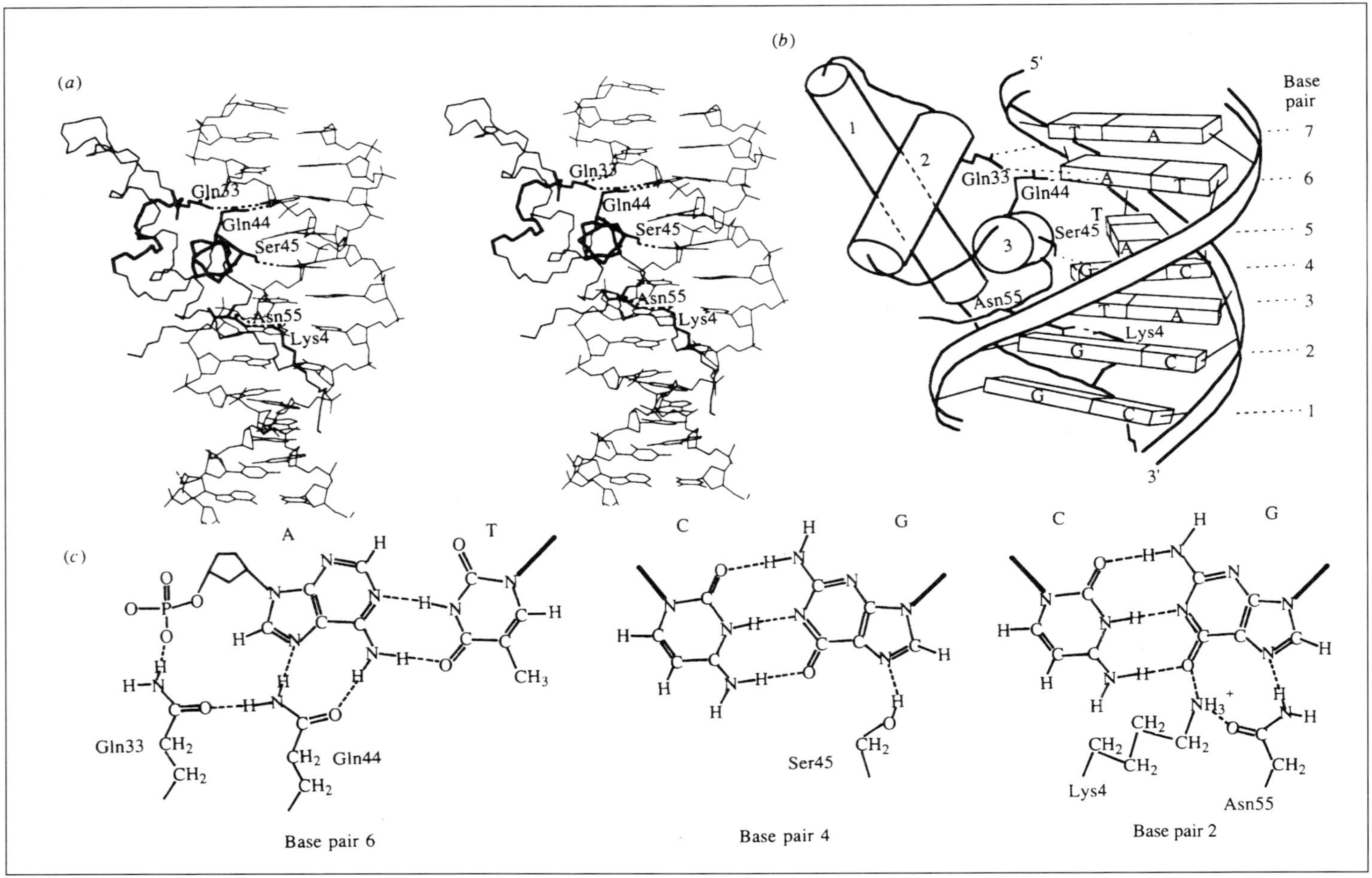

Fig. 9 The details of the λ*cI* complex interaction. The figure shows the crystal structure of one half of the λ*cI* repressor–operator complex. Only the region of the helix–turn–helix from the protein is included, as a stick model in the stereo view (top left) or as a cartoon cylinder with side-chains (top right). Below is shown the chemical detail of the interactions with base pairs in the operator. Note that side-chains interact both with base pair edges and with the phosphodiester backbone, as well as with each other, forming a network of hydrogen bond interactions (26).

dimers, resulting in efficient operator saturation for relatively small changes in the overall concentration of the protein (22).

The details of the sequence-specific interactions are instructive. Polar and charged side-chains participate to contact hydrogen bond donors and acceptors at the edges of a subset of base pairs within the operator. Not all of the available donor and acceptor groups on the base pairs contacted make interactions, and at some sites only one such contact is made. Several side-chains make multiple contacts to the DNA and to each other, helping to form an appropriate interaction interface. Although only a subset of the base pairs within the operator is contacted directly by the protein, genetic experiments show that repression *in vivo* is sensitive to the sequence at each position within the site (52). This is presumably due to sequence-dependent effects on the local DNA conformation. Indeed, in the crystal structure the operator was encompassed on a 20 bp fragment and this is distorted from linear B-form duplex, especially towards the ends of the fragment where it is bent towards the protein, presumably to maximize the interaction energy. Helical twist angles throughout the complex also vary significantly from pure B-duplex values (21–47° compared to 36°) and there are some high propeller twist values. All of these distortions may account for the observed sequence preferences.

Similar situations have now been observed in complexes of λ*cro* and *434cI*. These do, however, show some significant differences from the λ*cI* situation, especially in the degree of DNA distortion present in the 434 complex (53–56). In that case there is a significant overwinding of the DNA duplex in the region of the central A.T-rich operator sequence, leading in places to distinct narrowing of the width of the minor groove (8.8 Å versus the expected 11.5 Å for B-form helices). These A.T base pairs show high propeller twists stabilized by the inter-base pair hydrogen bonding described above (21), and also by water molecules bound in the minor groove. These data suggest that even in homologous systems, whilst the overall nature of protein–DNA interactions remains similar, there will be a good deal of variation in the structural details.

2.1.2 *E. coli* TrpR, MetJ, and LacI

TrpR: For *cI* the absolute concentration of the protein in the host cell determines the function of the repressor. It is an example of an irreversible genetic switch. In more general situations, genes need to be switched on and off in response to cellular conditions. A very elegant example of such physiological coupling via protein transcriptional regulators has been uncovered by structural studies of the *E. coli* tryptophan repressor protein, TrpR. This was the first system in which crystallographic data provided an explanation of such coupling (14, 57, 58). As well as binding DNA, the repressor also binds the end product of the metabolic pathway whose genes it contains, L-tryptophan. TrpR is an example of another small, dimeric protein containing an HTH DNA-binding protein. However, in the absence of L-tryptophan the molecule has a very low affinity for its operator sites. Crystal structures of the repressor in the presence or absence of L-Trp, and the ternary complex with bound operator fragments have revealed the molecular basis of this control. In the absence

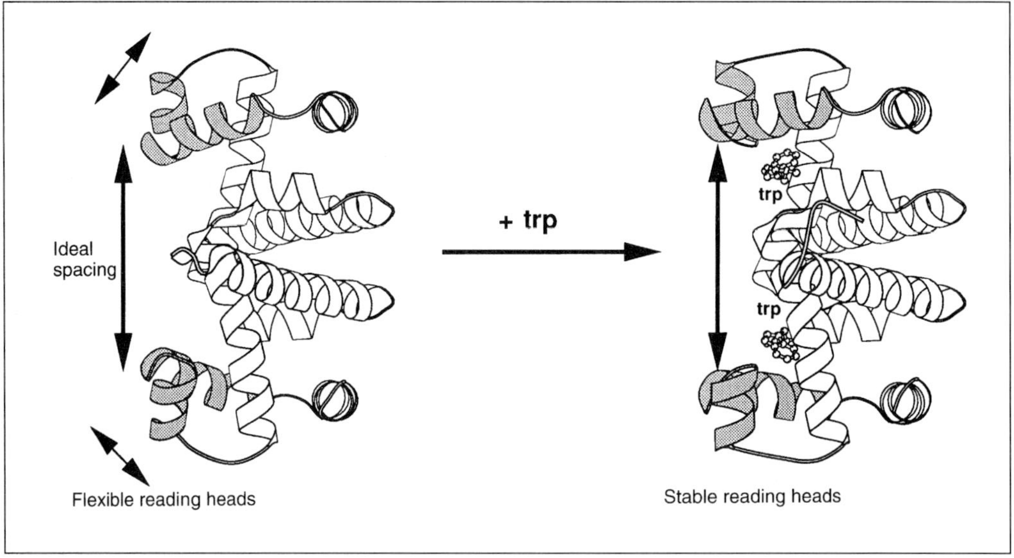

Fig. 10 Activation of TrpR. The figure illustrates the allosteric induction of DNA-binding for TrpR based on the three-dimensional structures of the apo, halo, and operator complexes (14, 57, 58). In the presence of L-Trp, the HTH motifs (helices D and E) move further apart enabling them to fit into adjacent major grooves along one face of a B-form helix.

of bound L-Trp, the relative separation of the HTH motifs in each monomer of the TrpR dimer is significantly less than the distance between successive major grooves along one face of a B-DNA duplex. In this conformation the repressor simply will not fit into the operator site. Binding of L-Trp triggers a coordinated conformational change within the molecule leading to an increase in the relative separation of the HTH motifs, enabling the molecule to fit into the DNA target, increasing the affinity for DNA by a factor of roughly a thousand (Fig. 10). Thus the physiological activity of TrpR is controlled by an induced fit mechanism. NMR data suggest that the conformation of apo-TrpR in solution is also somewhat denatured in the region of the recognition helix, which would also lower the affinity for the operator site.

As well as providing an explanation for physiological coupling of operator binding, the TrpR system also extended our understanding of DNA recognition in two additional ways. The crystal structures in this system were determined to very high resolution for macromolecular complexes (1.8 Å). This allowed the vital role played by water molecules to be observed clearly for the first time. Water molecules are ideal hydrogen bonding ligands and if two of their three possible sites for interaction are saturated the result is a uniquely orientated third position able to make a specific contact with another ligand. In effect, water molecules on the surfaces of proteins or nucleic acid molecules represent an extension to the recognition interface. This is well illustrated in the case of TrpR, where the operator–repressor complex crystal structure revealed that essentially all of the sequence specificity of the interaction is achieved by water-mediated contacts. Thus, the interaction is very sensitive to the

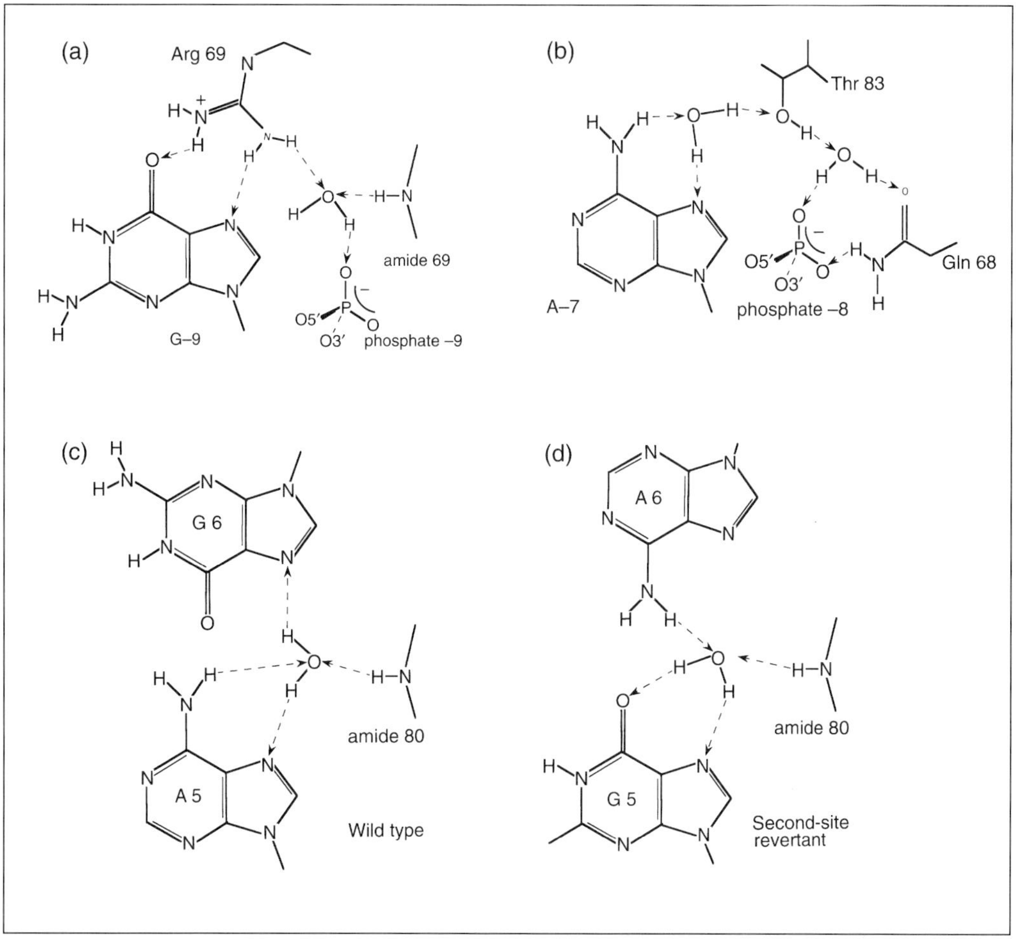

Fig. 11 The role of water in DNA binding. The role of water molecules in specifying particular nucleotide sequences in the TrpR:operator complex is illustrated. Hydrogen bonds are shown as arrows pointing away from donor groups. In (a) the water participates with an amino acid side-chain to contact a base, a situation reminiscent of the λ*cI* complex (see Fig. 9). In (b) a pair of water molecules form a network of contacts that probably specify adenine. In (c) a water molecule is contacted by two bases and the repressor backbone. Mutation of these bases results in large reductions of repressor affinity, except when the order of the purines is reversed allowing a similar set of contacts to be made (59). Figure based on Reference 12.

precise conformation of the DNA fragment and this led to the proposal of 'indirect readout' described above. Figure 2.11 shows some of the water-mediated contacts which have been identified and confirmed by mutagenesis experiments (59). Reference to other high resolution X-ray crystal structures of protein–nucleic acid complexes suggests that water is commonly present at the interfaces, although its importance in TrpR is extreme. In general, water molecules cannot be placed accurately in electron density maps unless the resolution is below 2.8 Å, and so structures at lower resolution may be missing important water-mediated contacts.

A second important feature of the TrpR system is that it binds to its operator sites in tandem, allowing protein–protein contacts between neighbouring TrpR dimers to produce cooperativity with respect to protein concentration (8). This was first postulated on the basis of the sequence repeats within the natural *trp* operator sites (60, 61) and various footprinting experiments. It has subsequently been confirmed by completion of a crystal structure for the tandemly bound complex. The major difference between this structure and those with a single dimer bound lies in the conformation of the N-terminal arm of the protein, which is disordered in solution and in the single complexes. In the tandemly bound complex, the arms form a stabilizing protein–protein interface, thus explaining the data showing that the arms were responsible for a 50-fold increase in affinity (62).

MetJ: The *E. coli* methionine repressor, MetJ, provides an interesting comparison to the TrpR system. Like TrpR it is a small dimeric molecule in solution (Mol. Wt. 11 999 Da), which is activated for operator binding by the non-cooperative binding of a small co-repressor, in this case *S*-adenosyl methionine, AdoMet (63–65). *Met* operator sites contain tandem repetitions of an 8 bp recognition target, the met box (consensus: 5′-dAGACGTCT-3′), and repressors bind cooperatively to such sites with respect to protein concentration; the minimum stable complex containing at least two repressor dimers (60). The consensus met box contains a 50 per cent identity to the consensus *trp* operator repeats. Indeed, the tandem binding of TrpR to its operator sites was first predicted by analogy to the *met* repressor system (61, 66).

Although there are these similarities between MetJ and TrpR, the molecular details of the interactions with their operators and the nature of the co-repressor effect in each case are very different. MetJ belongs to a different structural class of DNA-binding proteins, which contain a β-ribbon–helix–helix (RHH) tertiary fold, the β-ribbon being the motif which inserts into the major groove of the DNA duplex (36, 67–69). Amino acid side-chains from residues within this ribbon (Lys23 and Thr25) make sequence-specific hydrogen bonds with the edges of operator base pairs, an example of direct readout (Fig. 12).

Further interaction specificity is achieved due to sequence-dependent distortions of the DNA duplex from the standard B-form conformation, examples of indirect readout (68). Thus at the centre of each 8 bp met box, i.e. at the C.G dinucleotide step, the helical axis bends roughly 25° towards the repressor, which has the result of significantly increasing the structural complementarity between the two molecular surfaces and ensuring that similar contacts can be made to the operator by both protein subunits within a dimer. A second sequence-dependent distortion occurs at the junction between adjoining met boxes, namely at the sequence 5′-CTAG-3′ (Fig. 13). Pyrimidine–purine base-pair steps were predicted to be conformationally flexible due to weak stacking interactions between neighbouring bases along one strand of the DNA. In consequence, the bases tend to stack preferentially with their 5′ or 3′ neighbours creating a site which is over-twisted relative to B-form duplexes. In *met* operators, the helical twist at the T.A step is 44° as opposed to 36° for canonical B-duplex. The result of this over-twist is that the phosphodiester 3′ to the purine base,

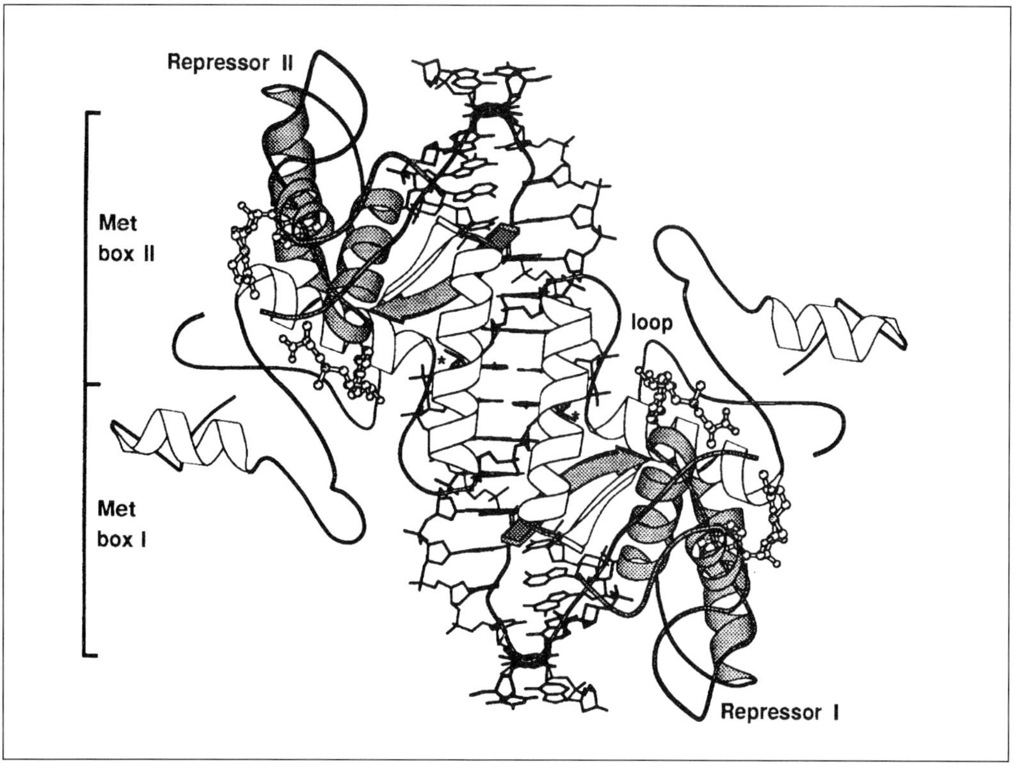

Fig. 12 Structure of the MetJ:operator complex. Cartoon representation of the X-ray crystal structure of the minimal MetJ repression complex (68). MetJ dimers are shown as ribbon representations, the two met box operator as a framework model, and AdoMet as ball and stick models.

i.e. between the A.G dinucleotide, becomes displaced outwards by ~2 Å. The protein makes a number of hydrogen bond contacts to this displaced phosphate, thus indirectly checking the sequence at the junction between met boxes. Interestingly, a recent crystal structure of a DNA oligonucleotide encompassing the CTAG sequence showed the same phosphate displacement in the absence of any bound protein (70), implying that in this case the recognition is more lock and key than induced fit. Reversing the order of the central dinucleotide between met boxes results in a fall in apparent affinity by over threefold (71), presumably reflecting the increased free energy cost of making the same conformational changes in a non-pyrimidine–purine site. The sequences of natural *met* operators vary considerably and there are relatively few examples of T.A steps at this point, however, the vast majority are pyrimidine–purine.

Crystal structures are available for apo-MetJ, the holo complex with AdoMet bound (67), and for the ternary complex with two consensus met box DNA operators (68). In all of these, there are no obvious conformational changes in the protein which could account for the observed co-repressor effect. Indeed, the AdoMet molecules are bound to the opposite faces of the repressor proteins from the face which binds DNA (Fig. 12). This suggested that the co-repressor effect is not caused by allosteric changes in protein conformation like in TrpR, although it should be noted that the MetJ:operator crystal structure is only available at 2.8 Å resolution and it may therefore not be revealing important roles for water molecules in stabilizing the interaction.

There are several possible explanations for the co-repressor effect without conformational change (72). The one which is currently favoured is based on the electrostatic effects of binding two AdoMet molecules per protein dimer. AdoMet contains a tertiary sulphur atom and hence carries a net positive charge. Experiments with the uncharged but almost isostructural adenosyl homocysteine suggest that it is the charge that is critical for the co-repressor effect. Calculation of the electrostatic potential on the DNA-binding surface, plus and minus bound AdoMet (72), suggested that in the presence of the positive charges there was a positive potential which would overlap the phosphodiester backbone of the operator, thus leading to favourable electrostatic interactions. This potential collapses in the absence of positively charged AdoMet molecules. Support for this 'electric genetic switch' model comes from direct measurements of the affinity of several repressor complexes carrying AdoMet homologues with titratable functional groups (73). These results are consistent with other systems in which it appears that long range electrostatic effects can mediate DNA–protein interactions. In particular, a number of 'super-repressors' have been isolated in TrpR (74) and lambda *cI* (75), which all involve amino acid substitutions to more basic residues at sites distal to the DNA-binding face .

As mentioned above, the 8-bp met box consensus sequence shares 50 per cent identity with a similar 8-bp repeat in *trp* operators (60, 61). Operator sequence variation and *in-vitro* selection experiments suggest that MetJ discriminates more strongly against non-consensus operators with T in position 2, i.e. the TrpR consensus base (76). This suggests very strongly that *in vivo* the two systems have evolved in each others presence and have adapted to minimize cross-talk between operators. The range of DNA sequences that can be contacted specifically by proteins of the RHH motif has been studied intensively, especially in the phage P22 proteins Mnt and Arc (77). Model building has been used to predict the effects of mutational changes at the interacting side-chains in the β-ribbon motif (78). However, *in vitro*-selection experiments with MetJ mutants suggest that, as with other DNA-binding motifs, there is no simple set of rules for defining DNA target sequences on the basis of protein sequences. Indeed, if the interacting amino acid residues have long side-chains, such as arginine, it appears that they can make contact with different base pairs depending on the flanking sequence context (79).

LacI: *E. coli* lactose repressor, LacI, provides yet another form of cooperative interaction. The repressor subunits are considerably larger than those of MetJ and TrpR, encompassing 360 amino acids, and the apo-repressor is the active form of the protein. Derepression (induction) is achieved by binding allolactose (or IPTG), the

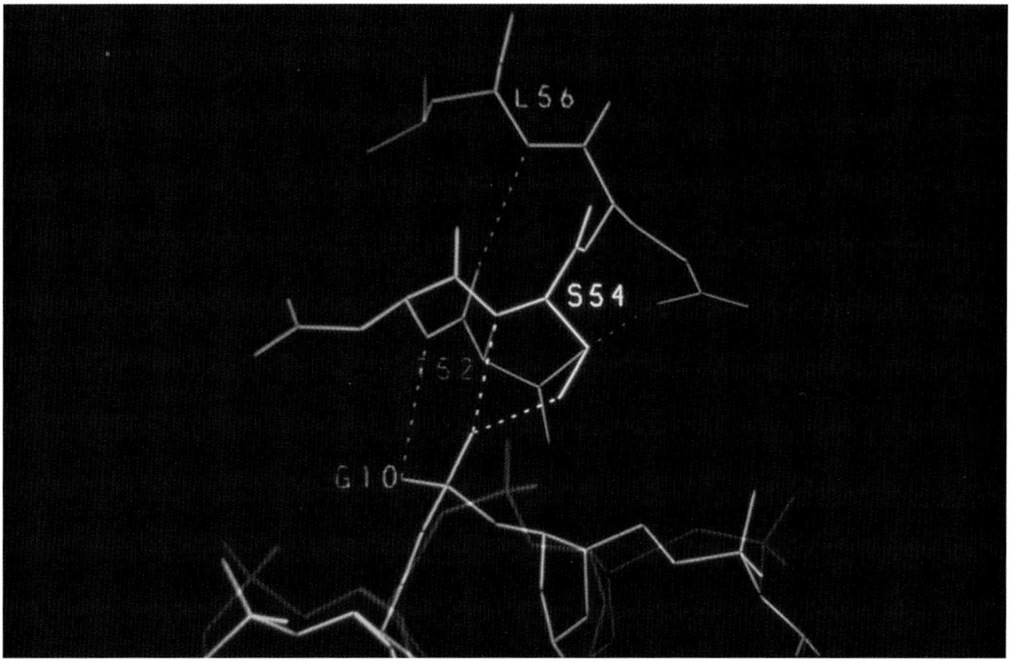

Fig. 13 Indirect readout in the MetJ:operator complex. Molecular graphics representation of the DNA–protein contacts at the centre of a minimal MetJ repression complex (Fig. 12). The sequence at the junction of the two met boxes (5'-CTAG-3') is distorted due to the over-twisting of the T.A dinucleotide step. This results in displacement of an adjacent phosphate group (green model) by 2 Å from its expected position in a B-form duplex (red model). The displaced phosphate is contacted by a series of hydrogen bonds from the N-terminus of the B helix in the repressor (blue model), which forms a rigid interaction surface unable to reach the undistorted phosphate. Thus, the protein specifies T.A (or pyrimidine;purine) without contacting the bases directly.

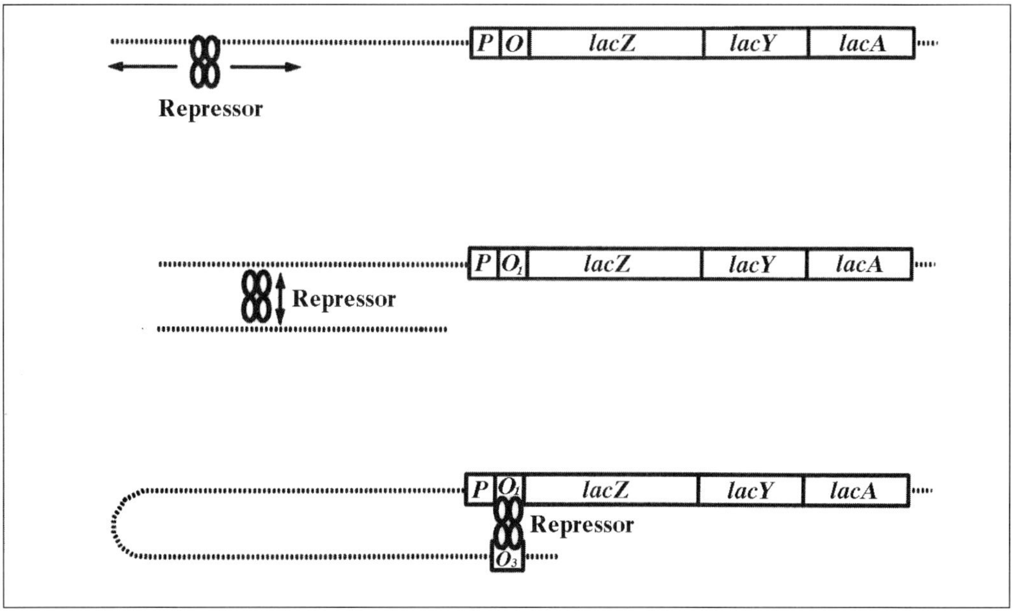

Fig. 14 Formation of higher-order complexes by the LacI repressor. The tetrameric repressor is shown as ovals, the *lac* promoter and primary operator by *P* and O_1, respectively, the upstream minor operator by O_3, and the structural gene as *lacZ*.

holo-repressor having a significantly lower affinity for operator sites. The repressor exists largely as a tetramer in solution at physiological concentrations, with each subunit containing an N-terminal HTH DNA-binding domain (80, 81). Operator sites contain dyad symmetric sequences which can be bound by just two of the available HTH motifs, allowing the other two to interact with the DNA at a distal secondary operator site, thus forming a repression loop (Fig. 14). The *lac* operator contains a primary site (highest affinity), situated in the 5′ promoter region of the operon, and two secondary sites (lower affinity) situated both up- and down-stream of the primary site (82). Thus three distinct looped complexes can be formed at this operator when repressor binds. Müller-Hill's group have shown that all three loops are required to form, in proportion to their respective affinities, in order to generate the wild-type levels of repression at the operon (83). As expected this depends on the multimerization state of the protein, and mutants which produce dimers are less effective repressors. (Note looped complexes can escape detection in the standard nitrocellulose filter-binding assay, presumably because they present primarily DNA surfaces to the filter (84).)

Although this system has been a paradigm for transcriptional repression for decades, it is only very recently that the X-ray crystal structure of the operator-bound form of the protein has been determined (85, 86). The large protein subunits are divided into five distinct regions; the N-terminal DNA-binding domains, a hinge helix, a core region containing two subdomains, and a C-terminal multimerization

Fig. 15 The X-ray structure of the *lac* repressor and its complex with DNA.Three-dimensional structure of the LacI protein (ribbons). The tetramer is bound to two separate DNA fragments (stick models) with each dimeric unit tethered to its neighbour via the C-helical bundle (taken from [86]).

helix (Fig. 15). The functional tetramer is a dimer of two tighter dimers, interacting extensively via the core region. Tetramer formation is mediated by formation of a four helix bundle by the C-terminal helices. The inducer binding site lies between the core subdomains and appears to operate by an allosteric mechanism. In a reverse of the situation in TrpR, IPTG binding results in increased separation of the amino-terminal subdomain of the core, which in turn increases the relative separation of the DNA-binding domains, resulting in the HTH motifs being too far apart to interact simultaneously with the operator half-sites. This mechanism is supported by the mapping of over 4000 single amino acid substitutions onto the three-dimensional structure (87, 86).

The overall DNA affinity is the result of interactions at both sets of DNA-binding modules, i.e. the two HTH motifs of a dimer. Effects of binding to DNA at a second site are seen even if the DNA is not an operator. Thus it appears that the binding energy from non-sequence-specifically bound DNA can make a positive contribution to the overall interaction energy (88). It also leads to effects on affinity when multiple proteins are binding, since looped complexes require DNA to bend. Proteins which bind and produce the appropriate DNA bend to be made can facilitate a second series of DNA–protein interactions, in effect making the two events appear cooperative.

Finally, it must be borne in mind that small molecular weight effectors (co-repressors/activators) can have effects on the cooperativity of binding. Thus, the TrpR apo-repressor binds two molecules of L-tryptophan, apparently non-cooperatively, resulting in both an increased stability and relative separation of the HTH DNA-binding motifs in the dimer. This concerted conformational change allows the HTH motifs to slot into adjoining major grooves of the DNA, resulting in a significant increase in the affinity for operator. Similar effects are seen in reverse with allolactose (or its analogues such as IPTG) and the LacI repressor. Effector molecules can also function by altering the position of the multimerization equilibrium in oligomeric systems, e.g. the hexameric arginine repressors (89).

A diagnostic test for cooperative behaviour is that binding reaches saturation at lower ligand concentrations than would be expected for a simple system. Carey (90) has shown that if a binding curve goes from 10 to 90 per cent saturation for a log concentration change of the varying ligand of 1.9 or less, the system under study is exhibiting positive cooperativity with respect to ligand binding. More sophisticated transformations of binding data are often used to construct Scatchard or Hill plots (91). Care should be exercised with such analyses, however, since they can easily be misinterpreted. Non-linear Scatchard plots in particular are often used to support interpretations of cooperative interactions. However, it is often difficult to obtain binding data over a wide enough range of ligand concentrations to assess the linearity or otherwise of such plots.

2.2 Sequence-specific RNA–protein interactions

2.2.1 tRNA synthetase complexes

A key step in the faithful decoding of the genetic message is the correct charging of tRNA molecules by their cognate synthetase enzymes, and their interactions have long been studied as examples of RNA–protein sequence specificity (92). Exquisite discrimination is possible, indeed is genetically essential, even though the RNA ligands in these complexes have highly conserved three-dimensional structures. Although tRNAs vary from 75 to 90 nt in length they all fold into a similar clover-leaf secondary structure, which in turn is folded to form the {L}-shaped tertiary structure having the anti-codon loop at one end of the {L} and the amino–acyl acceptor stem at the other (93–95). Presumably, this conserved fold is required for the similar interactions they must all make with the tRNA processing enzymes and the ribosomal components. Within this structural motif the tRNA synthetases must be able to identify the precise tRNA sequence such that each anti-codon uniquely encodes a single amino acid.

The folded tRNA motif illustrates the complexity and flexibility of RNA structure in general and the many differences from duplex DNA. As well as the standard four nucleotide bases, tRNAs contain large numbers of variant bases, some of which are complex modifications of the standard groupings. Base pairing and stacking throughout the molecule contribute to the fold but non-Watson–Crick base pairs, such as purine–purine, Hoogsteen, and reverse Hoogsteen pairs are common. There are also examples of triple base interactions. The overall structure is also stabilized by binding to a number of divalent metal ions (1). Interestingly, despite their frequency and sophistication there appears to be no absolutely critical role(s) for the modified bases and they can be replaced *in vitro* by standard residues with only minor effects on translational efficiency and tRNA charging. It may be that the modifications exist to help prevent misrecognition by non-cognate synthetases rather than promote cognate interactions (96), or to provide additional stability to the folded structure (97). Extensive sequence variation experiments have been carried out *in vitro* to isolate sequence-specificity determinants. These have characterized a set of identity determinants amongst the tRNAs. There appears to be no consistent set of rules for specifying identities, rather the interactions with synthetases can be thought of as idiosyncratic. Identity elements range from sections of the anti-codon, both loop and stem through the 'elbow' of the {L} into the acceptor stem. Figure 2.16 shows an example of this complexity in $tRNA^{Gln}$.

tRNA synthetases, perhaps reflecting their need for precise discrimination of the RNA targets, are a very diverse group of enzymes, having a range of quaternary structures from dimers to tetramers (98). Broadly, however, they fall into two structural groups, Class I and Class II (99, 100). Class I enzymes have two characteristic sequence motifs, KMSKS (one letter code) and HIGH, and also contain the conserved nucleotide binding fold, the Rossman fold. Class II proteins do not have the Rossman fold but do have an anti-parallel β-sheet. Figure 2.17 shows a schematic

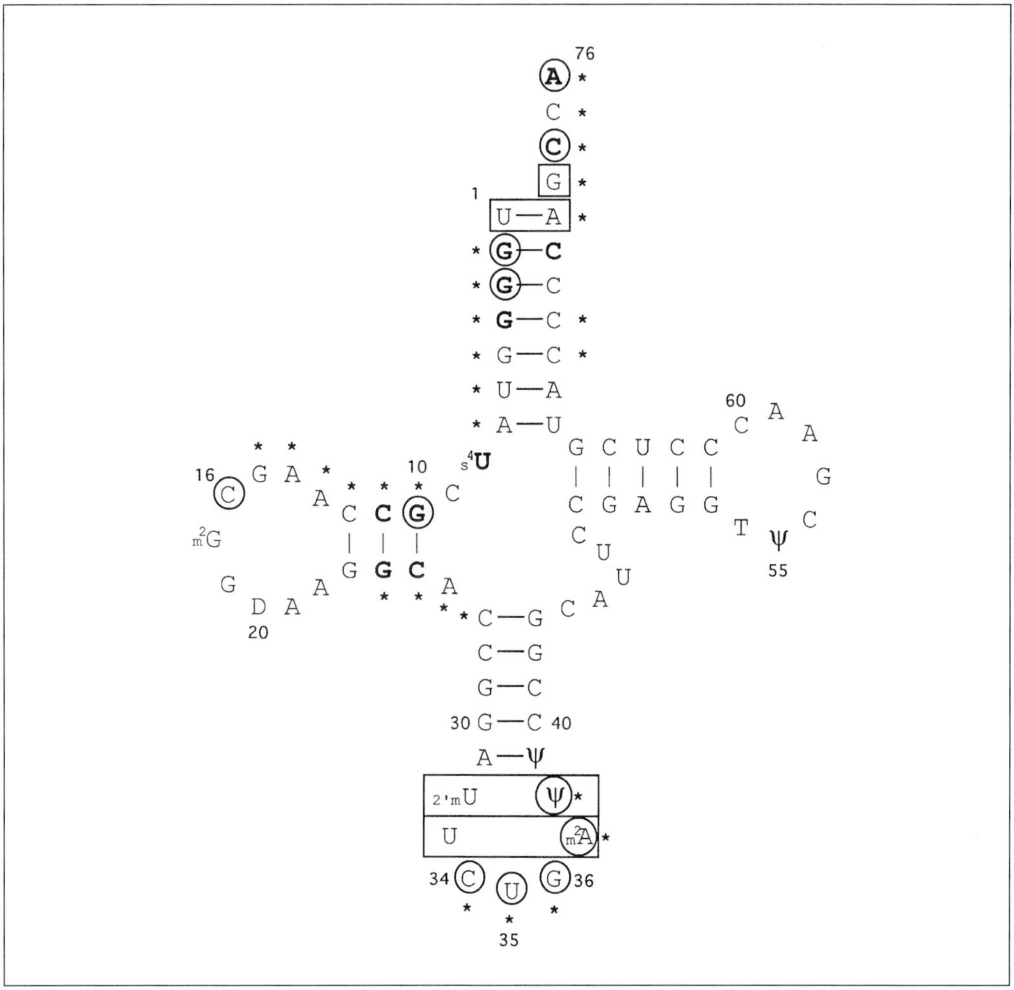

Fig. 16 tRNA synthetase:tRNA interactions. The figure shows a clover-leaf representation of tRNA^{Gln}. Bases in the molecule interacting directly with the cognate synthetase are circled and those making water-mediated contacts are shown in bold. Backbone interactions are shown as *. Taken from Reference 92.

of these two general folds. There are also differences in their molecular mechanisms of action etc., but here we will concentrate on their RNA recognition abilities.

The glutaminyl tRNA synthetase:tRNA^{Gln} complex: This complex from *E. coli* has been reported as a 2.5-Å resolution crystal structure (101). The enzyme is a Class I synthetase, being a monomer of over 550 amino acids (Fig. 17). There are two major protein domains; the active site containing the Rossman fold and an anti-codon binding region comprising two β-barrel structures. The tRNA lies along the surface of the protein with its variable loop on the outside of the complex. There is an

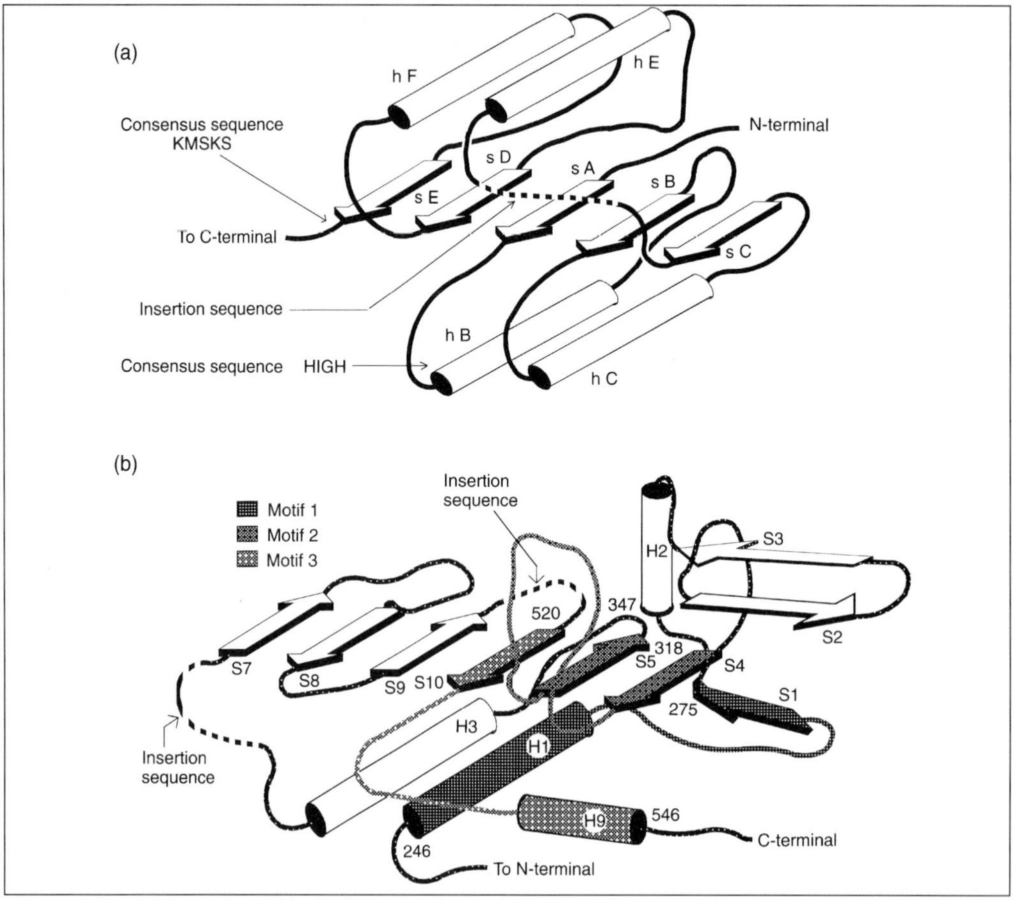

Fig. 17 Class I and Class II tRNA synthetases. The figure shows schematic representations of the tertiary folding of Class I (a) and Class II (b) tRNA synthetases. Taken from Reference 92.

extended RNA–protein interface as suggested by the number of contacted residues (Fig. 16), and some 2700 Å^2 of surface become buried on complex formation, similar to the upper values reported for DNA–protein complexes. There are a large number of direct and water-mediated base contacts with the protein. There are also contacts throughout the ribose–phosphate backbone of the RNA. Interaction with the anti-codon, which contains many identity elements in this case, appears particularly intense. Each base of the anti-codon is bound by a separate pocket on the protein. The anti-codon stem is extended by the formation of non-Watson–Crick base pairs (Figs 16 and 18). Base binding pockets consist of five or six amino acid residues, one of which is positively charged and makes a contact to the adjacent RNA phosphate group. The bases are recognized by direct hydrogen bonding and there is close to saturation coverage of the available hydrogen bonding sites. This is distinct from the

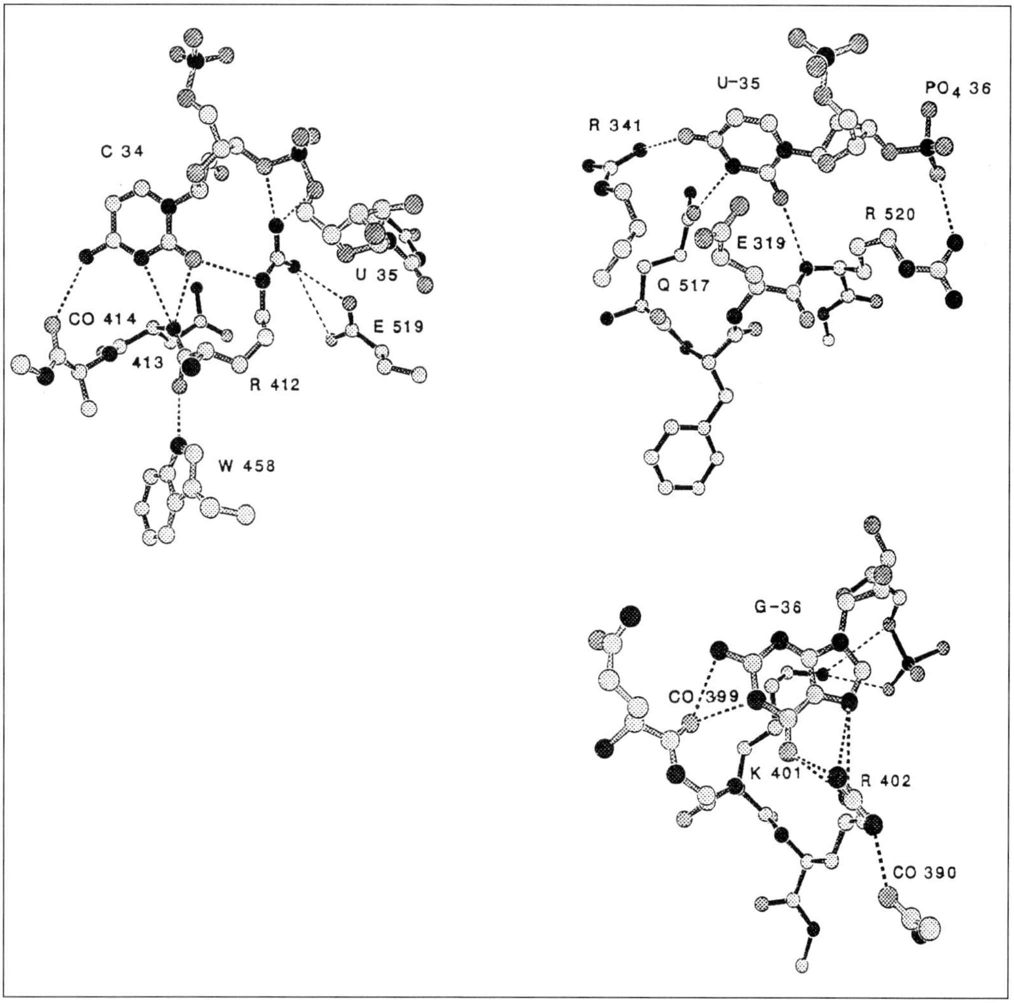

Fig. 18 Recognition of the anti-codon. Ball and stick models showing the RNA–protein interactions at the $tRNA^{Gln}$ anti-codon. Hydrogen bonds are shown as dotted lines. There is essentially saturation coverage of the hydrogen bond donors and acceptors at each base. Letters and numbers refer to nucleotide and amino acid identities and positions, respectively.

types of interaction discussed for DNA-binding proteins earlier, where only a sub-set of all possible contacts are made.

The aspartyl tRNA synthetase:tRNA^{Asp} complex: This is an example of a Class II protein, which forms and binds two tRNA molecules as a dimer (Fig. 19). Again each subunit is divided into two domains, one responsible for binding the acceptor stem and one the anti-codon. A major contrast with the glutaminyl/Class I systems is the orientation of the tRNA, which binds with its variable loop towards the protein. Sequence-specific base contacts are also made in the major groove rather than the minor

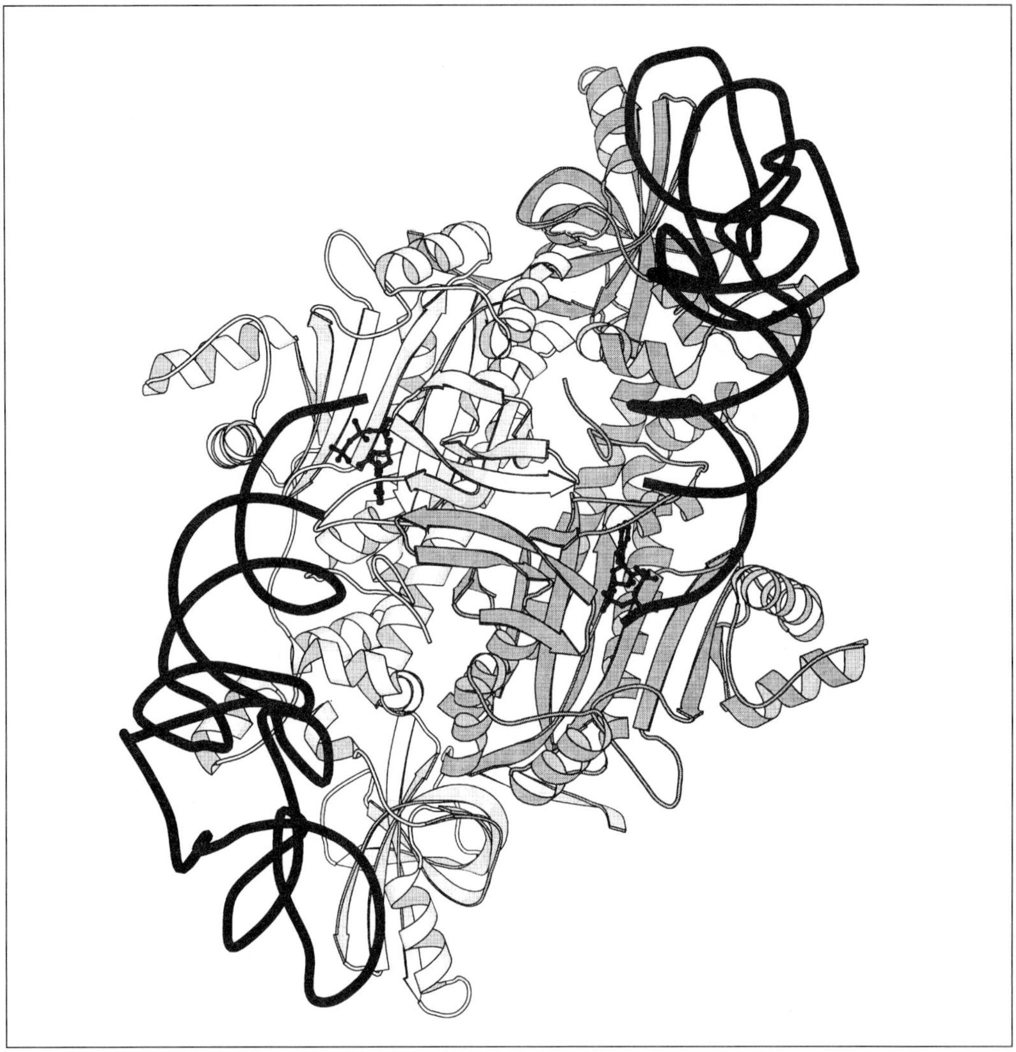

Fig. 19 Aspartyl tRNA synthetase complex. The diagram shows the tRNA complex formed with aspartyl tRNA synthetase, a dimeric, Class II enzyme. The protein subunits are shown as ribbons with one monomer shaded. The tRNAs are shown as continuous black ribbons. Note that the tRNA binds in the opposite orientation with respect to the protein compared to the $tRNA^{Gln}$ shown above (Fig. 16). Taken from Reference 92.

groove. Once again there is intensive interaction in the anti-codon loop with many of the hydrogen bonding sites saturated, although interactions occur with the surface of the protein rather than in individual binding pockets. The unstacked conformation of the anti-codon–protein complex is stabilized by an intra-RNA hydrogen bond contact from an exo-cyclic amino group to a phosphate, a similar interaction is seen in the RNA phage operators (see below).

For both types of tRNA–protein complex it is possible to compare the structures of the bound and free (tRNA Phe is the comparator for tRNA Gln) forms of the RNA ligand. This suggests that RNA conformational flexibility is a major component to the interactions. Base pairs are made or broken by complex formation and there are major rearrangements in the anti-codon loops to allow the intensive recognition interactions to occur.

2.2.2 RNA phage translational repression complex

The translational repression complex which forms between bacteriophage MS2 coat proteins (and their close relatives) and an RNA stem-loop of 19 nt (Fig. 20) has long been the paradigm for sequence-specific recognition of simple RNA target sites (102, 103). The phage RNA is the infective species, and serves as a polycistronic mRNA once inside target bacteria. However, the requirements for the differing translational products are vastly different, and the phage has evolved elegant regulatory mechanisms to ensure that protein production is closely tailored to these requirements. Some ten minutes post infection there is a switch in gene expression from 'early' to 'late', signalled by cessation of translation of the replicase subunit and self-assembly of progeny phage particles. These are subsequently released from the cell by the action of a phage encoded lysis protein. Replicase translation is controlled by sequence-specific binding of coat protein subunits to the intergenic region between the 3′ end of the coat protein subunit and the 5′ end of the replicase cistron. This sequesters both the Shine–Delgano sequence and the initiation codon (Fig. 20) of the

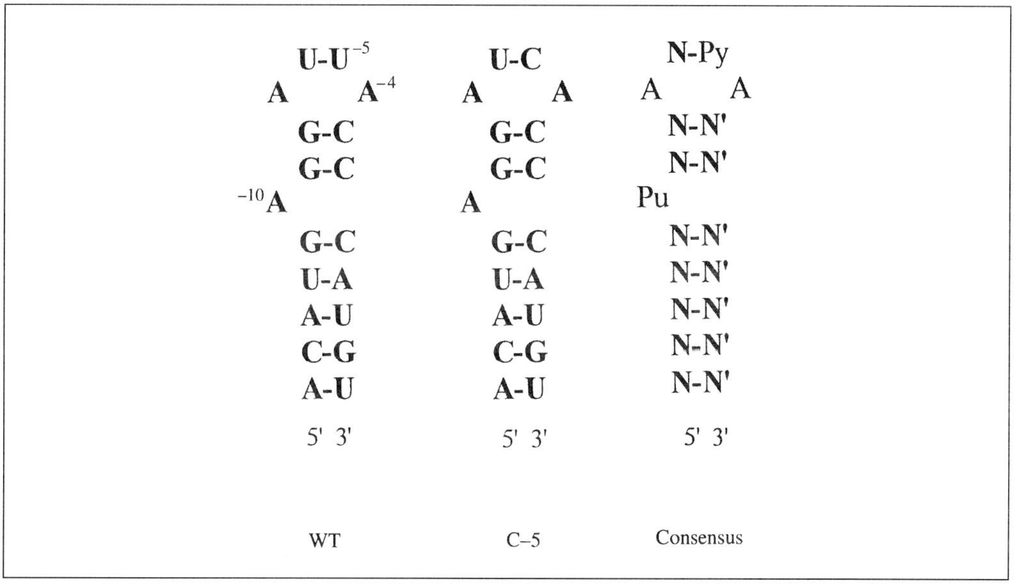

Fig. 20 Sequence of the RNA bacteriophage translational operator. Secondary structures of the MS2 operator sequences are shown. The consensus is shown, together with sequences for wild-type (WT) and the C–5 variant. Numbering is relative to the start of the replicase cistron for MS2. Py = pyrimidine, Pu = purine, and N–N′ represents any complementary Watson–Crick base pair.

replicase; an example of translational repression. The RNA–coat protein complex formed also marks the RNA for specific self-assembly into new phage capsids by serving as an assembly initiator.

Our understanding of this interaction at the molecular level has recently made major advances due to the determination by X-ray crystallography of three-dimensional structures for the wild-type, *T*=*3* phage particle (104, 105); RNA-free capsids, produced by over-expression of a recombinant coat protein gene in the absence of genomic phage RNA (106); a non-assembling coat protein dimer (107); and, most importantly, the structures of several operator capsids in which the 19 nt operator fragment has been soaked into crystals of RNA-free capsids, where it makes sequence-specific contacts to coat protein dimers in the structure (40, 41). Solution studies with chemically variant oligonucleotides, in which specific functional groups involved in the RNA–protein interaction have been substituted, leave little doubt that the complexes seen in the capsid crystals are representative of the repression complex which forms initially in solution between a coat protein dimer and the RNA operator (108–110). There is also very good correlation between the intermolecular contacts seen in the operator capsid crystals (Fig. 21) and coat protein mutations which give rise to reduced RNA binding (111–113).

The canonical operator fragment consists of a seven base-pair stem, interrupted by a bulged adenine at position –10 (see Fig. 20 for numbering system), closed at one end by a four base single-stranded loop. Extensive sequence variation experiments (102) showed that the identity of only four nucleotide positions of the 19-residue fragment were important for coat protein recognition. These are the adenines at positions –4, –7 and, –10 and the pyrimidine at –5. Interaction with the coat protein is preserved provided that the Watson–Crick base pairing of the stem is maintained and recognition is insensitive to the identity of the base at –6. Solution studies suggested that all three important adenines were recognized differentially (108, 109), and this has been confirmed by the X-ray structure of the operator complex (40, 41). The orientation of the RNA at A/B dimers within the *T*=*3* capsid is unique and allows the molecular details of the interaction to be modelled in the resulting electron density maps at 2.7–3.0 Å (Fig. 21).

Adenine residues at –4 and –10 interact with the protein in hydrophobic pockets in a roughly twofold symmetric arrangement across each coat protein dimer. The orientation of the base and the functional groups contacted at each site differ. Remarkably, the base at –7 is not contacted directly but is involved in an extended stacking interaction with the pyrimidine at –5, which in turn is stacked against Tyr85 from the protein. Solution structures of the RNA operator calculated from NMR data (114) suggest that the species bound by the protein is at best only a minor component of a set of conformers in equilibrium, confirming earlier predictions based on chemical reactivity of the various functional groups (115). It is clear that complex formation with coat protein must involve significant conformational changes from the dominant form of the RNA fragment in solution in which the A–10 is intercalated in the base-paired stem. Operator variants having a cytidine at position –5 (C–5) bind the coat protein much more tightly than the wild-type sequence. The X-ray crystal

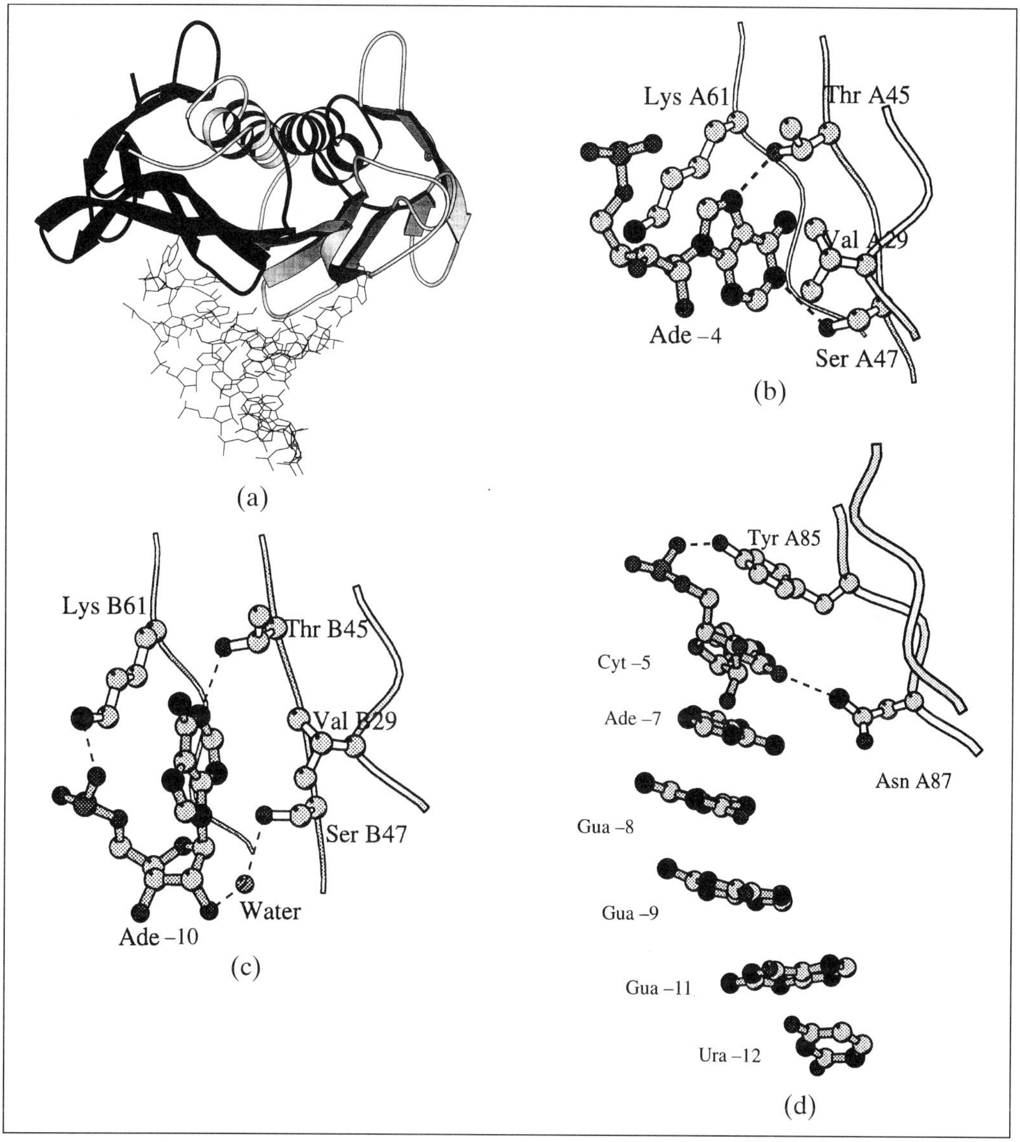

Fig. 21 The structure of the MS2 translational repression complex. (a) Structure of the operator (framework model) bound to an A/B coat protein dimer (ribbons) within a capsid. Note that the RNA forms a crescent which is complementary to the convex surface of the β-sheet in the protein. (b) The detailed sequence-specific RNA–protein contacts at the A–4 binding pocket, hydrogen bonds are shown as dotted lines. (c) The equivalent interaction to (b) on the B subunit with A–10. Note that although the adenine binding pocket consists of the same amino acids, the orientation of the base is different from A–4 and hence the detailed hydrogen bonding is different. (d) The roles of the –5 pyrimidine and A–7 are indicated. A–7 is not contacted by the protein directly but participates in an extended base stacking contact with G–8 and C–5, which in turn is stacked against Tyr A85 on the protein. G is too large to occupy the A–7 position and so A is specified indirectly. C–5 both stacks on the tyrosine and makes a hydrogen bond from O2 to the side-chain of Asn A87. This contact does not discriminate between U and C but the amino group of C makes an additional intra-RNA contact to the phosphate backbone (not shown).

structure of this operator variant suggests that this is due to an additional intra-RNA hydrogen bond between the exo-cyclic amino group of C–5 and the phosphodiester backbone, presumably stabilizing the conformers in solution closer to those bound by protein (41).

Recently, the first X-ray crystal structures of RNA aptamers selected by binding to the coat protein have been determined (116–117). These include fragments which would not have been expected to be tight binders on the basis of the previous operator variation experiments. In particular, a fragment with a three nucleotide loop and a three base gap to the bulged adenosine in the stem; a fragment lacking the bulged adenosine; and a further fragment lacking a fully base-paired stem all bind to the same site on the protein (Fig. 21). These studies emphasize the role of proteins in determining the RNA conformation which is bound and suggest that it will be even more difficult to define rules for sequence recognition in the RNA field than with DNA.

Postscript

We have discussed a few examples of protein–DNA and protein–RNA complexes which regulate gene expression in prokaryotes. Progress in this area in structure determination is so rapid that we could have chosen to illustrate a number of other systems. However, whilst each system has idiosyncratic features, there are many aspects of each intermolecular complex which seem familiar because they are simply variations on themes seen already elsewhere. Of these, the modularity of the interactions, i.e. the use of only domains of proteins or RNA molecules, appears a common trend. It is also clear that interactions are multivalent and that overall complex stability depends on a very large number of individual intermolecular contacts. This in turn makes predicting the nature of an interacting partner from the structure or sequence of one half of a complex difficult, if not impossible. However, the wealth of information likely to emerge in the next few years may change this situation for the better. It is clear that protein–nucleic acid interactions will remain a paradigm for molecular recognition studies and will continue to teach us how molecules take life and death decisions for the cells which produce them.

Acknowledgements

I am grateful to Prof. Simon Phillips (Leeds) and Dr Lars Liljas (Uppsala) for their assistance in the preparation of many of the figures shown in this chapter. Work in the author's laboratory on the MS2 bacteriophage and the *E. coli* MetJ repressor is funded by the UK BBSRC, The Wellcome Trust, and The Leverhulme Trust.

References

1. Saenger, W. (1984) *Principles of nucleic acid structure*. Springer-Verlag, New York.
2. Blackburn, G. M. and Gait, M. J. (1990) *Nucleic acids in chemistry and biology*. IRL Press, OUP, Oxford.

3. Calladine, C. R., Drew, H. R., and McCall, M. J. (1988) The intrinsic curvature of DNA in solution. *J. Mol. Biol.*, **210**, 127–137.
4. Dickerson, R. E., Drew, H. R., Conner, B. N., Wing, R. M., Fratini, A. V., and Kopka, M. L. (1982) The anatomy of A-, B- and Z-DNA. *Science*, **216**, 475–485.
5. Ha, J. H., Spolar, R. S., and Record, M. T. (1989) Role of the hydrophobic effect in stability of site-specific protein–DNA complexes. *J. Mol. Biol.*, **209**, 801–816.
6. Spolar, R. S. and Record, M. T. (1994) Coupling of local folding to site-specific binding of proteins to DNA. *Science*,. **263**, 777–784.
7. Luisi, B. F., Xu, W. X., Otwinowski, Z., Freedman, L. P., Yamamoto, K. R., and Sigler, P. B. (1991) Crystallographic analysis of the interaction of the glucocortocoid receptor with DNA. *Nature*, **352**, 497–505.
8. Lawson, C. L. and Carey, J. (1993) Tandem binding in crystals of a trp repressor/operator half-site complex. *Nature*, **366**, 178–182.
9. Davies, D., Sheriff, S., and Padlan, E. (1988) Antibody–antigen complexes. *J. Biol. Chem.*, **263**, 10541–10544.
10. Stormo, G. D. and Fields, D. S. (1998) Specificity, free energy and information content in DNA-protein interactions. *TIBS*, **23**, 109–113.
11. Seeman, N. C., Rosenberg, J. M., and Rich, A. (1976) Sequence-specific recognition of double helical nucleic acids by proteins. *PNAS (USA)*, **73**, 804–808.
12. Lilley, D. M. J. (ed.) (1995) *DNA–protein structural interactions*. IRL Press, OUP, Oxford.
13. Steitz, T. A. (1990) Structural studies of protein–nucleic acid interaction: the sources of sequence-specific binding. *Q. Rev. Biophys.*, **23**, 205–280.
14. Otwinowski, Z., Schevitz, R. W., Zhang, R. G., Lawson, C. L., Joachimiak, A., Marmorstein, R. Q., Luisi, B. F., and Sigler, P. B. (1988) Crystal structure of *trp* repressor/operator complex at atomic resolution. *Nature*, **335**, 321–329.
15. Uhlenbeck, O. C. (1987) A small catalytic oligoribonucleotide. *Nature*, **328**, 596–600.
16. Cech, T. R. and Bass, B. L. (1986) Biological catalysis by RNA. *Annu. Rev. Biochem.*, **55**, 599–629.
17. Strobel, S. A. and Doudna, J. A. (1997) RNA seeing double: close-packing of helices in RNA tertiary structure. *TIBS*, **22**, 262- 266.
18. Draper, D. E. (1995). Protein–RNA recognition. *Annu. Rev. Biochem.*, **64**, 593–620.
19. Varani, G. (1997) RNA–protein intermolecular recognition. *Acc. Chem. Res.*, **30**, 189–195.
20. Crick, F. H. C. and Klug, A. (1975) Kinky helix. *Nature*, **255**, 530–533.
21. Nelson, H. C. M., Finch, J. T., Luisi, B. F., and Klug, A. (1987) The structure of an oligo(dA).oligo(dT) tract and its biological implications. *Nature*, **330**, 221–226.
22. Ptashne, M. (1986) *A genetic switch*. Cell Press, Cambridge, MA.
23. von Hippel, P. H. and Berg, O. G. (1989). Facilitated target location in biological systems. *J. Biol. Chem.*, **264**, 675–678.
24. McKay, D. B. and Steitz, T. A. (1981) Structure of catabolite gene activator protein at 2.9 Å resolution suggests binding to left-handed B-DNA. *Nature*, **290**, 744–749.
25. Anderson, W. F., Ohlendorf, D. H., Takeda, Y., and Matthews, B. W. (1981) Structure of the *cro* repressor from bacteriophage λ and its interaction with DNA. *Nature*, **290**, 754–758.
26. Pabo, C. O. and Lewis, M. (1982) The operator-binding domain of λ repressor: structure and DNA recognition. *Nature*, **298**, 443–447.
27. Wharton, R. P., Brown, E. L., and Ptashne, M. (1984) Substituting an α-helix switches the sequence specific DNA interactions of a repressor. *Cell*, **38**, 361–369.
28. Anderson, W. F., Takeda, Y ., Ohlendorf, D. H., and Matthews, B. W. (1982) Proposed α-helical super-secondary structure associated with protein–DNA recognition. *J. Mol .Biol.*, **159**, 745–751.

29. Harrison, S. C. (1991) A structural taxonomy of DNA-binding domains. *Nature*, **353**, 715–719.
30. Aggarwaal, A. K. and Harrison, S. C. (1990) DNA recognition by proteins with the helix-turn-helix motif. *Annu. Rev. Biochem.*, **59**, 933–969.
31. Jordan, S. R. and Pabo, C. O. (1988) Structure of the lambda complex at 2.5 Å resolution: details of the repressor-operator interactions. *Science*, **242**, 893–899.
32. Beamer, L. J. and Pabo, C. O. (1992) Refined 1.8 Å crystal structure of the lambda repressor–operator complex. *J. Mol. Biol.*, **227**, 177–196.
33. Brennan, R. G., Roderick, S. L., Takeda, Y., and Matthews, B. W. (1990). Protein–DNA conformational changes in the crystal structure of a λ Cro-operator complex. *PNAS (USA)*, **87**, 8165–8169.
34. Schultz, S. C., Shields, G. C., and Steitz, T. A. (1991). Crystal structure of a CAP–DNA complex: the DNA is bent by 90°. *Science*, **253**, 1001–1007.
35. Luisi, B. F. (1992) DNA transcription. Zinc standard for economy. *Nature*, **356**, 379–380.
36. Raumann, B. E., Brown, B. M., and Sauer, R. T., (1994) Major groove DNA recognition by β-sheets: the ribbon–helix–helix family of gene regulatory proteins. *Curr. Opinion Struct. Biol.*, **4**, 36–43.
37. Patikoglou, G. and Burley, S. K. (1997) Eukaryotic transcription factor–DNA complexes. *Annu. Rev. Biophys. Biomol. Struct.*, **26**, 289–325.
38. Wigley, D. B., Davies, G. J., Dodson, E. J., Maxwell, A., and Dodson, G. (1991) Crystal structure of an N-terminal fragment of the DNA gyrase B protein. *Nature*, **351**, 624–629.
39. Cusack, S. (1995). 11 down and 9 to go. *Nature Struct. Biol.*, **2**, 824–831.
40. Valegård, K., Murray, J. B., Stockley, P. G., Stonehouse, N. J., and Liljas, L. (1994). Crystal structure of an RNA bacteriophage coat protein–operator complex. *Nature*, **371**, 623–626.
41. Valegård, K., Murray, J. B., Stonehouse, N. J.; Van den Worm, S. H. E., Stockley, P. G., and Liljas, L. (1997). The three-dimensional structures of two complexes between recombinant MS2 capsids and RNA operator fragments reveal sequence-specific protein–RNA interactions. *J. Mol. Biol.*, **271**, 724–738.
42. Oubridge, C., Ito, N., Evans, P. R., Teo, C. H., and Nagai, K. (1994).Crystal structure at 1.92 Å resolution of the RNA-binding domain of the U1A spliceosomal protein complexed with the RNA hairpin. *Nature*, **372**, 432–438.
43. Nagai, K. (1996) RNA–protein complexes. *Curr. Op. Struct. Biol.*, **6**, 53–61 .
44. Rossmann, M. G. and Johnson, J. E. (1989) Icosahedral RNA virus structure. *Annu. Rev. Biochem.*, **58**, 533–573
45. Valegård, K., Liljas, L., Fridborg, K., and Unge, T. (1990). The three-dimensional structure of the bacterial virus MS2. *Nature*, **345**, 36–41.
46. St. Johnston, D., Brown, N. H., Gall, J. G., and Jantsch, M. (1992) A conserved double-stranded RNA-binding domain. *PNAS (USA)*, **89**, 10979–10983.
47. Bycroft, M., Grünert, S., Murzin, A. G., Proctor, M., and St. Johnston, D. (1995) NMR solution structure of a dsRNA binding domain from *Drosophila staufen* protein reveals homology to the N-terminal domain of ribosomal protein S5. *EMBO J.*, **14**, 3563–3571.
48. Kharrat, A., Macias, M. J., Gibson, T. J., Nilges, M., and Pastore, A. (1995) Structure of the dsRNA binding domain of *E. coli* RNase III. *EMBO J.*, **14**, 3572–3584.
49. Musco, G., Stier, G., Joseph, C., Castiglione Morelli, M. A., Nilges, M., Gibson, T. J., and Pastore, A. (1996) Three-dimensional structure and stability of the KH domain: molecular insights into the fragilexsyndrome. *Cell*, **85**, 237–245.
50. Nagai, K., Oubridge, C., Jessen, T. H., Li, J., and Evans, P. R. (1990) Crystal structure of the RNA-binding domain of the U1 small nuclear ribonucleoprotein A. *Nature*, **348**, 515–520.

51. Avis, J. M., Allain, F. H-T., Howe, P. W. A., Varani, G., Neuhaus, D., and Nagai, K. (1996) Solution structure of the N-terminal RNP domain of U1A protein: the role of C-terminal residues in structure stability and RNA binding., *J. Mol. Biol.*, **257**, 398–411.
52. Sauer, R. T., Jordan, S. R., and Pabo, C. O. (1990). λ repressor: A model system for understanding protein–DNA interactions and protein stability. *Adv. Prot. Chem.*, **40**, 1–61.
53. Rogers, D. W. and Harrison, S. C. (1993) The complex between phage 434 repressor DNA-binding domain and operator site OR3: structural differences between consensus and non-consensus half-sites. *Structure*, **1**, 227–240.
54. Mondragon, A. and Harrison, S. C. (1991) The phage 434 Cro/O_R1 complex at 2.5 Å resolution. *J. Mol. Biol.*, **219**, 321–334.
55. Koudelka, G. B., Harbury, P., Harrison, S. C., and Ptashne, M. (1988) DNA twisting and the affinity of bacteriophage 434 operator for bacteriophage 434 repressor. *PNAS (USA)*, **85**, 4633- 4637.
56. Koudelka, G. B. and Carlson, P. (1992) DNA twisting and the effects of non-contacted bases on affinity of 434 operator for 434 repressor. *Nature*, **355**, 886–889.
57. Schevitz, R. W., Otwinowski, Z., Joachimiak, A., Lawson, C. L., and Sigler, P. B. (1985). The three-dimensional structure of *trp* repressor. *Nature*, **317**, 782–786.
58. Zhang, R-G., Joachimiak, A., Lawson, C. L., Schevitz, R. W., Otwinowski, Z., and Sigler, P. G. (1987) The crystal structure of *trp* aporepressor at 1.8 Å shows how binding tryptophan enhances DNA affinity. *Nature*, **327**, 591–597.
59. Joachimiak, A., Haran, T. E., and Sigler, P. B. (1994) Mutagenesis supports water mediated recognition in the trp repressor–operator system. *EMBO J.*, **13**, 367–372.
60. Phillips, S. E. V., Manfield, I., Parsons, I., Davidson, B. E., Rafferty, J. B., Somers, W. S., Margarita, D., Cohen, G. N., Saint-Girons, I., and Stockley, P. G. (1989) Co-operative tandem binding of *met* repressor of *Escherichia coli. Nature*, **341**, 711–715.
61. Phillips , S. E. V. and Stockley, P. G. (1994) Similarity of *E. coli met* and *trp* repressors. *Nature*, **368**, 106.
62. Carey, J., Lewis, D. E. A., Lavoie, T. A., and Yang, J. (1991). How does *trp* repressor bind to its operator? *J. Biol. Chem.*, **266**, 24509–24513.
63. Old, I. G., Phillips, S. E. V., Stockley, P. G., and Saint Girons, I. (1991) Regulation of methionine biosynthesis in the Enterobacteriaceae. *Prog. Biophys. Molec. Biol.*, **56**, 145–185.
64. Saint-Girons, I., Belfaiza, J., Guillou, Y., Perrin, D., Guiso, N., Barzu, O., and Cohen, G. N. (1986) Interactions of the *E. coli* methionine repressor with the *metF* operator and with its co-repressor, *S*-adenosylmethionine. *J. Biol. Chem.*, **261**, 10936–10940.
65. Saint-Girons, I., Parsot, C., Zakin, M. M., Barzu, O., and Cohen, G. N. (1988) Methionine biosynthesis in Enterobacteriaceae: Biochemical, regulatory, and evolutionary aspects. *CRC Critical Reviews in Biochemistry* (CRC Press), **23**, S1-S42.
66. Phillips, S. E. V. and Stockley, P. G. (1996) Structure and function of *Escherichia coli met* repressor: similarities and contrasts with *trp* repressor. *Phil. Trans. R. Soc. London* B, **351**, 527–535.
67. Rafferty, J. B., Somers, W. S., Saint-Girons, I., and Phillips, S. E. V. (1989) Three dimensional crystal structures of the *Escherichia coli* Met repressor with and without co-repressor. *Nature*, **341**, 705–710.
68. Somers, W. S. and Phillips, S. E. V. (1992) Crystal structure of the *met* repressor–operator complex at 2.8 Å resolution: DNA recognition by β-strands. *Nature*, **359**, 387–393.
69. He, Y. Y., McNally, T., Manfield, I., Navratil, O., Old, I. G., Phillips, S. E. V., Saint-Girons, I., and Stockley, P. G. (1992). Probing *met* repressor–operator recognition in solution. *Nature*, **359**, 431–433.

70. Urpi, L., Tereshko, V., Malinina, L., Huynh-Dinh, T., and Subirana, J. A. (1996). Structural comparison between the d(CTAG) sequence in oligonucleotides and trp and met repressor–operator complexes. *Nature Struct. Biol.*, **3**, 325–328.
71. Wild, C. M., McNally, T., Phillips, S. E. V., and Stockley, P. G. (1996) Effects of systematic variation of the minimal *Escherichia coli met* consensus operator site: *In vivo* and *in vitro met* repressor binding. *Mol. Micro.*, **21**, 1125–1135.
72. Phillips, K. and Phillips, S. E. V., (1994) Electrostatic activation of *E.coli* methionine repressor. *Structure*, **2**, 309–316.
73. Parsons, I. D., Persson, B., Mekhalfia, A., Blackburn, G. M., and Stockley, P. G. (1995) Probing the molecular mechanism of action of co-repressor in the *E.coli* methionine repressor–operator complex using surface plasmon resonance. *Nucleic Acids Res.*, **23**, 211–216.
74. Klig, L. S. and Yanofsky, C. (1988) Increased binding of operator DNA by *trp* super-repressor EK49. *J. Biol. Chem.*, **263**, 243–246.
75. Nelson, H. C. M. and Sauer, R. T. (1985). Lambda repressor mutations that increase the affinity and specificity of operator binding. *Cell*, **42**, 549–558.
76. He, Y-Y., Stockley, P. G., and Gold, L. (1996) *In vitro* evolution of the DNA binding sites of *Escherichia coli* methionine repressor, MetJ. *J. Mol. Biol.*, **255**, 55–66.
77. Brown, B. M., Milla, M. E., Smith, T. L., and Sauer, R. T. (1994) Scanning mutagenesis of the Arc repressor as a functional probe of operator recognition. *Nature Struct. Biol.*, **1**, 164–168.
78. Suzuki, M. (1995). DNA recognition by a β-sheet. *Prot. Eng.*, **8**, 1–4.
79. Elworthy, S., Ellison, A. L., and Stockley, P. G. A MetJ variant has specificity for particular tandem arrays of two distinct 8 bp DNA binding sites. In preparation.
80. Miller, J. H. (1978) The *lacI* gene: its role in *lac* operon control and its use as a genetic system. In *The operon* (ed. J. H. Miller and W. S. Reznikoff). Cold Spring Harbor Laboratory, Cold Spring Harbor, New York.
81. Ogata, R. T. and Gilbert, W. (1979) DNA-binding site of lac repressor probed by dimethylsulphate methylation of *lac* operon. *J. Mol. Biol.*, **132**, 709–728.
82. Besse, M., Von Wilcken-Bergmann, B., and Müller-Hill, B. (1986). Synthetic lac operator mediates repression through lac repressor when introduced upstream and downstream from lac repressor. *EMBO J.*, **5**, 1377–1381.
83. Oehler, S., Eismann, E. R., Krämer, H., and Müller-Hill, B. (1990) The three operators of the *lac* operon cooperate in repression, *EMBO J*, **9**, 973–979.
84. Fickert, R. and Müller-Hill, B. (1992) How Lac repressor finds lac operator *in vitro*. *J. Mol. Biol.*, **226**, 59–68.
85. Lewis, M., Chang, G, Horton, N. C., Kercher, M. A., Pace, H. C., Schumacher, M. A., Brennan, R. G., and Lu, P. (1996) Crystal structure of the lactose operon repressor and its complexes with DNA and inducer. *Science*, **271**, 1247–1254.
86. Pace, H. C., Kercher, M. A., Lu, P., Markiewicz, P., Miller, J. H., Chang, G., and Lewis, M. (1997) Lac repressor genetic map in real space. *TIBS*, **22**, 334- 339.
87. Markiewicz, P., Kleina, L. G., Cruz, C., Ehret, S., and Miller, J. H. (1994) Genetic studies of the lac repressor. XIV. Analysis of 4000 altered *Escherichia coli lac* repressors reveals essential and non-essential residues, as well as 'spacers' which do not require a specific sequence. *J. Mol. Biol.*, **240**, 421–433.
88. Levandoski, M. M., Tsodikov, O. V., Frank, D. E., Melcher, S. E., Saeker, R. M., and Record, M. T. (1996) Co-operative and anti-co-operative effects in binding of the first and second plasmid Osym operators to a LacI tetramer: evidence for contributions of non-operator DNA binding by wrapping and looping. *J. Mol. Biol.*, **260**, 697–717.

89. Van Duyne, G. D., Ghosh, G., Maas, W. K., and Sigler, P. B. (1996) Structure of the oligomerization and L-arginine binding domain of the arginine repressor of *Escherichia coli*. *J. Mol. Biol.*, **256**, 377–391.
90. Carey, J. (1991). Gel retardation. *Meth. Enzymol.*, **208**, 103–117.
91. Wyman, J. and Gill, S. J. (1990) in *Binding and linkage: functional chemistry of biological macromolecules*. University Science Books, Mill Valley, California.
92. Nagai, K. and Mattaj, I. W. (1994) *RNA–protein interactions*. IRL Press, OUP, Oxford.
93. Robertus, J. D., Ladner, J. E., Finch, J. T., Rhodes, D., Brown, R. S., Clark, B. F. C., and Klug, A. (1974) Structure of yeast phenylalanine tRNA at 3 Å resolution. *Nature*, **250**, 546–551.
94. Kim, S. H., Suddath, F. L., Quigley, G. J., McPherson, A., Sussman, J. L., Wang, A. H. J., Seeman, N. C., and Rich, A. (1974). Three-dimensional tertiary structure of yeast phenylalanine transfer RNA. *Science*, **185**, 435–440.
95. Moras, D., Comarmond, M. B., Fischer, J., Thierry, J. C., Ebel, J. P., and Giegé, R. (1980). Crystal structure of $tRNA^{Asp}$. *Nature*, **288**, 669–674.
96. Muramatsu, T., Nishikawa, K., Nemoto, F., Kuchino, Y., Nishimura, S., Miyazawa, T., and Yokoyama, S. (1988). Codon and amino acid specificities of a transfer RNA are both converted by a single post-transcriptional modification. *Nature*, **336**, 179–181.
97. Sampson, J. R. and Uhlenbeck, O. C. (1988). Biochemical and physical characterisation of an unmodified yeast phenylalanine transfer RNA transcribed *in vitro*. *PNAS (USA)*, **85**, 1033–1037.
98. Schimmel, P. and Söll, D. (1979). Aminoacyl-tRNA synthetases: general features and recognition of transfer RNAs. *Ann. Rev. Biochem.*, **48**, 601–648.
99. Eriani, G., Delarue, M., Poch, O., Gangloff, J., and Moras, D. (1990). Partition of tRNA synthetases into two classes based on mutually exclusive sets of sequence motifs. *Nature*, **347**, 203–206.
100. Delarue, M. and Moras, D. (1993). The aminoacyl-tRNA synthetase family: modules at work. *BioEssays*, **15**, 675–687.
101. Rould, M. A., Perona, J. J., and Steitz, T. A. (1991). Structural basis for transfer RNA recognition by glutaminyl-tRNA synthetase. *Nature*, **352**, 213–218.
102. Witherell, G. W., Gott, J. M., and Uhlenbeck, O. C. (1991). Specific interaction between RNA phage coat proteins and RNA. *Progr. Nucl. Acid Res. Mol. Biol.*, **40**, 185–220.
103. Stockley, P. G., Stonehouse, N. J., and Valegård, K. (1994) The molecular mechanism of RNA phage morphogenesis. *Int. J. Biochem.*, **26**, 1249–1260.
104. Valegård, K., Liljas, L., Fridborg, K., and Unge, T. (1990) The three-dimensional structure of the bacterial virus MS2 capsids. *Nature*, **345**, 36–41.
105. Golmohammadi, R., Valegård, K., Fridborg, K., and Liljas, L. (1993). The refined structure of bacteriophage MS2 at 2.8 Å resolution. *J. Mol. Biol.*, **234**, 620–639.
106. Stonehouse, N. J, Valegård, K., Golmohammadi, R., van den Worm, S., Walton, C., Stockley, P. G., and Liljas, L. (1996). Crystal Structures of MS2 Capsids with mutations in the subunit FG loop. *J. Mol. Biol.*, **256**, 330–339.
107. Ni, C.-Z., Syed, R., Kodandapani, R., Wickersham, R., Peabody, D. S., and Ely, K. R. (1995). Crystal structure of the MS2 coat protein dimer: implications for RNA binding and virus assembly. *Structure*, **3**, 255–263.
108. Talbot, S. J., Goodman, S. T. S., Bates, S. R. E., Fishwick, C. W. G., and Stockley, P. G. (1990). Use of synthetic oligoribonucleotides to probe RNA–protein interactions in the MS2 translational operator complex, *Nucl. Acids Res.*, **18**, 3521–3528.
109. Stockley, P. G., Stonehouse, N. J., Murray, J. B., Goodman, S. T. S., Talbot, S. J., Adams, C.

J., Liljas, L., and Valegård, K. (1995) Probing sequence-specific RNA recognition by the bacteriophage MS2 coat protein. *Nucl. Acids Res.*, **23**, 2512–2518.

110. Baidya, N. and Uhlenbeck, O. C. (1995). The role of 2′ hydroxyl groups is an RNA–protein interaction. *Biochemistry*, **34**, 12363–12368.
111. Stockley, P. G., Stonehouse, N. J., Walton, C., Walters, D. A., Medina, G., Macedo, J. M. B., Hill, H. R., Goodman, S. T. S., Talbot, S. J., Tewary, H. K., Golmohammadi, R., Liljas, L., and Valegård, K. (1993). Molecular mechanism of RNA-phage morphogenesis. *Biochem. Soc. Trans.*, **21**, 627–633.
112. Peabody, D. S. (1993). The RNA binding site of bacteriophage MS2 coat protein. *EMBO J.* **12**, 595–600.
113. Peabody, D. S. and Lim. F. (1996). Complementation of RNA binding site mutations in MS2 coat protein heterodimers. *Nucl. Acids Res.*, **24**, 2352–2359.
114. Borer, P. N., Lin, Y., Wang, S., Roggenbuck, M. W, Gott, J. M., Uhlenbeck, O. C., and Pelezer, I. (1995). Proton NMR and structural features of a 24-nucleotide RNA hairpin. *Biochemistry*, **34**, 6488–6503.
115. Talbot, S. J., Medina, G., Fishwick, C. W. G., Haneef, I., and Stockley, P. G. (1991) Hyper-reactivity of adenines and conformational flexibility of a translational repression site. *FEBS Letts.*, **283**, 159–164.
116. Convery, M. A., Rowsell, S., Stonehouse, N. J., Ellington, A. D., Hirao, I., Murray, J. B., Peabody, D. S., Phillips ,S. E. V., and Stockley, P. G. (1998) The crystal structure of an RNA aptamer protein complex to 2.8 Å resolution: defining the rules of RNA–protein recognition. *Nature Struct. Biol.* **5**, 133–139.
117. Rowsell, S., Stonehouse, N. J., Convery, M. A., Adams, C. J., Ellington, A. D., Hirao, I., Peabody, D. S., Stockley, P. G., and Phillips, S. E. V. (1998) Crystal structures of a series of RNA aptamers complexed to the same protein target. *Nature Struct. Biol.* **5**, 970–975.

3 Promoters, sigma factors, and variant RNA polymerases

JOHN D. HELMANN

1. Introduction

Transcription, the copying of genetic information from its DNA repository into functional RNA molecules, is catalyzed by RNA polymerase (RNAP). In this chapter we review the structure and function of bacterial RNAP, discuss the DNA determinants of promoter strength, and consider several examples of alterations to RNAP which regulate gene expression.

2. Conserved structure of RNAP

RNAP has a complex architecture that reflects the evolutionary history of its source organism. All life on earth can be divided into three domains, the Bacteria, the Archaea, and Eucarya, based on molecular relatedness, as originally inferred from sequencing of conserved macromolecules such as ribosomal RNA (1). This division of life into three domains is also reflected in the structure of RNAP. RNAP in all free-living organisms is a multisubunit enzyme with a complex structure that differs between the three domains. In contrast, some bacteriophage, notably T7, and eukaryotic mitochondria contain a structurally distinct, single-subunit RNAP.

Bacterial RNAP, particularly that from *Escherichia coli*, is among the best understood multisubunit RNAP and shares amino acid similarity with enzymes from eukaryotes and the Archae (Fig. 1). In most species, RNAP contains a catalytically active core component with subunit composition $\beta\beta'\alpha_2$ (2). Although the core enzyme can synthesize RNA, it is not capable of selective transcription initiation due to an inability to recognize promoter sites (3–5). Association of core with a selectivity factor, known as σ, forms a holoenzyme with the ability to recognize promoter sites and thereby direct the transcription of specific genes. Most promoters require the association of core with the major σ factor, known as σ^{70} in *E. coli* and σ^{A} in *Bacillus subtilis*. However, core can also associate with many alternative σ subunits which

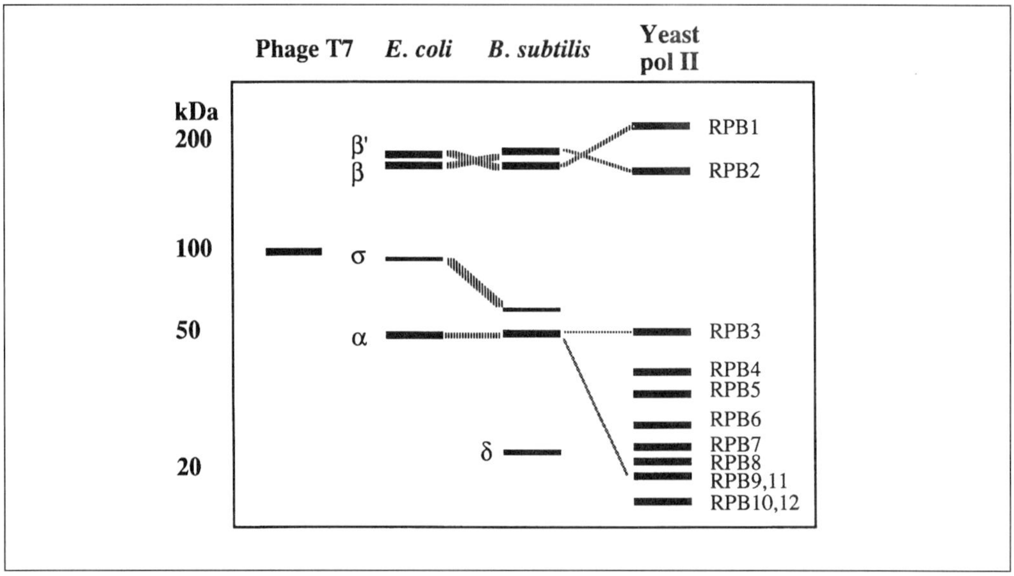

Fig. 3.1 Subunit composition of representative RNAPs as displayed by SDS-PAGE. Sequence similarities indicative of an evolutionary relationship (homology) are indicated by the dashed lines. RNAP from phage T7 is a single polypeptide chain of 99 kDa (17). Most eubacterial RNAP have a conserved core structure of $\beta\beta'\alpha_2$ associated with a family of specificity proteins known as σ factors. In addition, RNAP often contains other co-purifying polypeptides of uncertain function such as the *B. subtilis* δ protein (132, 133) and the ω polypeptides (not shown). RNAP II from *S. cerevisiae* is a 12 subunit enzyme resolvable into 10 discrete bands by SDS-PAGE (8). Each protein is designated RPB for RNA polymerase B subunit.

control specialized subsets of genes (Table 1). In *B. subtilis*, for example, one recent review lists 11 distinct σ factors known to direct gene expression under various growth conditions (6). In fact, at least eight additional σ factors have been identified in the course of sequencing the *B. subtilis* genome. Although most alternative σ factors are structurally and functionally related to σ^{70}, the σ^{54} family proteins are unrelated and will be discussed separately.

RNAP purified from eukaryotes and from the Archaea has an even more complex subunit structure (Fig. 1) (7–9). In eukaryotes there are three nuclear RNAPs, designated RNApol I, pol II, and pol III, together with organellar RNAP active in the mitochondria and chloroplasts. The nuclear enzymes have two large subunits, similar in sequence to β and β′, and two smaller subunits that each share limited similarity to the bacterial α subunit (9, 10). Interestingly, several subunits are shared by two or even all three of the nuclear RNAPs. In the Archaea, the β and/or β′ subunits are sometimes split into separate polypeptides encoded by separate genes. The similarities between eukaryotic and archaeal RNAP extend to the general transcription factors as well. For example, transcription initiation in Archaea requires factors with similarity to the eukaryotic factors TATA-binding protein (TBP) and TFIIB (11–14).

Mitochondria are generally thought to have evolved from endosymbiotic bacteria

Table 1 σ factors and their functions

Organism	σ	Gene	Function
E. coli	σ^{70} (σ^{A})	*rpoD* (*sigA*)	Transcription of housekeeping genes
	σ^{32} (σ^{H})	*rpoH* (*htpR*)	Heat shock
	σ^{24} (σ^{E})	*rpo*	Extreme heat shock, periplasmic stress
	σ^{28} (σ^{F})	*fliA*	Flagellar-based motility
	σ^{38} (σ^{S})	*rpoS* (*katF*)	Stationary phase genes
	σ^{54} (σ^{N})	*rpoN*, *glnF*	Nitrogen regulated genes
	σ^{FecI}	*fecI*	Ferric citrate uptake
B. subtilis	σ^{A}	*sigA*	Transcription of housekeeping genes
	σ^{B}	*sigB*	General stress response
	σ^{C}	Unknown	Unknown
	σ^{D}	*sigD*	Flagellar-based motility
	σ^{E}	*sigE* (*spoIIGB*)	Sporulation, early mother cell
	σ^{F}	*sigF* (*spoIIAC*)	Sporulation, early forespore
	σ^{G}	*sigG* (*spoIIG*)	Sporulation, late forespore
	σ^{H}	*sigH* (*spo0H*)	Competence and early sporulation
	σ^{K}	*sigK*	Sporulation, late mother cell
	σ^{L}	*sigL*	Degradative enzymes
	ECF σs	*sigV sigW*, *sigX* *sigY sigZ*, etc.	Unknown

related to the α subgroup of the proteobacteria. The mitochondrial RNAP is relatively simple and contains a single large polypeptide chain associated with a specificity factor (15, 16). Intriguingly, the mitochondrial enzyme is most similar to the single subunit RNAP encoded by coliphages of the T7 family (17), while the specificity factor shares sequence and functional similarity with bacterial σ factors (16, 18). Unexpectedly, the DNA sequence of a primitive mitochondrial genome reveals the presence of genes for potential β, β′, α, and σ subunits for a bacterial type RNAP (19). It is not yet known whether the encoded RNAP is actually functional, nor is it understood why this bacteria-like enzyme was replaced during evolution by the single subunit enzyme of the T7 family.

Like mitochondria, chloroplasts are presumed to have evolved from endosymbiotic bacteria. The chloroplast RNAP is most similar to bacterial RNAP in subunit composition, except that the β′ subunit is split into β′ and β″ subunits (20). Sequencing of the complete chloroplast genome from several organisms has revealed genes for the β, β′, β″, and α-like subunits of the chloroplast RNAP. However, no gene for a σ-like function was found, despite biochemical evidence that chloroplast RNAP contained σ-like polypeptides. Recently, the gene for a chloroplast σ factor was identified in the nucleus of the unicellular organism, *Cyanidium caldarium* (21, 22). It is likely that chloroplasts may contain more than one type of RNAP with different promoter selectivities. In fact, the *rpoB* gene in tobacco chloroplasts can be deleted and the resulting organelle, while no longer capable of photosynthesis, is maintained (23). The nature and role of this second transcription system are not yet understood.

2.1 Structure and function of *E. coli* RNA polymerase

The Eσ^{70} RNA polymerase from *E. coli* is the best characterized bacterial transcription enzyme. This enzyme contains a conserved core, comprised of β, β′, and two α subunits, that is reversibly associated with the primary σ subunit, σ^{70}. Assembly of RNAP follows a pathway involving initial dimerization of α to α_2, addition of β to form $\alpha_2\beta$, and then addition of β′ to form the core enzyme (24). Each of the subunits of RNAP has been overproduced to high level in *E. coli*, and can be purified for *in-vitro* reconstitution of core and holoenzymes containing defined genetic alterations (e.g. truncations or point mutations) in one or more subunits (24, 25). This has allowed researchers to investigate the *in vitro* transcription properties of RNAP containing alterations that would be lethal *in vivo*, and therefore inaccessible using classical genetic approaches.

2.1.1 Roles of the β and β′ subunits

β and β′ together account for nearly half of the mass of RNAP holoenzyme and contain the catalytic center of the enzyme. Classically, the β subunit is the site of mutations leading to resistance to the antibiotic rifampicin (*rif*^{r} mutants). In contrast, mutations to streptolydigin resistance map to both the β and β′ subunits (26). Biochemical experiments, including chemical crosslinking, have mapped the binding site for the initiating nucleoside triphosphates and the growing RNA chain to the β and β′ subunits (27–31). The ability of RNAP to clamp tightly to the DNA template during the elongation phase also resides in the β and β′ subunits, which are thought to contribute to a high affinity, duplex-DNA binding site in front of the catalytic center and a single-strand specific DNA-binding site just upstream of the active site (32). It is also likely that the ability of elongating RNAP to recognize pause signals and factor-independent termination signals, typically GC-rich hairpins followed by a U-rich sequence in the nascent transcript, is mediated by interactions of the nascent RNA with the β and/or β′ subunits (33–35). Finally, β′ has been shown to be a target for transcription activation by the bacteriophage N4 single-strand binding protein at N4 late promoters (36).

2.1.2 Roles of the α subunit

The α subunit of RNAP is the initiator of RNAP assembly and makes important interactions with DNA, transcription factors, and RNA during the initiation and elongation processes (37–40). The α subunit has two separable protein domains, an amino-terminal domain (NTD) and a carboxyl-terminal domain (CTD), linked by a flexible linker peptide (41, 42). Sites important for dimerization and assembly with β and β′ are located in the NTD (43). The sites of DNA, RNA, and protein activator interaction have been primarily found in the CTD, although some activators also make contact with α NTD (44, 45). The α CTD is currently the only portion of the core RNAP for which high resolution structural information is available (46, 47).

The α CTD can bind to DNA in the region upstream of the –35 region contacts made by the σ subunit. Some specific DNA sequences, located between –40 and –65, offer a particularly tight binding site for the α CTD and can thereby increase the affinity of RNAP for the promoter. These sequences, known as upstream promoter (UP) elements, play an important role in expression of highly expressed genes such as ribosomal RNA operons (38) and flagellin (48). In recent experiments, it has been possible to dissect the contributions of the individual α subunits to UP element binding at the *rrnB* P1 UP element. In these studies, RNAP was purified from a strain expressing wild-type α and bearing an affinity tag (hexahistidine) on a subunit also containing a point mutation that blocks the interaction of α with β, but not with β′. As a result, the only RNAP molecules recovered by affinity chromatography contain a wild-type α subunit in the position that contacts β, and a mutant α in the other position (so-called heterodimeric RNAP). By attaching a DNA cleaving agent to one of the two α subunits, it was then possible to determine that the upstream portion of the UP element is contacted by the β′-associated α subunit (49).

The α CTD is a target for several transcription activator proteins that bind DNA in the region upstream of the promoter element. The cyclic-AMP receptor protein (CRP) interactions with RNAP have been particularly well studied and are explored in more detail in Chapter 4. Remarkably, the flexible linker region between the α NTD and α CTD allows an incredible diversity of interactions between upstream bound activator proteins and RNAP bound at a promoter (50).

2.1.3 Roles of the σ subunit

The σ subunit of RNAP determines promoter selectivity, helps catalyze the DNA strand opening near the start site of transcription, and is a target for one class of transcription activators (51–53). In general, σ subunits seem to have two folded domains corresponding, roughly, to conserved regions designated 2 and 4 (54, 55). Region 2 determines –10 recognition and participates in DNA melting, presumably by forming an ssDNA-specific binding site (56, 57). In fact, region 2 alone imparts upon core enzyme the ability to bind tightly and selectively to single-stranded DNA sequences representing the non-template strand present within an open complex (58). This binding reaction is strongly affected by mutations in σ known to impair promoter melting (59), suggesting that σ does indeed contact and stabilize the single-stranded DNA within the open complex. Region 2 also appears to contain important determinants of core binding (60, 61). The high resolution structure of region 2 from *E. coli* σ^{70} has recently been determined and there is excellent agreement between previous biochemical inferences and the structure of this region (61). For example, all amino acid side-chains implicated in –10 recognition and DNA melting are found on one face of an exposed α-helical element.

Region 4 contains the portion of σ responsible for recognition of the –35 promoter region and is a target for interaction with activator proteins. Recognition of the –35 region is thought to be mediated by a helix–turn–helix type (HTH) DNA-binding motif in the carboxyl terminal region (region 4.2) of factors of the σ^{70} family (51, 52). Assuming that this DNA–protein interaction is similar to those described for other HTH proteins (Chapter 2), selective binding likely involves the formation of specific and complementary hydrogen bonds between the edges of the –35 region base pairs and amino acid side-chains of the HTH recognition helix.

Though less conserved than regions 2 and 4, regions 1 and 3 also play roles in σ factor function. Region 1, in those σ factors that retain this region, prevents the interaction of σ factor with DNA unless σ is associated with core enzyme (62, 63). Derivatives of σ^{70}, for example, in which region 1 is removed can now bind to promoter DNA with sequence selectivity, albeit with very low affinity, even in the absence of core RNAP (62, 64, 65). Genetic studies demonstrate that alterations in region 3 can affect core-binding (66) and can also influence the process of initiation as judged by alterations in the relative production of various aborted RNA products (67).

3. The bacterial transcription cycle

The process of RNA synthesis can be described as a cycle beginning with the localization of promoter sites and ending with the release of the completed transcript and core RNAP (Fig. 2). The reconstitution of the released core enzyme with σ factor and subsequent promoter recognition completes the cycle (68). Understanding the precise molecular details of each step in the transcription cycle will greatly aid our ability to interpret the effects of activators and repressors on gene expression (Chapter 4).

3.1 Promoter localization

Promoter recognition encompasses the events preceding formation of the first phosphodiester bond (transcript initiation). The first step in promoter recognition is the

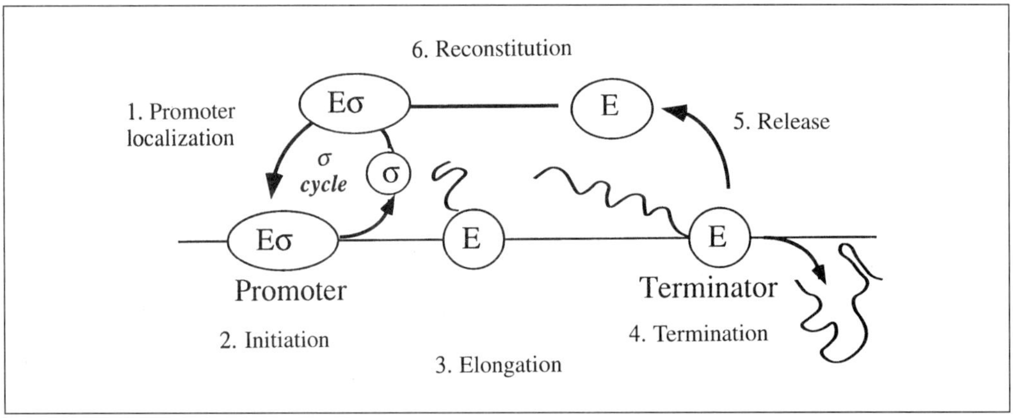

Fig. 2 The bacterial transcription cycle. The core RNAP ($\beta\beta'\alpha_2$) is represented as E. In the first step, RNAP holoenzyme (Eσ) binds the template DNA (represented by a thin line) and locates specific promoter sequences. Transcript initiation occurs leading to the synthesis of short, aborted products and eventual escape from the promoter region ('promoter clearance') to form the stable elongation complex. These 'ternary' complexes contain RNAP, the DNA template, and a nascent RNA transcript (wavy line) but generally lack the σ subunit. Termination of transcription, release of the transcript and template, and reconstitution of the liberated core enzyme (E) with σ factor complete the cycle.

non-specific association of RNAP with the DNA template through predominantly electrostatic interactions (69). This loosely associated enzyme then diffuses randomly along the contour of the DNA, with frequent dissociation and reassociation (depending on the reaction conditions), until a promoter site is encountered (70, 71). A more stable recognition complex is then formed in which RNAP makes specific interactions with the promoter DNA (5, 69, 72–74). This initial recognition complex is called a 'closed complex' since the DNA strands are not yet melted. In this step, promoter selectivity is determined by sequence specific contacts between the –35, and possibly the –10, hexamers and the associated σ subunit (51, 52, 56).

The role of the –10 region in the early stages of promoter localization is not completely clear. The ability of RNAP holoenzyme to bind specifically to –10 region ssDNA (58, 59), and the observation that only the non-template strand is important for recognition of certain key bases with the –10 region (75), suggests that much, or perhaps all, of –10 region sequence recognition may occur subsequent to DNA melting. According to this model, the closed complex is localized primarily or entirely by sequence recognition of the –35 region and perhaps upstream components of the –10 region (e.g. the extended –10 region, see below), and the complex becomes firmly associated with the promoter only after strand separation and sequence-specific binding of the resulting non-template ssDNA to a site on holoenzyme, including key contacts with σ region 2.3.

Since the synthesis of RNA requires the use of single-stranded DNA template, separation of the two DNA strands is a prerequisite for the initiation reaction (76). Strand separation is a complex reaction in which RNAP stabilizes melting of the DNA over a 12 to 16 base-pair region spanning the start site of transcription to form an 'open complex.' Early studies suggested that strand separation was an all-or-none reaction with a concerted melting across the region of the 'transcription bubble' (69, 77, 78). More recently it has become appreciated that this is not always the case: at some promoters, RNAP may melt the DNA in a stepwise fashion (76, 79). For example, at phage λ P_R and the T7 A1 promoters, RNAP forms a complex with a small (≤12 bp) region of strand separation which only extends to form the fully melted (≥14 bp) open complex upon binding of three Mg(II) ions (80–82). At an *E. coli* ribosomal promoter, *rrnB* P1, RNAP does not stably melt the DNA in the absence of bound NTP substrates (83, 84). An intermediate partially melted (or 'nucleated') complex has been observed at a *B. subtilis* tRNA operon promoter, P_{trnS} (85) and at T7 A1 (82). Most dramatically, several distinct partially melted intermediates have been documented for the flagellin promoter (79).

The overall structure of RNAP–promoter complexes can be inferred, at least at low resolution, from a comparison of the three-dimensional structure of RNAP core and holoenzyme (86) and the visualization of promoter complexes by electron and atomic force microscopy methods (87). In the open promoter complex, it is estimated that 70 to 80 bp of largely duplex DNA is wrapped over the surface of RNAP with a resulting net bend angle of >50°. The presence of a large groove on the surface of RNAP suggests the presence of a template-binding channel that can be either in an open or a closed conformation, depending, at least in part, on the presence or absence of the σ subunit.

3.2 Transcript initiation and promoter clearance

The open complex binds to the two appropriate initiating NTP substrates, as specified by the template strand, and the process of transcription commences when the first two substrates are polymerized to a dinucleotide. The start sites chosen for initiation are determined primarily by distance from the conserved –10 element and the preference of RNAP for initiating with a purine rather than a pyrimidine (88, 89). It is known that the concentration of a given NTP required to initiate transcription is much higher than the concentration required during elongation. Evolution has exploited this fact to directly regulate gene expression in response to intracellular NTP levels. For example, at the *E. coli pyrC* promoter, high levels of CTP allow RNAP to initiate at the preferred distance from the –10 region, but the resulting transcripts contain a stable RNA hairpin that blocks access of ribosomes to the ribosome-binding site and consequently shuts off translation (89). If CTP levels are low, RNAP stretches one base further downstream and initiates instead with a preferred substrate (GTP) at a non-preferred distance (9 bp) from the –10 region. In this case, the resulting transcripts are effectively translated thereby allowing synthesis of PyrC, an important enzyme for the *de novo* synthesis of pyrimidine nucleotides. In this system, RNAP acts to directly sense the level of an important metabolite within the cell (CTP) by its effect on start site selection.

In the early stages of RNA synthesis short oligonucleotide products, of lengths generally less than 15 nucleotides, are repeatedly synthesized and released from the enzyme in a process known as 'abortive synthesis' (90, 91). During abortive synthesis the σ subunit is retained and RNAP maintains extensive contact with the promoter region DNA. At some point, triggered by undefined cues, a large conformational change occurs leading to the formation of a highly processive elongation complex (ternary complex) and the concomitant ejection of the σ subunit (92). At this step the extent of contact between RNAP and DNA is reduced substantially as the 60–80 bp DNaseI footprint characteristic of open complexes is transformed into the much smaller 25 to 40 bp footprint characteristic of elongation complexes (92). This transformation is often referred to as promoter clearance and can limit the rate of chain initiation both *in vivo* and *in vitro* (93–95). Since it takes RNAP at least 1 second to clear the promoter region, even the strongest promoters (such as ribosomal RNA promoters) can initiate no more than one RNA chain per second (96).

The ability of RNAP to escape from the abortive phase of transcription and enter into a productive elongation phase is strongly affected by the DNA sequence in the early transcribed region (35). In general, RNAP initiates transcription many times at each promoter before ultimately entering successfully into a productive elongation mode (35, 97). The relative proportion of abortive and productive products can also be affected by the tendency of RNAP to enter into a non-productive slippage synthesis while transcribing through certain DNA sequences, particularly stretches of three or more non-template T residues (encoding U). In these regions, RNAP may synthesize long transcripts containing repeated U residues and fail to transition into a productive elongation mode (93, 98–101).

The *E. coli codBA* operon illustrates how slippage synthesis can regulate transcription (102). At this promoter, start site selection is controlled by NTP levels, much as at *pyrC*. However, in this case, it is the concentration of the second NTP that determines start site selection within the sequence GATTTTTTG. If UTP levels are high, initiation begins with the synthesis of an AU dinucleotide, and RNAP then enters into a non-productive cycle of slippage synthesis and makes long transcripts of sequence AU_n. In contrast, if UTP levels are low, RNAP shifts to initiation with a GA dinucleotide and, for reasons that are not well understood, the resulting transcripts are much less prone to slippage synthesis. Since *codBA* encodes proteins involved in cytosine uptake and utilization, it makes sense to derepress expression only if pyrimidine levels are low in the cell. The examples of the *pyrC* and *codBA* operons illustrate how RNAP can sense intracellular concentrations of NTP substrates, and use this information to coordinate the expression of operons encoding nucleotide metabolism functions.

3.3 Elongation, pausing, and the 'inchworm' model

Although the template-directed synthesis of an RNA chain is highly processive, it is not uniform. On average, between 20 and 50 nucleotides are added to the chain per second under physiological conditions, but the rate of chain extension is quite variable (33, 103). Elongating RNAP frequently pauses for periods of several seconds before continuing chain elongation (35). These pauses can be triggered by specific features of the transcript including, but not limited to, the potential for hairpin formation (104). In some cases, notably amino acid biosynthetic operons, pauses play an important physiological role by allowing translation to be closely coupled to transcription (105). In the vicinity of the promoter, some pauses may be elicited by interactions between the still present σ subunit and sequences in the early transcribed region. As RNAP progresses down the template, the exposure of regions on the non-template strand that resemble the –10 element of the promoter may allow σ, or at least region 2 of σ, to bind to this DNA and stall the elongating RNAP (106).

The elongation of RNA chains was originally thought of as a monotonic process in which the addition of a base to the growing RNA chain coincided with the downstream translocation of a rigid RNAP by one base on the template (107). However, structural characterization of defined RNAP elongation complexes revealed an unexpected variability and called into question this simple model (108, 109). Remarkably, the downstream boundary of RNAP ternary complexes (so-called because they contain DNA, RNAP, and RNA) remains stationary for several cycles of nucleotide addition and then, upon addition of a single nucleotide, moves forward by as many as 8 or 10 bases (109–111). These observations led to an 'inchworm model' for transcriptional elongation (112). In this model, binding of RNAP to the template is mediated by two DNA-binding sites which alternate between fixed and mobile modes of interaction, much like the anterior and posterior ends of the namesake caterpillar (35, 112). As nucleotides are added to the growing RNA chain, the RNAP footprint on the DNA initially contracts and the distance between the active center of RNAP and the downstream boundary of RNAP–DNA contacts decreases from 18 to as few as 10 bp (113, 114). Then the fixed downstream DNA-binding site becomes mobile and translocates by 8 to 10 bp along the template to complete the cycle. A refinement of this model proposes that RNAP moves monotonically through some regions of the DNA template and then, in response to unknown signals, can enter into one or more cycles of 'inchworming' (110).

In addition to the apparent structural differences in RNAP complexes at different template positions, there is considerable variability in stability. While most elongation complexes are very stable, and are active even when stored for weeks at room temperature, at some sites the complex becomes either reversibly or irreversibly altered to an inactive state (115). Originally, these complexes were defined as 'dead-end' complexes, and were shown to still contain RNAP, RNA, and DNA and were thus distinguished from 'release sites' at which termination occurs. Recent studies indicate that some of these complexes are still capable of elongation when incubated for extended periods of time with NTPs and these have been defined as reversibly 'arrested' complexes, whereas at other sites RNAP becomes irreversibly 'arrested' (116, 117). In general, these arrested complexes correlate with a disengagement of the RNA 3′-end from the enzyme active site, and concomitant backward translocation of the active center to a site positioned as many as 10 or 12 nucleotides from the RNA 3′-end. These complexes can be reactivated by transcript cleavage to create a new RNA 3′-end positioned in the active site (118). The resulting 3′-end fragment is then released from the enzyme and RNAP can try once again to elongate successfully through the region. These transcript cleavage reactions are strongly stimulated by protein factors. In *E. coli*, these factors are GreA and GreB (119). In eukaryotes, functionally related proteins have been well characterized (35).

A new model has emerged that links the observations of structural heterogeneity during elongation, reversible and irreversible arrest, and active site translocation (116, 117). According to these studies, RNAP contacts a relatively constant length of DNA during elongation. The apparent structural heterogeneity that led to the original formulation of the inchworm model is actually due to the rapid equilibration of RNAP between different positions on the template. The RNAP can translocate backwards on the DNA with extrusion of the 3′-end of the RNA from the front of the enzyme, and this can account for the apparent shortening of the DNase I footprint of some elongation complexes. Since these complexes are still in equilibrium with RNAP properly positioned on the template, and with the RNA 3′-end associated with the catalytic center, they can still be chased to form full-length RNA products by NTP addition. An important implication of all these models is that the structure of RNAP ternary complexes can vary dramatically at different positions along the DNA template. As a consequence, the susceptibility of RNA to regulatory modifications may also vary (118, 120).

It is likely that pausing and termination are also affected by the conformational variability of RNAP during elongation (34). It has been proposed that signals in the DNA template, or perhaps in the nascent transcript, cause RNAP to enter a cycle of 'inchworming', with the subsequent downstream translocation leading to pausing of the polymerase or release of the RNA (113, 114). These sites of pausing and possible transcript release may thus correspond to sequences where the probability of backward RNAP translocation is greater than the probability of successful nucleotide addition and elongation.

The notion that transcript elongation is in competition with the potential backward translocation of RNAP along the DNA has been accompanied by equally drastic changes in our perception of the role of RNA in ternary complexes. Formerly, the ternary complex was thought to be stabilized by an extended RNA–DNA hybrid, with each cycle of nucleotide addition extending the leading edge of the hybrid by one base with concomitant melting of a base-pair at the trailing edge (107). Indeed, it has been shown that the sites at which RNA is released (termination sites, see below) are predicted by sequence-dependent variations in the stability of this putative RNA–DNA hybrid (121, 122). However, several recent results have bolstered support for an alternative model which proposes that the transcript is held in the ternary complex by two RNA-binding sites, with the DNA–RNA hybrid used primarily for indexing the transcript to maintain the proper register for chain elongation (35, 123). Dramatic support for this notion comes from the observation that DNA polymerases can replicate DNA containing RNAP ternary complexes without displacement of the RNAP or the transcript (124, 125). Moreover, RNAP can faithfully complete the nascent RNA chains after the passage of DNA polymerase, even under conditions where a hybrid could not be consistently maintained. Therefore, a DNA–RNA hybrid cannot be the primary determinant of RNAP ternary complex stability. In summary, the process of transcript elongation is much more complicated than envisioned even just a few years ago, and the implications of these various models for gene regulation are only beginning to emerge.

3.4 Termination

The termination of RNA synthesis occurs at specific sites where the RNA transcript has a high probability of being released (122, 126). In *E. coli*, there are at least two classes of termination sites designated rho-dependent and rho-independent (122). In general, most rho-independent termination sites are at the end of operons or in regions between genes, while rho-dependent termination sites can be found along the length of a transcription unit (127). Importantly, transcript termination is a primary control point for regulating gene expression in numerous operons (128).

Experimentally, it is clear that some sites where RNAP ceases elongation *in vitro* are not physiologically relevant release sites: some sites may be long pauses and RNAP may eventually resume elongation (chaseable sites), while others may be stable ternary complexes that have not released RNA but cannot be further elongated ('dead-end' or 'arrested' complexes) (108, 115, 118). In addition, it is important to distinguish true release sites from subsequent processing events which can alter the 3'-end of the message (129).

3.4.1 Rho-independent termination

Rho-independent release sites are frequently associated with a GC-rich hairpin followed by a short U-rich region at the 3′-end of the RNA transcript. The roles of the hairpin and U-rich region in the pausing of RNAP and the release of the RNA transcript are not well understood and are the topics of active study (33, 130). Dissociation of the RNAP–RNA–DNA complex into its macromolecular components could occur by several pathways: the RNA chain may be released first, leaving core RNAP bound to DNA, or the template may be released first, leaving an RNAP–RNA complex(131). Alternatively, both DNA and RNA may be released from core simultaneously. The order and rate of these dissociation events may determine the efficiency with which RNAP is able to rebind σ and subsequently initiate a new chain (recycling) (131). Indeed, in *B. subtilis* an auxiliary RNAP subunit, designated δ (Fig. 1), releases RNA from RNAP–RNA complexes and thereby enhances the rate of recycling observed *in vitro* (132, 133).

3.4.2 Rho-dependent termination and polarity

Rho-dependent sites allow the release of RNA in response to Rho factor, an RNA-dependent helicase (134). In *E. coli*, transcription and translation are frequently coupled; ribosomes load on nascent messages as they are being transcribed. Events which lead to an uncoupling of transcription and translation often result in premature transcription termination, a phenomenon known as 'polarity' (127). Polarity can result in a failure to express downstream genes in an operon when translation of an upstream gene is interrupted by a nonsense mutation. Polarity can also result if translation is slowed by amino acid limitation: when cells are starved of isoleucine the majority of nascent *lacZ* transcripts are aborted prior to reaching the end of the gene (103).

When transcription and translation are uncoupled, large stretches of single-stranded RNA are exposed which can then bind the hexameric Rho protein. Subsequent ATP hydrolysis fuels the unidirectional translocation of Rho along the transcript, where, upon encounter with RNAP, the Rho protein triggers RNA release (135). Clearly, RNA molecules which are normally not translated, such as rRNA and tRNA, must lack Rho-target sites or RNAP must be suitably modified to reduce its susceptibility to Rho-mediated termination (136, 137).

4. Promoter structure

Promoters are defined as the DNA sequences required for the accurate initiation of transcription. In early experiments, it was demonstrated that RNAP is able to selectively transcribe some regions of a duplex DNA molecule while ignoring others (4). Before the availability of recombinant DNA methods, the selectivity of transcription was often measured using the DNA genomes of bacteriophage. These DNA molecules could be prepared in substantial quantities, were relatively homogeneous, and contained strong transcription signals. On many phage DNAs, RNAP synthesizes RNA molecules which hybridize to only one of the two separated DNA strands. This strand selectivity provided a convenient assay for the selectivity of transcription. Selectivity experiments using T4 DNA, for example, led to the realization that RNAP contained two separable components, a core enzyme capable of non-selective RNA synthesis, and a σ factor necessary for selectivity (3).

4.1 Discovery of promoter core elements: the –35 and –10 regions

The advent of DNA sequencing technology (first chemical sequencing and, later, enzymatic methods) led to the realization that many strong promoters for *E. coli* RNAP contained similar DNA sequences positioned upstream of the start point of transcription. These conserved regions contain two sets of 6 bp each centered near –35 and –10 relative to the start site (designated +1). It is rare for a naturally occurring promoter to match the derived consensus at all 12 positions: an average bacterial promoter contains between eight and nine matches to consensus (138, 139). Most mutations which alter promoter strength occur within these consensus elements: alterations which bring the promoter closer to consensus (by changing a non-consensus base to consensus) typically increase strength while those which decrease the fit to consensus often decrease strength (5, 140). These observations led to the notion that the sequence of the –35 and –10 elements determines promoter strength.

4.2 A modern view of promoter structure

In recent years it has become increasingly clear that the above simple view of promoters does not adequately convey the complexity of the RNAP–promoter interaction (96). Although it is true that the adage 'consensus is best' has some predictive power, there are many exceptions to the rule. Indeed, there is only a modest correlation between fit to consensus and *in vivo* promoter strength (141), and a somewhat better correlation between fit to consensus and the rate at which RNAP binds a promoter *in vitro* (96). Furthermore, several examples have been discovered of mutations to consensus leading to decreases in promoter strength (142).

A more modern view of promoter structure acknowledges the contributions of the entire promoter region to RNAP binding and initiation (Fig. 3) (96). RNAP contacts between 60 and 90 bp of DNA when bound at a promoter. Numerous studies have described the details of these contacts, including the changes which occur as RNAP progresses from the initial closed complex, through intermediate states, to form the fully melted open complex (72, 74, 78, 81, 85, 92, 109, 143–146). The pattern which emerges is of an initial interaction limited to one face of the helix in the region between –60 and –10 (closed complex), extension of the downstream footprint to about +10 or +20 in the absence of extensive DNA melting (intermediate or nucleated complex), and melting of the transcription bubble with concomitant close association

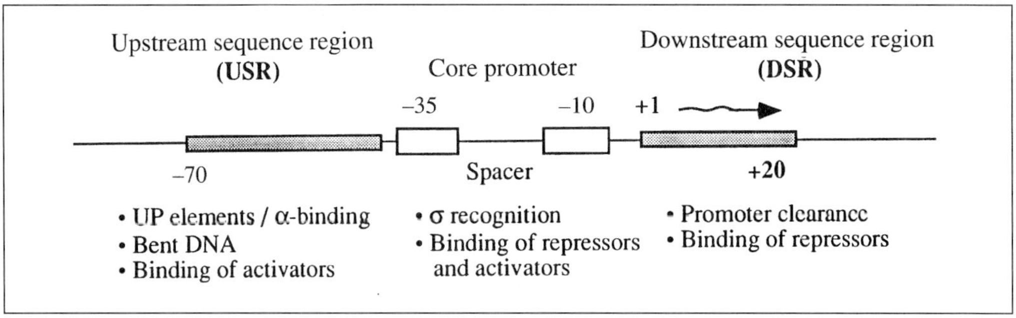

Fig. 3 Components of a bacterial promoter. The promoter can be divided into an 'upstream sequence region' (USR), the core promoter elements, and the 'downstream sequence region' (DSR) (36). Representative functions for each region are indicated.

of RNAP with the non-template strand (open complex). As noted above, this model needs to be modified to account for the multiple, partially melted intermediates observed in several recent studies. Perhaps not surprisingly, the DNA sequence throughout the large region contacted by RNAP can affect initiation.

4.2.1 Upstream sequence region: UP elements and bent DNA

A common technique for defining the boundaries of a promoter is to test the abilities of a nested set of deletions to promote transcription. Often, sequences extending well upstream of the –35 region are required for efficient transcription, but not for a low (basal) level of expression (48, 147–149). Although in many cases, these 'upstream activating sequences' include the binding sites of transcription factors (Chapter 4), several examples now exist of 'factor-independent activation'. Upstream activating sequences include UP elements, which bind the α subunits of core RNAP (38, 48), and intrinsically bent DNA (150). These sequences stimulate transcription even in a purified system containing only DNA, RNAP, and NTP substrates and should therefore be considered as part of the promoter (Fig. 3).

UP elements were first defined at *rrnB* P1, one of the strongest promoters in *E. coli* (38). This strength is due to several sequence features: both the –10 and –35 elements are close to consensus, there is a strong factor-independent activation region (UP element) between –40 and –65, and there are three binding sites for the DNA-bending protein FIS between –65 and –120 (38, 83, 151). The ability of the UP element to activate transcription *in vitro* depends on the α CTD: when RNAP reconstituted with C-terminally deleted α subunits was used for transcription the 30-fold stimulation characteristic of the UP element was lost as was the extended DNaseI footprint in this upstream region (38, 152). Indeed, the purified α proteins bind directly to the UP element, suggesting that contacts between the UP element and α mediate transcriptional activation (38). An UP element is also part of the strong σ^D-dependent promoter for flagellin in *B. subtilis* (48). Like *rrnB*, the flagellin UP element activates transcription 30-fold both *in vivo* and *in vitro* and interacts specifically with the isolated α subunits of RNAP. Interestingly, neither the *rrnB* nor the flagellin UP element appear to be intrinsically bent although the DNA may assume a bent conformation upon interaction with RNAP.

DNA sequences containing between four and six adenines in a row (A-tracts) adopt an altered structure which leads to a static bend at the junction of the A-tract and the adjacent B-form DNA (153). Although the magnitude of the bend is rather small (estimated at 18°), a repetition of A-tracts phased with the helical repeat of DNA (10 to 11 bp/turn) can lead to substantial net bending. DNA sequences associated with bending are often found in the USR of bacterial promoters (139, 154–157). In several cases, these sequence regions have been shown to activate transcription both *in vivo* and *in vitro* (150). It seems likely that the ability of bent DNA to activate transcription is due to improved contacts with the α subunits of RNAP, much as occurs when an UP element is present. Indeed, UP elements are characteristically AT-rich sequences. The distinction between transcription activation by UP elements (sequences which bind α) and intrinsic DNA bends (which may also bind α) may be largely a matter of nomenclature.

4.2.2 Functions of the core promoter elements and the spacer region

The highly conserved –35 and –10 hexamer elements play a critical role in determining both the selectivity of transcription and the frequency of initiation (promoter strength) (96). Both genetic and biochemical experiments argue that σ is responsible for most contacts between RNAP and the conserved bases of these core elements, but the details of these interactions are not clear.

The –35 and –10 elements are separated by a non-conserved DNA element known as the spacer region. In general, the length of the spacer is more important than its sequence for determining promoter strength. For *E. coli* σ^{70} holoenzyme, optimal activity is found for spacers of 17 bp, while increasing or decreasing the spacer length leads to decreased promoter strength (158, 159). Changing the spacer length alters both the linear and the rotational disposition of the core elements. It is hypothesized that during the process of open complex formation the –35 and –10 regions are rotated relative to one another leading to an untwisting of the spacer DNA which subsequently facilitates open complex formation (76, 160–162). Increasing or decreasing the length of the spacer may impair this facilitating mechanism. The relative rotational displacement of the –35 and –10 regions is determined by the product of the spacer length and the average twist per bp over the length of the spacer region. The local helical twist, in turn, is very sensitive to sequence (particularly nearest neighbor effects) and to DNA supercoiling (163) (see Chapter 6).

Although the sequence of the spacer region is often considered unimportant for promoter strength, this too is an oversimplification. The spacer region contains an important sequence element designated the –15 or 'extended –10' region (164). This motif has the consensus of TnTG and is separated from the –10 region by one base. The –15 region is fairly well conserved in σ^A-dependent promoters of *B. subtilis* (139), but is less well conserved in *E. coli* σ^{70} promoters (138). In *E. coli*, the –15 region is found associated with a class of promoters, the 'extended –10' promoters, which often lack a –35 element. Mutational studies have demonstrated that the –15 region is an important determinant of promoter strength and can strongly affect open complex formation (144, 165). Remarkably, an σ^{70} derivative completely lacking conserved region 4.2 suffices for transcription from an extended –10 promoter (166). Recent genetic studies suggest that –15 region recognition is mediated by a portion of σ factor just C-terminal to region 2.4, in a newly defined element called region 2.5 (167). In addition to the –15 region, there is statistical evidence that DNA sequences within the spacer region are not random: there is a correlation between the length of the spacer and the dinucleotide composition of the DNA (168).

4.2.3 The downstream sequence region (DSR) and clearance

Further complexities of relating promoter sequence to strength are apparent from analyses of strong phage promoters (94, 169). In the case of λ P_L, mutations of the –10 element which bring the promoter closer to consensus led to a decrease in promoter strength, although the mutant promoter could bind RNAP more rapidly *in vitro*. The deleterious effect of these –10 region mutations could be reversed by substitution of DNA sequences in the +3 to +20 early transcribed region (169). From these studies it was hypothesized that the downstream sequence region (DSR) could influence the rate of promoter clearance and thereby overcome a kinetic barrier imposed by having very strong interactions between RNAP and the core elements. This analysis is supported by the effects of strongly activating USR elements on promoter strength *in vivo*: the presence of phased poly(A) tracts in the USR stimulates transcription from a promoter with relatively poor core elements, but actually inhibits initiation from a promoter with near consensus core elements (170). Further, it can be demonstrated that in the latter case RNAP is stalling in the early transcribed region and is therefore limited at the step of promoter clearance (95). In addition to influencing the rate and efficiency of promoter clearance, changes in the DSR (and the early transcribed region) can also affect the efficiency with which RNAP recognizes downstream termination signals (171, 172). The basis of these effects is not understood. As noted above, the DSR is also key to the regulation of several *E. coli* pyrimidine metabolism genes.

In conclusion, the emerging picture of bacterial promoters is one of great complexity. RNAP interacts with a large region of DNA during promoter recognition, transcript initiation, abortive synthesis, and eventual promoter clearance. Sequences as far upstream as –70 and as far downstream as +20 or more bind RNAP, apparently by wrapping around the enzyme. These interactions involve several, if not all, subunits of RNAP and result in bending, untwisting, and localized melting of the DNA. It should not be too surprising, then, that the sequence determinants of these interactions are equally complex and distributed.

5. Alternative σ factors and their roles

The discovery that RNAP could be separated into a catalytically active core enzyme and a σ specificity factor immediately prompted the speculation that σ factor replacement would be a powerful mechanism for activating the transcription of new sets of genes (3). Studies of *B. subtilis* phage SP01 initially indicated that σ switching was, in fact, a mechanism of gene regulation (173, 174). Subsequently, multiple alternative σ factors were found to control processes as diverse as spore formation, flagellar-based motility, and stress responses (6). It is likely that all bacteria will contain multiple σ factors: *B. subtilis* has at least 19, *E. coli* has seven, and even the small 1.8 megabase genome of *Haemophilus influenzae* Rd (the first completely sequenced microorganism) reveals four σ factor genes (175).

5.1 Structural families of alternative sigmas

Based on sequence comparisons bacterial σ factors fall into two large families: *E. coli* σ^{70} together with the vast majority of the alternative σ factors form the σ^{70} family (Figs. 4 and 5), while the nitrogen regulation σ from *E. coli* and it relatives form the σ^{54} family (56, 176). Within the σ^{70} family, the primary σ factors (σ^{A} proteins) form a closely related group (group I), while the *E. coli* stationary phase σ, σ^{S} (see Chapter 7), and certain *Streptomyces* σ proteins form a closely related family of proteins which are, however, dispensable for growth (group II) (177). The group III alternative σ factors form several clusters: many of the *B. subtilis* factors important for endospore formation form a group, σ factors from diverse organisms important (in general) for flagellar-based motility form a second cluster, and an even more divergent group consists of σ factors that, in general, control *e*xtra*c*ytoplasmic *f*unctions (ECF subfamily) (178). In contrast with the σ^{70} family proteins, σ factors of the σ^{54} family are quite diverged in both structure and function (179) (Table 1, Fig. 4; see also Chapters 7 and 11).

The σ^{54} family of regulatory proteins recognize promoters with conserved core elements positioned near –12 and –24 (179). Unlike the σ^{70} holoenzyme, the σ^{54} holoenzyme is unable to form open complexes in the absence of an activator protein. σ^{54} itself forms a stable protein–DNA complex at some promoters and appears to be responsible for most of the DNA–protein interactions observed in σ^{54} holoenzyme–promoter complexes. Extensive mutational and biochemical analysis supports a two domain model for σ^{54} function (Fig. 4; 180–184): the carboxyl-terminal domain (III) contains DNA-binding and essential core-binding determinants and is linked to a glutamine-rich amino-terminal domain (I) by a non-conserved, but often highly acidic, intervening region (II). Region I functions to receive the signal from the activator protein and allosterically modulates the DNA-binding and promoter melting functions localized in region III (181). Indeed, mutations within this region can bypass the requirement for an activator protein (185, 186).

5.2 Regulation of σ factor activity

Since σ factors act catalytically to allow transcription initiation, the production of even small amounts of active σ factor can have profound consequences for gene

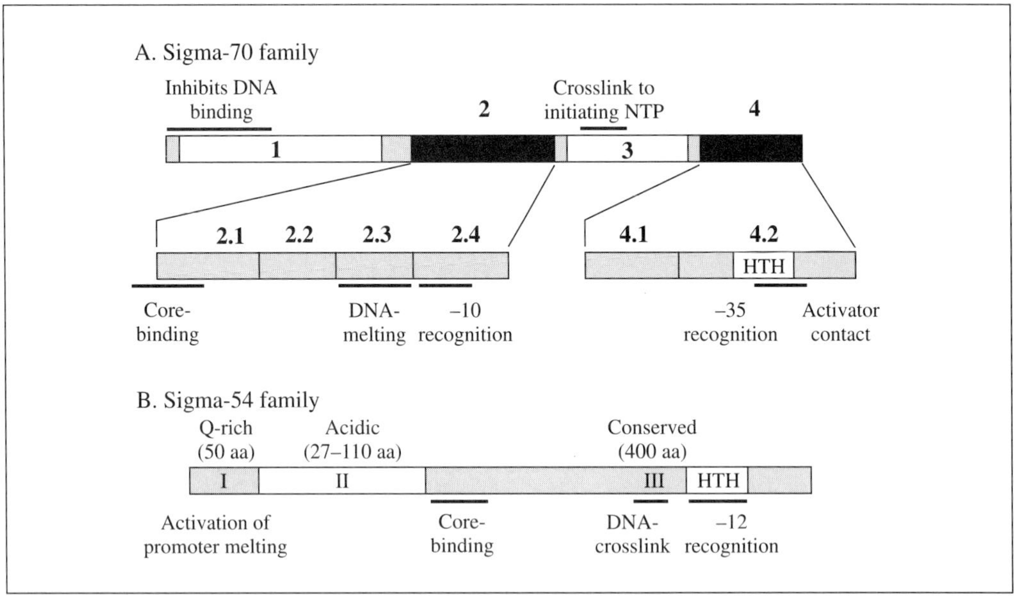

Fig. 4 Structure–function maps for bacterial σ factors. A. The σ^{70} family of proteins have four conserved regions with the indicated functional roles. Region 1 acts to prevent σ interaction with DNA in the absence of allosteric activation by core enzyme (63); region 2 participates in core-binding, promoter recognition, and DNA melting (51, 52); region 3 contains amino acids close to the active center of RNAP (201) and may contribute to core-binding (66); region 4 contains determinants of −35 recognition and provides a target for some activator proteins (53). B. The σ^{54} family of proteins have two conserved regions (I and III) separated by variable region II. Core- and DNA-binding determinants are largely localized in region III, while region I is important for the σ^{54} holoenzyme to be able to respond to activator proteins and thereby establish a productive open complex (179–184, 202).

expression. Therefore, the activity of σ factors is tightly regulated, often at multiple levels. Some of the common regulatory mechanisms include:

(1) transcriptional control of the σ factor gene;

(2) translational control of σ factor synthesis;

(3) synthesis of an inactive pro-σ requiring proteolytic activation; and

(4) the presence of an anti-σ factor (see also Chapter 11).

The complexity of regulation of σ factor activity is evident from studies of *E. coli* σ^{32} (σ^{H}) which controls the expression of several heat shock proteins (Fig. 6; 187). Transcription of the σ^{32} gene (*rpoH*) involves three different σ^{70}-dependent promoters and a promoter used at very high temperatures by the σ^{24} ECF sigma factor (188). Upon temperature upshift, the transcription of *rpoH* increases only slightly (about twofold) even though the levels of σ^{32} protein increase by at least tenfold (189). Therefore, much of the regulation of the heat shock response is post-transcriptional. Increased levels of σ^{32} are due to both a derepression of translation of the *rpoH* message and a transient increase in the stability of the σ^{32} protein (187).

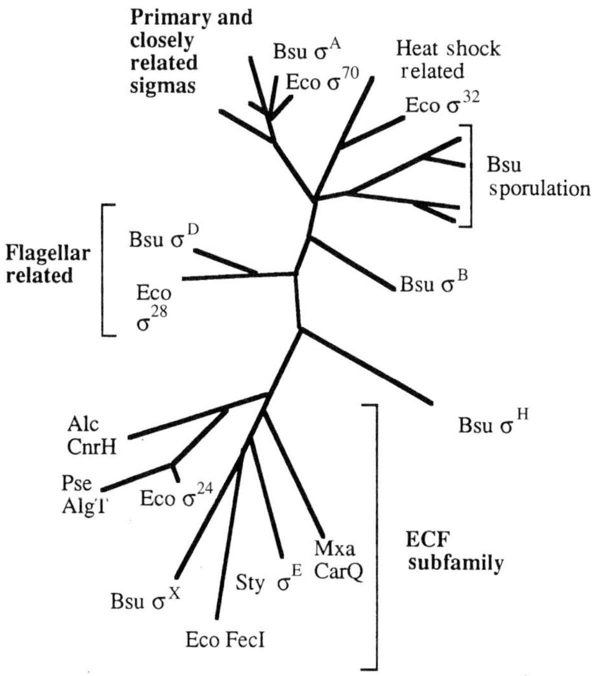

Fig. 5 Phylogenetic tree for bacterial σ^{70} family factors (adapted from Reference 178). Bacterial σ factors cluster into the closely related primary σ factors (group I proteins, represented by Bsu σ^A and Eco σ^{70}) and the non-essential alternative σ factors. The alternative σ factors include some which are closely related in sequence to σ^{70} (the group II proteins such as σ^S; not shown), and others which are more distantly related (group III). The group III proteins include a family of *B. subtilis* sporulation regulators, heat shock (σ^{32}) related proteins, and σ factors controlling flagellar motility and related functions. Recently, it has become appreciated that a large family of even more distantly related proteins also encode σ factors which, in general, control extracytoplasmic functions (ECF subfamily; see Reference 178 for further discussion).

Proteolysis of σ^{32} is governed by the HflB(FtsH) protease (which is itself controlled by σ^{32}) and is facilitated by the heat shock proteins DnaK, DnaJ, and GrpE (also under σ^{32} control) (187, 190, 191). The region of σ^{32} recognized by DnaK is located just C-terminal to region 2 and is highly conserved among σ^{32} homologs, but not among other σ factors (192). Thus, induction of the heat shock response leads to the synthesis of factors which degrade σ^{32} and thereby shut off the response.

6. Modification of RNAP and transcriptional control

At any given time, bacteria contain a heterogeneous population of RNAP. In growing cells, much of the enzyme is elongating RNA chains and is likely to be associated with elongation factors while only a small fraction is present as holoenzyme (5). The

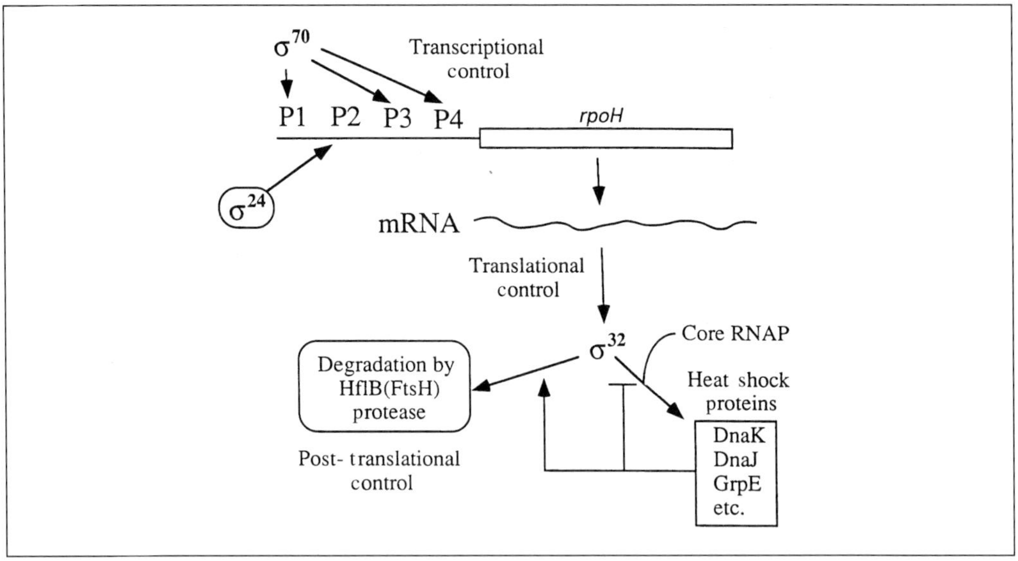

Fig. 6 Mechanisms of control of σ^{32} activity in *E. coli* (adapted from Reference 187). Transcription of the *rpoH* gene, encoding σ^{32}, is directed by three σ^{70}-dependent and one σ^{24}-dependent promoters. The σ^{24}-dependent promoter is most important under extreme stress conditions and at very high temperatures. The *rpoH* mRNA has a complex secondary structure which is likely to be important in the heat shock-dependent increase in translation. Once synthesized, σ^{32} directs the expression of numerous heat shock proteins including four proteins which are positive effectors for σ^{32} proteolysis; DnaK, DnaJ, GrpE, and HflB. By binding to σ^{32}, DnaK, DnaJ, and GrpE also decrease the ability of σ^{32} to bind core RNAP.

relative amount of each holoenzyme is determined by the abundance and relative affinity of the various σ factors, by the presence of σ factor antagonists (anti-σ factors), and by the availability of free core enzyme. Promoter activity, in turn, is controlled by the availability of the suitable holoenzyme, the presence of competing promoters within the same regulon, and the action of activator and repressor proteins. During growth transitions, stress responses, or developmental changes such as sporulation, the population of RNAP can vary greatly from that in growing cells, leading to a large-scale redirection of the transcriptional machinery.

Heterogeneity can also arise by modification of the core subunits. It has been demonstrated that RNAP purified from stationary phase *E. coli* contains covalently modified core subunits (193, 194). Although these modifications are not well understood, it is likely that they contribute (together with the synthesis of σ^S; Chapter 7) to the changes in gene expression characteristic of stationary phase *E. coli*.

Some of the best understood examples of RNAP modification occur during phage infection. These include the synthesis of new σ factors (phages T4gp55 and SPO1gp28), synthesis of RNAP binding proteins (phage T4 AsiA), modification of the core subunits (phage T4 Alt and Mod), and the synthesis of new phage specific RNAPs (e.g. phage T7 RNAP; Fig. 1). Phage T4 is one of the best understood bacteriophages and uses several of these strategies.

6.1 The complex genetic program of phage T4

Phage T4 has a large double-stranded DNA genome of 166 kb and a complex pattern of gene expression (195). Upon injection of the phage DNA into *E. coli*, σ^{70} holoenzyme initiates transcription from strong early promoters, many of which contain intrinsic DNA-bends in their USR. The phage-encoded Alt protein is injected into the cell with the DNA and subsequently ADP-ribosylates one of the two α subunits of core on a specific arginine (residue 265). Within two minutes of infection, synthesis of the phage-encoded Mod protein leads to ADP-ribosylation of the other α subunit.

The phage-mediated ADP-ribosylation of the α subunits may play a role in the shut-off of host transcription. The modification of the α subunits occurs within the carboxyl-terminal domain of α at a position known to be important for interaction with UP elements and with positive activator proteins, such as the cyclic-AMP dependent activator, CAP. It is therefore possible that these modifications contribute to the shut off of the very strong rRNA promoters which typically account for the majority of cellular transcription. This is predicted to liberate a large pool of RNAP for use by phage T4. Shut off of host transcription is further ensured by the action of Alc which causes premature transcription termination on DNA templates lacking the covalent modifications characteristic of T4 DNA (e.g. hydroxymethyl cytosine) (120).

The transition to middle gene expression requires the T4 encoded MotA protein which binds specific sequences located in the –30 region of the promoter (196). MotA activates transcription by a modified form of RNAP containing σ^{70} and the σ^{70}-binding protein AsiA (197). Although AsiA was originally described as an anti-σ, it is now thought to alter the conformation of the σ^{70} holoenzyme to allow proper interaction of RNAP with MotA (bound near the –35 region) while still allowing σ^{70}-dependent recognition of the –10 region (195–198).

Activation of late gene transcription is perhaps the most complex of all. First, late gene transcription requires the presence of a phage-encoded alternative σ factor, the product of gene 55 (σ^{gp55}). Second, the T4 late promoters are activated by interactions between mobile (diffusible) components of the replication machinery and the modified RNAP (199). This interaction does not require the carboxyl-terminal domain of the α subunit, and the region(s) of RNAP contacted by the replication proteins are not yet known (200).

7. Conclusions

Transcription is the most frequent control point in bacterial gene expression. Consequently, it is important to develop a familiarity with the complexities of bacterial RNAP structure and function. Perhaps because of its central role in gene expression, RNAP has evolved a complex subunit architecture. The β and β' subunits play key roles in catalysis, binding of template DNA and NTP substrates, and interaction with regulatory molecules. The α subunits form a structural scaffold for the assembly of RNAP, interact with upstream regions of promoter DNA, and are the target of many transcription activators. The σ subunits determine promoter selectivity, facilitate the

DNA-melting step of open complex formation, and are the target of some transcription activators. Together, these subunits form a complex macromolecular machine that accurately copies DNA into RNA and is able to sense and integrate numerous regulatory signals.

Acknowledgement

Work in the author's laboratory on RNAP is supported by NIH grant GM47446.

References

1. Woese, C. R., Kandler, O., and Wheelis, M. L. (1990) Towards a natural system of organisms: Proposal for the domains Archaea, Bacteria, and Eucarya. *Proc. Natl. Acad. Sci. USA*, **87**, 4576–4579.
2. Rowland, G. C. and Glass, R. E. (1990) Conservation of RNA polymerase. *Bioessays*, **12**, 343–346.
3. Burgess, R. R., Travers, A. A., Dunn, J. J., and Bautz, E. (1969) Factor stimulating transcription by RNA polymerase. *Nature*, **221**, 43–46.
4. Chamberlin, M. J. (1974) The selectivity of transcription. *Ann. Rev. Biochem.*, **43**, 721–775.
5. McClure, W. R. (1985) Mechanism and control of transcription initiation in prokaryotes. *Ann. Rev. Biochem.*, **54**, 171–204.
6. Haldenwang, W. G. (1995) The sigma factors of *Bacillus subtilis*. *Microbiol. Rev.*, **59**, 1–30.
7. Young, R. A. (1991) RNA polymerase II. *Ann. Rev. Biochem.*, **60**, 689–715.
8. Woychik, N. A. and Young, R. A. (1994) Exploring RNA polymerase II structure and function. In *Transcription: mechanisms and regulation* (ed. R. C. Conaway and J. Conaway), pp. 227–242. Raven Press, New York.
9. Archambault, J. and Friesen, J. D. (1993) Genetics of eukaryotic RNA polymerase I, II, and III. *Microbiol. Rev.*, **57**, 703–724.
10. Larkin, R. M. and Guilfoyle, T. J. (1997) Reconstitution of yeast and *Arabidopsis* RNA polymerase α-like subunit heterodimers. *J. Biol. Chem.*, **272**, 12824–12830.
11. Kosa, P. F., Ghosh, G., Dedecker, B. S., and Sigler, P. B. (1997) The 2.1-Å crystal structure of an archaeal preinitiation complex: TATA-box-binding protein-transcription factor (II)B core-TATA-box. *Proc. Natl. Acad. Sci. USA*, **94**, 6042–6047.
12. Hausner, W., Wettach, J., Hethke, C., and Thomm, M. (1996) Two transcription factors related with the eucaryal transcription factors TATA-binding protein and transcription factor IIB direct promoter recognition by an archaeal RNA polymerase. *J. Biol. Chem.*, **271**, 30144–30148.
13. Qureshi, S. A., Bell, S. D., and Jackson, S. P. (1997) Factor requirements for transcription in the Archaeon *Sulfolobus shibatae*. *EMBO J.*, **16**, 2927–2936.
14. Thomm, M. (1996) Archaeal transcription factors and their role in transcription initiation. *FEMS Microbiol. Rev.*, **18**, 159–171.
15. Shadel, G. S. and Clayton, D. A. (1993) Mitochondrial transcription initiation: Variation and conservation. *J. Biol. Chem.*, **268**, 16083–16086.
16. Jaehning, J. A. (1993) Mitochondrial transcription: is a pattern emerging? *Mol. Microbiol.*, **8**, 1–4.
17. McAllister, W. T. and Raskin, C. A. (1993) The phage RNA polymerases are related to DNA polymerases and reverse transcriptases. *Mol. Microbiol.*, **10**, 1–6.

18. Shadel, G. S. and Clayton, D. A. (1995) A *Saccharomyces cerevisiae* mitochondrial transcription factor, sc-mtTFB, shares features with sigma factors but is functionally distinct. *Mol. Cell. Biol.*, **15,** 2101–2108.
19. Lang, B. F., Burger, G., O'Kelly, C. J., Cedergren, R., Golding, G. B., Lemieux, C., Sankoff, D., Turmel, M., and Gray-M., W. (1997) An ancestral mitochondrial DNA resembling a eubacterial genome in miniature. *Nature*, **387,** 493–497.
20. Igloi, G. L. and Koessel, H. (1992) The transcriptional apparatus of chloroplasts. *CRC Critical Reviews in Plant Sciences*, **10,** 525–558.
21. Liu, B. and Troxler, R. F. (1996) Molecular characterization of a positively photoregulated nuclear gene for a chloroplast RNA polymerase sigma factor in *Cyanidium caldarium. Proc. Natl. Acad. Sci. USA*, **93,** 3313–3318.
22. Tanaka, K., Oikawa, K., Ohta, N., Kuroiwa, H., Kuroiwa, T., and Takahashi, H. (1996) Nuclear encoding of a chloroplast RNA polymerase sigma subunit in a red alga. *Science*, **272,** (1932–1935.
23. Allison, L. A., Simon, L. D., and Maliga, P. (1996) Deletion of *rpoB* reveals a second distinct transcription system in plastids of higher plants. *EMBO J.*, **15,** 2802–2809.
24. Fujita, N. and Ishihama, A. (1996) Reconstitution of RNA polymerase. *Methods Enzymol.*, **273**, 121–130.
25. Tang, H., Kim, Y., Severinov, K., Goldfarb, A., and Ebright, R. H. (1996) *Escherichia coli* RNA polymerase holoenzyme: Rapid reconstitution from recombinant alpha, beta, beta', and sigma subunits. *Methods Enzymol.*, **273**, 130–134.
26. Jin, D. J. and Zhou, Y. N. (1996) Mutational analysis of structure–function relationship of RNA polymerase in *Escherichia coli. Methods Enzymol.*, **273**, 300–319.
27. Markovtsov, V., Mustaev, A., and Goldfarb, A. (1996) Protein–RNA interactions in the active center of transcription elongation complex. *Proc. Natl. Acad. Sci. USA*, **93,** 3221–3226.
28. Mustaev, A., Kashlev, M., Lee, J., Polyakov, A., Lebedev, A., Zalenskaya, K., Grachev, M., Goldfarb, A., and Nikiforov, V. (1991) Mapping of the priming substrate contacts in the active center of *Escherichia coli* RNA polymerase. *J. Biol. Chem.*, **266,** 23927–23931.
29. Borukhov, S., Lee, J., and Goldfarb, A. (1991) Mapping of a contact for the RNA 3′ terminus in the largest subunit of RNA polymerase. *J. Biol. Chem.*, **266,** 23932–23935.
30. Hanna, M. M. (1996) Photochemical cross-linking analysis of protein–nucleic acid interactions in *Escherichia coli* transcription complexes from lambda P-R′ promoter. *Methods Enzymol.*, 403–418.
31. Zaychikov, E., Martin, E., Denissova, L., Kozlov, M., Markovtsov, V., Kashlev, M., Heumann, H., Nikiforov, V., Goldfarb, A., and Mustaev, A. (1996) Mapping of catalytic residues in the RNA polymerase active center. *Science*, **273,** 107–109.
32. Nudler, E., Avetissova, E., Markovtsov, V., and Goldfarb, A. (1996) Transcription processivity: protein–DNA interactions holding together the elongation complex. *Science*, **273,** 211–217.
33. Chan, C. L. and Landick, R. (1994) New perspectives on RNA chain elongation and termination by *E. coli* RNA polymerase. In *Transcription: Mechanisms and regulation* (ed. R. C. Conaway and J. Conaway), pp. 297–321. Raven Press, New York.
34. Landick, R. (1997) RNA polymerase slides home: Pause and termination site recognition. *Cell*, **88,** 741–744.
35. Uptain, S. M., Kane, C. M., and Chamberlin, M. J. (1997) Basic mechanisms of transcript elongation and its regulation. *Ann. Rev. Biochem.*, **66,** 117–172.
36. Miller, A., Wood, V, Ebright, R. H., and Rothman-Denes, L. B. (1997) RNA polymerase beta′ subunit: A target of DNA binding-independent activation. *Science*, **275,** 1655–1657.

37. Schauer, A. T., Cheng, S. W. C., Zheng, C., St.Pierre, L., Alessi, D., Hidayetoglu, D. L., Costantino, N., Court, D. L., and Friedman, D. I. (1996) The alpha subunit of RNA polymerase and transcription antitermination. *Mol. Microbiol.*, **21,** 839–851.
38. Ross, W., Gosink, K. K., Salomon, J., Igarishi, K., Zou, C., Ishihama, A., Severinov, K., and. Gourse, R. L (1993) A third recognition element in bacterial promoters: DNA binding by the α subunit of RNA polymerase. *Science*, **262,** 1407–1413.
39. Liu, K., Zhang, Y., Severinov, K., Das, A., and Hanna, M. M. (1996) Role of *Escherichia coli* RNA polymerase alpha subunit in modulation of pausing, termination of anti-termination by the transcription elongation factor NusA. *EMBO J.*, **15,** 150–161.
40. Ishihama, A. (1992) Role of the RNA polymerase α subunit in transcription activation. *Mol. Microbiol.*, **6,** 3283–3288.
41. Jeon, Y. H., Yamazaki, T., Otomo, T., Ishihama, A., and Kyogoku, Y. (1997) Flexible linker in the RNA polymerase alpha subunit facilitates the independent motion of the C-terminal activator contact domain. *J. Mol. Biol.*, **267,** 953–962.
42. Blatter, E. E., Ross, W., Tang, H., Gourse, R. L., and Ebright, R. H. (1994) Domain organization of RNA polymerase alpha subunit: C-terminal 85 amino acids constitute a domain capable of dimerization and DNA binding. *Cell*, **78,** 889–896.
43. Heyduk, T., Heyduk, E., Severinov, K., Tang, H., and Ebright, R. H. (1996) Determinants of RNA polymerase alpha subunit for interaction with beta, beta', and sigma subunits: Hydroxyl-radical protein footprinting. *Proc. Natl. Acad. Sci. USA*, **93,** 10162–10166.
44. Busby, S. and Ebright, R. H. (1997) Transcription activation at class II CAP-dependent promoters. *Mol. Microbiol.*, **23,** 853–859.
45. Busby, S. and Ebright, R. H. (1994) Promoter structure, promoter recognition, and transcription activation in prokaryotes. *Cell*, **79,** 743–746.
46. Gaal, T., Ross, W., Blatter, E. E., Tang, H., Jia, X., Krishnan, V. V., Assa-Munt, N., Ebright, R. H., and Gourse, R. L. (1996) DNA-binding determinants of the α subunit of RNA polymerase: Novel DNA-binding domain architecture. *Genes and Development*, **10,** 16–26.
47. Jeon, Y. H., Negishi, T., Shirakawa, M., Yamazaki, T., Fujita, N., Ishihama, A., and Kyogoku, Y. (1995) Solution structure of the activator contact domain of the RNA polymerase alpha subunit. *Science*, **270,** 1495–1497.
48. Fredrick, K., Caramori, T., Chen, Y. F., Galizzi,V, and Helmann, J. D. (1995) Promoter architecture in the flagellar regulon of *Bacillus subtilis*: High-level expression of flagellin by the σ^D RNA polymerase requires an upstream promoter element. *Proc. Natl. Acad. Sci. USA*, **92,** 2582–2586.
49. Murakami, K., Kimura, M., Owens, J. T., Meares, C. F., and Ishihama, A. (1997) The two alpha subunits of *Escherichia coli* RNA polymerase are asymmetrically arranged and contact different halves of the DNA upstream element. *Proc. Natl. Acad. Sci. USA*, **94,** 1709–1714.
50. Belyaeva, T. A., Brown, J. A., Fujita, N., Ishihama, A., and Busby, S. J. W. (1996) Location of the C-terminal domain of the RNA polymerase alpha subunit in different open complexes at the *Escherichia coli* galactose operon regulatory region. *Nucl. Acids Res.*, **24,** 2243–2251.
51. Gross, C. A., Lonetto, M., and Losick, R. (1992) Bacterial sigma factors. In *Transcriptional regulation* (ed. S. L. McKnight and K. R. Yamamoto), pp. 129–176. Cold Spring Harbor Press, Cold Spring Harbor, NY.
52. Helmann, J. D. (1994) Bacterial sigma factors. In *Transcription: Mechanisms and regulation* (ed. R. C. Conaway and J. Conaway), pp. 1–17. Raven Press, New York.
53. Li, M., Moyle, H., and Susskind, M. M. (1994) Target of the transcriptional activation function of phage λ cI protein. *Science*, **263,** 75–77.

54. Chen, Y.-F. and Helmann, J. D. (1995) The *Bacillus subtilis* flagellar regulatory protein σ^D: overproduction, domain analysis and DNA-binding properties. *J. Mol. Biol.*, **249,** 743–753.
55. Severinova, E., Severinov, K., Fenyö, D., Marr, M., Brody, E. N., Roberts, J. W., Chait, B. T., and Darst, S. A. (1996) Domain organization of the *Escherichia coli* RNA polymerase σ^{70} subunit. *J. Mol. Biol.*, **263,** 637–647.
56. Helmann, J. D. and Chamberlin, M. J. (1988) Structure and function of bacterial σ factors. *Ann. Rev. Biochem.*, **57,** 839–872.
57. Juang, Y. L. and Helmann, J. D. (1994) A promoter melting region in the primary σ factor of *Bacillus subtilis*: Identification of functionally important aromatic amino acids. *J. Mol. Biol.*, **235,** 1470–1488.
58. Marr, M. T. and Roberts, J. W. (1997) Promoter recognition as measured by binding of polymerase to nontemplate strand oligonucleotide. *Science*, **276,** 1258–1260.
59. Huang, X., Lopez de Saro, F. J., and Helmann, J. D. (1997) σ factor mutations affecting the sequence selective interaction of RNA polymerase with –10 region single-stranded DNA. *Nucl. Acids Res.*, **25,** 2603–2609.
60. Joo, D. M., Ng, N., and Calendar, R. (1997) A sigma-32 mutant with a single amino acid change in the highly conserved region 2.2 exhibits reduced core RNA polymerase affinity. *Proc. Natl. Acad. Sci. USA*, **94,** 4907–4912.
61. Malhotra, A., Severinova, E., and Darst, S. A. (1996) Crystal structure of a σ^{70} subunit fragment from *E. coli* RNA polymerase. *Cell*, **87,** 127–136.
62. Dombroski, A. J., Walter, W. A., Record Jr., M. T., Siegele, D. A., and Gross, C. A. (1992) Polypeptides containing highly conserved regions of transcription initiation factor σ^{70} exhibit specificity of binding to promoter DNA. *Cell*, **70,** 501–512.
63. Dombroski, A. J., Walter, W. A., and Gross, C. A. (1993) Amino-terminal amino acids modulate σ-factor DNA-binding activity. *Genes and Development*, **7,** 2446–2455.
64. Dombroski, A. J., Johnson, B. D., Lonetto, M., and Gross, C. A. (1996) The sigma subunit of *Escherichia coli* RNA polymerase senses promoter spacing. *Proc. Natl. Acad. Sci. USA*, **93,** 8858–8862.
65. Dombroski, A. J. (1997) Recognition of the –10 promoter sequence by a partial polypeptide of sigma-70 *in vitro*. *J. Biol. Chem.*, **272,** 3487–3494.
66. Zhou, Y. N., Walter, W. A., and Gross, C. A. (1992) A mutant σ^{32} with a small deletion in conserved region 3 of σ has reduced affinity for core RNA polymerase. *J. Bacteriol.*, **174,** 5005–5012.
67. Hernandez, V. J., Hsu, L. M., and Cashel, M. (1996) Conserved region 3 of *Escherichia coli* sigma-70 is implicated in the process of abortive transcription. *J. Biol. Chem.*, **271,** 18775–18779.
68. Travers, A. A. and Burgess, R. R. (1969) Cyclic re-use of the RNA polymerase sigma factor. *Nature*, **222,** 537–540.
69. Leirmo, S. and Record Jr, M. T. (1990) Structural, thermodynamic and kinetic studies of the interaction of $E\sigma^{70}$ RNA polymerase with promoter DNA. In *Nucleic acids and molecular biology* (ed. F. Eckstein and D. M. J. Lilley), pp. 123–151. Springer–Verlag, Berlin, Heidelberg.
70. von Hippel, P. H. and Berg, O. G. (1989) Facilitated target location in biological systems. *J. Biol. Chem.*, **264,** 675–678.
71. Kabata, H., Kurosawa, O., Arai, I., Washizu, M., Margarson, S. A., Glass, R. E., and Shimamoto, N. (1993) Visualization of single molecules of RNA polymerase sliding along DNA. *Science*, **262,** 1561–1563.

72. Kovacic, R. T. (1987) The 0_C closed complexes between *Escherichia coli* RNA polymerase and two promoters, T7-A3 and *lac* UV5. *J. Biol. Chem.*, **262,** 13654–13661.
73. Roe, J. H., Burgess, R. R., and Record, Jr, M. T. (1985) Temperature dependence of the rate constants of the *Escherichia coli* RNA polymerase-λ P_R promoter interaction: Assignment of the kinetic steps corresponding to protein conformational change and DNA opening. *J. Mol. Biol.*, **184,** 441–453.
74. Schickor, P., Metzger, W., Werel, W., Lederer, H., and Heumann, H. (1990) Topography of intermediates in transcription initiation of *E. coli*. *EMBO J.*, **9,** 2215–2220.
75. Roberts, C. W. and Roberts, J. W. (1996) Base-specific recognition of the nontemplate strand of promoter DNA by *E. coli* RNA polymerase. *Cell*, **86,** 495–501.
76. deHaseth, P. and Helmann, J. D. (1995) Open complex formation by *Escherichia coli* RNA polymerase: the mechanism of polymerase-induced strand separation of double helical DNA. *Mol. Microbiol.*, **16,** 817–824.
77. Buc, H. and McClure, W. R. (1985) Kinetics of open complex formation between *Escherichia coli* RNA polymerase and the *lacUV5* promoter: Evidence for a sequential mechanism involving three steps. *Biochemistry*, **24,** 2712.
78. Kirkegaard, K., Buc, H., Spassky, A., and Wang, J. C. (1983) Mapping of single-stranded regions in duplex DNA at the sequence level: Single-strand-specific cytosine methylation in RNA polymerase-promoter complexes. *Proc. Natl. Acad. Sci. USA*, **80,** 2544–2548.
79. Chen, Y. F. and Helmann, J. D. (1997) DNA-melting at the *Bacillus subtilis* flagellin promoter nucleates near –10 and expands unidirectionally. *J. Mol. Biol.*, **267,** 47–59.
80. Suh, W. C., Leirmo, S., and Record, Jr, M. T. (1992) Roles of Mg(II) in the mechanism of formation and dissociation of open complexes between *Escherichia coli* RNA polymerase and the λP_R promoter: Kinetic evidence for a second open complex requiring Mg(II). *Biochemistry*, **31,** 7815–7825.
81. Suh, W.-C., Ross, W., and Record, Jr, M. T. (1993) Two open complexes and a requirement for Mg^{2+} to open the λP_R transcription start site. *Science*, **259,** 358–361.
82. Zaychikov, E., Denissova, L., Meier, T., Goette, M., and Heumann, H. (1997) Influence of Mg^{2+} and temperature on formation of the transcription bubble. *J. Biol. Chem.*, **272,** 2259–2267.
83. Newlands, J. T., Ross, W., Gosink, K. K., and Gourse, R. L. (1991) Factor-independent activation of *Escherichia coli* rRNA transcription. II. Characterization of complexes of *rrnB* P1 promoters containing or lacking the upstream activator region with *Escherichia coli* RNA polymerase. *J. Mol. Biol.*, **220,** 569–583.
84. Ohlsen, K. L. and Gralla, J. D. (1992) DNA melting within stable closed complexes at the *Escherichia coli rrnB* P1 promoter. *J. Biol. Chem.*, **267,** 19813–19818.
85. Juang, Y.-L. and Helmann, J. D. (1995) Pathway of promoter melting by *Bacillus subtilis* RNA polymerase at a stable RNA promoter: Effects of temperature, delta protein, and sigma factor mutations. *Biochemistry*, **34,** 8465–8473.
86. Polyakov, A., Severinova, E., and Darst, S. A. (1995) Three-dimensional structure of *E. coli* core RNA polymerase: promoter binding and elongation conformations of the enzyme. *Cell*, **83,** 365–373.
87. Rees, W. A., Keller, R. W., Vesenka, J. P., Yang, G., and Bustamante, C. (1993) Evidence of DNA bending in transcription complexes imaged by scanning force microscopy. *Science*, **260,** 1646–1649.
88. Fredrick, K. and Helmann, J. D. (1997) RNA polymerase sigma factor determines start-site selection but is not required for upstream promoter element activation on heteroduplex (bubble) templates. *Proc. Natl. Acad. Sci. USA*, **94,** 4982–4987.

89. Liu, J. and Turnbough, Jr, C. L. (1994) Effects of transcriptional start site sequence and position on nucleotide-sensitive selection of alternative start sites at the *pyrC* promoter in *Escherichia coli*. *J. Bacteriol.*, **176,** 2938–2945.
90. Carpousis, A. J. and Gralla, J. D. (1980) Cycling of ribonucleic acid polymerase to produce oligonucleotides during initiation *in vitro* at the *lac* UV5 promoter. *Biochemistry,* **19,** 3245–3253.
91. Munson, L. and Reznikoff, W. S. (1981) abortive initiation and long ribonucleic acid synthesis. *Biochemistry,* **20,** 2181–2085.
92. Krummel, B. and Chamberlin, M. J. (1989) RNA chain initiation by *Escherichia coli* RNA polymerase. Structural transitions of the enzyme in early ternary complexes. *Biochemistry,* **28,** 7829–7842.
93. Jin, D.-J. and Turnbough Jr, C. L. (1994) An *Escherichia coli* RNA polymerase defective in transcription due to its overproduction of abortive initiation products. *J. Mol. Biol.,* **236,** 72–80.
94. Kammerer, W., Deuschle, U., Gentz, R., and Bujard, H. (1986) Functional dissection of *Escherichia coli* promoters: information in the transcribed region is involved in late steps in the overall process. *EMBO J.,* **5,** 2995–3000.
95. Ellinger, T., Behnke, D., Bujard, H., and Gralla, J. D. (1994) Stalling of *Escherichia coli* RNA polymerase in the +6 to +12 region *in vivo* is associated with tight binding to consensus promoter elements. *J. Mol. Biol.,* **239,** 455–465.
96. Knaus, R. and Bujard, H. (1990) Principles governing the activity of *E. coli* promoters. In *Nucleic acids and molecular biology* (ed. F. Eckstein and D. M. J. Lilley), pp. 110–122. Springer–Verlag, Heidelberg.
97. Hsu, L. M. (1996) Quantitative parameters for promoter clearance. *Methods Enzymol.* **273,** 59–71.
98. Jacques, J.-P. and Susskind, M. M. (1990) Psudeo-templated transcription by *Escherichia coli* RNA polymerase at a mutant promoter. *Genes and Development,* **4,** 1801–1810.
99. Jin, D. J. (1994) Slippage synthesis at the *galP2* promoter of *Escherichia coli* and its regulation by UTP concentration and cAMP-cAMP receptor protein. *J. Biol. Chem.,* **269,** 17221–17227.
100. Jin, D. J. (1996) A mutant RNA polymerase reveals a kinetic mechanism for the switch between nonproductive stuttering synthesis and productive initiation during promoter clearance. *J. Biol. Chem.,* **271,** 11659–11667.
101. Xiong, X. F. and Reznikoff, W. S. (1993) Transcriptional slippage during the transcription initiation process at a mutant *lac* promoter *in vivo*. *J. Mol. Biol.,* **231,** 569–580.
102. Qi, F. and Turnbough, J. C. L. (1995) Regulation of *codBA* operon expression in *Escherichia coli* by UTP-dependent reiterative transcription and UTP-sensitive transcriptional start site switching. *J. Mol. Biol.,* **254,** 552–565.
103. Vogel, U., Sørensen, M., Pedersen, S., Jensen, V. F., and Kilstrup, M. (1992) Decreasing transcription elongation rate in *Escherichia coli* exposed to amino acid elongation. *Mol. Microbiol.,* **6,** 2191–2200.
104. Chan, C. L. and Landick, R. (1993) Dissection of the *his* leader pause site by base substitution reveals a multipartite signal that includes a pause RNA hairpin. *J. Mol. Biol.,* **233,** 25–42.
105. Yanofsky, C. (1988) Transcription attenuation. *J. Biol. Chem.,* **263,** 609–612.
106. Ring, B. Z., Yarnell, W. S., and Roberts, J. W. (1996) Function of *E. coli* RNA polymerase sigma factor sigma-70 in promoter-proximal pausing. *Cell,* **86,** 485–493.
107. von Hippel, P. H., Bear, D. G., Morgan, W. D., and McSwiggen, J. A. (1984) Protein–

nucleic acid interactions in transcription: A molecular analysis. *Ann. Rev. Biochem.*, **53,** 389–446.

108. Krummel, B. and Chamberlin, M. J. (1992) Structural analysis of ternary complexes of *Escherichia coli* RNA polymerase: Individual complexes halted along different transcription units have distinct and unexpected biochemical properties. *J. Mol. Biol.* **225,** 221–237.
109. Krummel, B. and Chamberlin, M. J. (1992) Structural analysis of ternary complexes of *Escherichia coli* RNA polymerase: Deoxyribonuclease I footprinting of defined complexes. *J. Mol. Biol.*, **225,** 239–250.
110. Nudler, E., Goldfarb, A., and Kashlev, M. (1994) Discontinuous mechanism of transcription elongation. *Science*, **265,** 793–796.
111. Zaychikov, E., Denissova, L., and Heumann, H. (1995) Translocation of the *Escherichia coli* transcription complex observed in the registers 11 to 20: 'Jumping' of RNA polymerase and asymmetric expansion and contraction of the 'transcription bubble'. *Proc. Natl. Acad. Sci. USA*, **92,** 1739–1743.
112. Chamberlin, M. J. (1995) New models for the mechanism of transcription elongation and its regulation. *Harvey Lect.*, **88,** 1–21.
113. Nudler, E., Kashlev, M., Nikiforov, V., and Goldfarb, A. (1995) Coupling between transcription termination and RNA polymerase inchworming. *Cell*, **81,** 351–357.
114. Wang, D., Meier, T. I., Chan, C. L., Feng, G., Lee, D. N., and Landick, R. (1995) Discontinuous movements of DNA and RNA in RNA polymerase accompany formation of a paused transcription complex. *Cell*, **81,** 341–350.
115. Arndt, K. M. and Chamberlin, M. J. (1990) RNA chain elongation by *Escherichia coli* RNA polymerase: Factors affecting the stability of elongating ternary complexes. *J. Mol. Biol.*, **213,** 79–108.
116. Komissarova, N. and Kashlev, M. (1997) RNA polymerase switches between inactivated and activated states by translocating back and forth along the DNA and the RNA. *J. Biol. Chem.*, **272,** 15329–15338.
117. Komissarova, N. and Kashlev, M. (1997) Transcriptional arrest: *Escherichia coli* RNA polymerase translocates backward, leaving the 3′ end of the RNA intact and extruded. *Proc. Natl. Acad. Sci. USA*, **94,** 1755–1760.
118. Erie, D. A., Hajiseyedjavadi, O., Young, M. C., and von Hippel, P. H. (1993) Multiple RNA polymerase conformations and GreA: Control of the fidelity of transcription. *Science*, **262,** 867–873.
119. Koulich, D., Orlova, M., Malhotra, A., Sali, A., Darst, S. A., and Borukhov, S. (1997) Domain organization of *Escherichia coli* transcript cleavage factors GreA and GreB. *J. Biol. Chem.*, **272,** 7201–7210.
120. Kashlev, M., Nudler, E., Goldfarb, A., White, T., and Kutter, E. (1993) Bacteriophage Alc protein: A transcription termination factor sensing local modification of DNA. *Cell*, **75,** 147–154.
121. Yager, T. D. and von Hippel, P. H. (1991) A thermodynamic analysis of RNA transcript elongation and termination in *Escherichia coli*. *Biochemistry*, **30,** 1097–1118.
122. von Hippel, P. H. and Yager, T. D. (1992) The elongation–termination decision in transcription. *Science*, **255,** 809–812.
123. Nudler, E., Mustaev, A., Lukhtanov, E., and Goldfarb, A. (1997) The RNA–DNA hybrid maintains the register of transcription by preventing backtracking of RNA polymerase. *Cell*, **89,** 33–41.
124. Liu, B., Wong, M. L., Tinker, R. L., Geiduschek, E. P., and Alberts, B. M. (1993) The DNA

replication fork can pass RNA polymerase without displacing the nascent transcript. *Nature*, **366**, 33–39.

125. Liu, B. and Alberts, B. M. (1995) Head-on collision between a DNA replication apparatus and RNA polymerase transcription complex. *Science*, **267**, 1131–1137.
126. Das, A. (1993) Control of transcription termination by RNA-binding proteins. *Ann. Rev. Biochem.*, **62**, 893–930.
127. Richardson, J. P. (1991) Preventing the synthesis of unused transcripts by Rho factor. *Cell*, **64**, 1047–1049.
128. Henkin, T. M. (1996) Control of transcription termination in prokaryotes. *Ann. Rev. Genet.*, **30**, 35–57.
129. Surratt, C. K., Milan, S. C., and Chamberlin, M. J. (1991) Spontaneous cleavage of RNA in ternary complexes of *Escherichia coli* RNA polymerase and its significance for the mechanism of transcription. *Proc. Natl. Acad. Sci. USA*, **88**, 7983–7987.
130. Reynolds, R. and Chamberlin, M. J. (1992) Parameters affecting transcription termination by *Escherichia coli* RNA Polymerase. II. Construction and analysis of hybrid terminators. *J. Mol. Biol.*, **224**, 53–63.
131. Arndt, K. M. and Chamberlin, M. J. (1988) transcription termination in *Escherichia coli*: Measurement of the rate of enzyme release from Rho-independent terminators. *J. Mol. Biol.*, **202**, 271–285.
132. Juang, Y. L. and Helmann, J. D. (1994) The δ subunit of *Bacillus subtilis* RNA polymerase: An allosteric effector of the initiation and core-recycling phases of transcription. *J. Mol. Biol.*, **239**, 1–14.
133. López de Saro, F., Woody, A.-Y. M., and Helmann, J. D. (1995) Structural analysis of the *Bacillus subtilis* δ factor: A protein polyanion which displaces RNA from RNA polymerase. *J. Mol. Biol.*, **252**, 189–202.
134. Platt, T. (1994) Rho and RNA: models for recognition and response. *Mol. Microbiol.*, **11**, 983–990.
135. Geiselmann, J., Wang, Y., Seifried, S. E., and von Hippel, P. H. (1993) A physical model for the translocation and helicase activities of *Escherichia coli* transcription termination protein Rho. *Proc. Natl. Acad. Sci. USA*, **90**, 7754–7758.
136. Greenblatt, J., Nodwell, J. R., and Mason, S. W. (1993) Transcriptional antitermination. *Nature*, **364**, 401–406.
137. Squires, C. L., Greenblatt, J., Li, J., Condon, C., and Squires, C. L. (1993) Ribosomal RNA antitermination *in vitro*: Requirement for Nus factors and one or more unidentified cellular components. *Proc. Natl. Acad. Sci. USA*, **90**, 970–974.
138. Hawley, D. K. and McClure, W. R. (1983) Compilation and analysis of *Escherichia coli* promoter sequences. *Nucl. Acids Res.*, **11**, 2237–2255.
139. Helmann, J. D. (1995) Compilation and analysis of *Bacillus subtilis* σ^A-dependent promoter sequences: Evidence for extended contact between RNA polymerase and upstream promoter DNA. *Nucl. Acids Res.*, **23**, 2351–2360.
140. Youderian, P., Bouvier, S., and Susskind, M. M. (1982) Sequence determinants of promoter activity. *Cell*, **30**, 843–853.
141. Mulligan, M. E., Hawley, D. K., Entriken, R., and McClure, W. R. (1984) *Escherichia coli* promoter sequences predict *in vitro* RNA polymerase selectivity. *Nucl. Acids Res.*, **12**, 789–800.
142. Graña, D., Gardella, T., and Susskind, M. M. (1988) The effects of mutations in the *ant* promoter of phage P22 depend on context. *Genetics*, **120**, 319–327.
143. Cowing, D. W., Mecsas, J., Record Jr, M. T., and Gross, C. A. (1989) Intermediates in the

formation of the open complex by RNA polymerase holoenzyme containing the sigma factor σ^{32} at the *groE* promoter. *J. Mol. Biol.*, **210,** 521–530.

144. Grimes, E., Busby, S., and Minchin, S. (1991) Different thermal energy requirement for open complex formation by *Escherichia coli* RNA polymerase at two related promoters. *Nucl. Acids Res*, **19,** 6113–6118.
145. Mecsas, J., Cowing, D. W., and Gross, C. A. (1991) Development of RNA polymerase–promoter contacts during open complex formation. *J. Mol. Biol.*, **220,** 585–597.
146. Simpson, R. B. (1979) The molecular topography of RNA polymerase–promoter interaction. *Cell*, **18,** 277–285.
147. Banner, C. D. B., Moran Jr, C. P., and Losick, R. (1983) Deletion analysis of a complex promoter for a developmentally regulated gene from *Bacillus subtilis. J. Mol. Biol.*, **168,** 351–365.
148. Hsu, L. M., Giannini, J. K., Leung, T. C., and Crosthwaite, J. C. (1991) Upstream sequence activation of *Escherichia coli argT* promoter *in vivo* and *in vitro. Biochemistry*, **30,** 813–822.
149. Leirmo, S. and Gourse, R. L. (1991) Factor-independent activation of *Escherichia coli* rRNA transcription. I. Kinetic analysis of the roles of the upstream activator region and supercoiling on transcription of the *rrnB* P1 promoter *in vitro. J. Mol. Biol.*, **220,** 555–568.
150. Pérez-Martín, J., Rojo, F., and DeLorenzo, V. (1994) Promoters responsive to DNA bending: a common theme in prokaryotic gene expression. *Microbiol. Rev.*, **58,** 268–290.
151. Ross, W., Thompson, J. F., Newlands, J. T., and Gourse, R. L. (1990) *E. coli* Fis protein activates ribosomal RNA transcription *in vitro* and *in vivo. EMBO J.*, **9,** 3733–3742.
152. Rao, L., Ross, W., Appleman, J. A., Gaal, T., Leirmo, S., Schlax, P. J., Record Jr, M. T., and Gourse, R. L. (1994) Factor independent activation of *rrnB* P1: An 'extended' promoter with an upstream element that dramatically increases promoter strength. *J. Mol. Biol.*, **235,** 1421–1435.
153. Crothers, D. M., Haran, T. E., and Nadeau, J. G. (1990) Intrinsically bent DNA. *J. Biol. Chem.*, **265,** 7093–7096.
154. Plaskon, R. R. and Wartell, R. M. (1987) Sequence distributions associated with DNA curvature are found upstream of strong *E. coli* promoters. *Nucl. Acids Res.*, **15,** 785–796.
155. VanWye, J. D., Bronson, E. C., and Anderson, J. N. (1991) Species-specific patterns of DNA bending and sequence. *Nucl. Acids Res.*, **19,** 5253–5261.
156. Espinosa-Urgel, M. and Tormo, A. (1993) σ^S-dependent promoters in *Escherichia coli* are located in DNA regions with intrinsic curvature. *Nucl. Acids Res.*, **21,** 3667–3670.
157. Tanaka, K.-I., Muramatsu, S., Yamada, H., and Mizuno, T. (1991) Systematic characterization of curved DNA segments randomly cloned from *Escherichia coli* and their functional significance. *Molecular and General Genetics*, **226,** 367–376.
158. Ayers, D. G., Auble, D. T., and deHaseth, P. L. (1989) Promoter recognition by *Escherichia coli* RNA polymerase: Role of the spacer DNA in functional complex formation. *J. Mol. Biol.*, **207,** 749–756.
159. Stefano, J. E. and Gralla, J. D. (1982) Spacer mutations in the *lac* p^s promoter. *Proc. Natl. Acad. Sci. USA*, **79,** 1069–1072.
160. Amouyal, M. and Buc, H. (1987) Topological unwinding of strong and weak promoters by RNA polymerase. A comparison between the *lac* wild-type and the UV5 sites of *Escherichia coli. J. Mol. Biol.*, **195,** 795–808.
161. Borowiec, J. A. and Gralla, J. D. (1987) All three elements of the *lac* p^s promoter mediate its transcriptional response to DNA supercoiling. *J. Mol. Biol.*, **195,** 89–97.
162. Travers, A. A. (1987) Structure and function of *E. coli* promoter DNA. *CRC Crit. Rev. Biochem.*, **22,** 181–219.

163. Wang, J. Y. and Syvanen, M. (1992) DNA twist as a transcriptional sensor for environmental changes. *Mol. Microbiol.*, **6,** 1861–1866.
164. Keilty, S. and Rosenberg, M. (1987) Constitutive function of a positively regulated promoter reveals new sequences essential for activity. *J. Biol. Chem.*, **262,** 6389–6395.
165. Burns, H. and Minchin, S. (1994) Thermal energy requirement for strand separation during transcription initiation: The effect of supercoiling and extended protein DNA contacts. *Nucl. Acids Res.*, **22,** 3840–3845.
166. Kumar, A., Malloch, R. A., Fujita, N., Smillie, D. A., Ishihama, A., and Hayward, R. S. (1993) The minus 35-recognition region of *Escherichia coli* sigma 70 is inessential for initiation of transcription at an 'extended minus 10' promoter. *J. Mol. Biol.*, **232,** 406–418.
167. Barne, K. A., Bown, J. A., Busby, S. J. W., and Minchin, S. D. (1997) Region 2.5 of the *Escherichia coli* RNA polymerase σ^{70} subunit is responsible for the recognition of the 'extended –10' motif at promoters. *EMBO J.*, **16,** 4034–4040.
168. Beutel, B. A. and Record Jr, M. T. (1990) *Escherichia coli* promoter spacer regions contain nonrandom sequences which correlate to spacer length. *Nucl. Acids Res.*, **18,** 3597–3604.
169. Knaus, R. and Bujard, H. (1988) P_L of coliphage lambda: an alternative solution for an efficient promoter. *EMBO J.*, **7,** 2919–2923.
170. Ellinger, T., Behnke, D., Knaus, R., Bujard, H., and Gralla, J. D. (1994) Context-dependent effects of upstream A-tracts: Stimulation or inhibition of *Escherichia coli* promoter function. *J. Mol. Biol.*, **239,** 466–475.
171. Goliger, J. A., Yang, X., Guo, H.-C., and Roberts, J. W. (1989) Early transcribed sequences affect termination efficiency of *Escherichia coli* RNA polymerase. *J. Mol. Biol.*, **205,** 331–341.
172. Telesnitsky, A. P. W. and Chamberlin, M. J. (1989) Sequences linked to prokaryotic promoters can affect the efficiency of downstream termination sites. *J. Mol. Biol.*, **205,** 315–330.
173. Tjian, R. and Pero, J. (1976) Bacteriophage SPO1 regulatory proteins directing late gene transcription *in vitro*. *Nature*, **262,** 753–757.
174. Pero, J., Nelson, J., and Fox, T. D. (1975) Highly asymmetric transcription by RNA polymerase containing phage-SP01-induced polypeptides and a new host protein. *Proc. Natl. Acad. Sci. USA*, **72,** 1589–1593.
175. Fleischmann, R. D., Adams, M. D., White, O., Clayton, R. A., Kirkness, E. F., Kerlavage, A. R., *et al.* (1995) Whole-genome random sequencing and assembly of *Haemophilus influenzae* Rd. *Science*, **269,** 496–512.
176. Gribskov, M. and Burgess, R. R. (1986) Sigma factors from *E. coli, B. subtilis,* phage SP01, and phage T4 are homologous proteins. *Nucl. Acids Res.*, **14,** 6745–6761.
177. Lonetto, M., Gribskov, M., and Gross, C. A. (1992) The σ^{70} family: Sequence conservation and evolutionary relationships. *J. Bacteriol.*, **174,** 3843–3849.
178. Lonetto, M. A., Brown, K. L., Rudd, K. E., and Buttner, M. J. (1994) Analysis of the *Streptomyces coelicolor sigE* gene reveals the existence of a subfamily of eubacterial σ factors involved in the regulation of extracytoplasmic functions. *Proc. Natl. Acad. Sci. USA*, **91,** 7573–7577.
179. Merrick, M. J. (1993) In a class of its own—the RNA polymerase sigma factor σ^{54} (σ^N). *Mol. Microbiol.*, **10,** 903–909.
180. Cannon, W., Austin, S., Moore, M., and Buck, M. (1995) Identification of close contacts between the sigma-N (sigma-54) protein and promoter DNA in closed promoter complexes. *Nucl. Acids Res.*, **23,** 351–356.

181. Cannon, W., Missailidi, S., Smith, C., Cottier, A., Austin, S., Moore, M., and Buck, M. (1995) Core RNA polymerase and promoter DNA interactions of purified domains of σ^N: Bipartite functions. *J. Mol. Biol.*, **248,** 781–803.
182. Merrick, M. and Chambers, S. (1992) The helix–turn–helix motif of σ^{54} is involved in recognition of the –13 promoter region. *J. Bacteriol.*, **174,** 7221–7226.
183. Wong, C., Tintut, Y., and Gralla, J. D. (1994) The domain structure of sigma 54 as determined by analysis of a set of deletion mutants. *J. Mol. Biol.*, **236,** 81–90.
184. Wong, C. and Gralla, J. D. (1992) A role for the acidic trimer repeat region of transcription factor σ^{54} in setting the rate and temperature dependence of promoter melting *in vivo*. *J. Biol. Chem.*, **267,** 24762–24768.
185. Syed, A. and Gralla, J. D. (1997) Isolation and properties of enhancer-bypass mutants of sigma 54. *Mol. Microbiol.*, **23,** 987–995.
186. Wang, J. T., Syed, A., Hsieh, M., and Gralla, J. D. (1995) Converting *Escherichia coli* RNA polymerase into an enhancer-responsive enzyme: role of an NH_2-terminal leucine patch in σ^{54}. *Science*, **270,** 992–994.
187. Bukau, B. (1993) Regulation of the *Escherichia coli* heat-shock response. *Mol. Microbiol.*, **9,** 671–680.
188. Gross, C. A. (1996) Function and regulation of the heat shock proteins. In *Escherichia coli and Salmonella* (ed. F. C. Neidhardt), pp. 1382–1399. ASM Press, Washington DC.
189. Straus, D. B., Walter, W. A., and Gross, C. A. (1987) The heat shock response of *E. coli* is regulated by changes in the concentration of σ^{32}. *Nature*, **329,** 348–351.
190. Tomoyasu, T., Gamer, J., Bukau, B., Kanemori, M., Mori, H., Rutman, A. J., Oppenheim, A. B., Yura, T., Yamanaka, K., Niki, H., Hiraga, S., and Ogura, T. (1995) *Escherichia coli* FtsH is a membrane-bound, ATP-dependent protease which degrades the heat-shock transcription factor σ^{32}. *EMBO J.*, **14,** 2551–2560.
191. Herman, C., Thévenet, D., D'Ari, R., and Bouloc, P. (1995) Degradation of σ^{32}, the heat shock regulator in *Escherichia coli*, is governed by HflB. *Proc. Natl. Acad. Sci. USA*, **92,** 3516–3520.
192. McCarty, J. S., Rüdiger, S., Schönfeld, H.-J., Schneider-Mergner, J., Nakahigashi, K., Yura, T., and Bukau, B. (1996) Regulatory region C of *E. coli* heat shock transcription factor, σ^{32}, constitutes a DnaK binding site and is conserved among Eubacteria. *J. Mol. Biol.*, **256,** 829–837.
193. Ozaki, M., Fujita, N., Wada, A., and Ishihama, A. (1992) Promoter selectivity of the stationary-phase forms of *Escherichia coli* RNA polymerase and conversion *in vitro* of the S1 form enzyme into a log-phase enzyme-like form. *Nucl. Acids Res.*, **20,** 257–261.
194. Ozaki, M., Wada, A., Fujita, N., and Ishihama, A. (1991) Growth phase-dependent modification of RNA polymerase in *Escherichia coli*. *Mol. Gen. Genet.*, **230,** 17–23.
195. Brody, E. N., Kassavetis, G. A., Ouhammouch, M., Sanders, G. M., Tinker, R. L., and Geiduschek, E. P. (1995) Old phage, new insights: Two recently recognized mechanisms of transcriptional regulation in bacteriophage T4 development. *FEMS Micro. Letts.*, **128,** 1–8.
196. March-Amegadzie, R. and Hinton, D. M. (1995) The bacteriophage T4 middle promoter P_{uvsX}: Analysis of regions important for binding of the T4 transcriptional activator MotA and for activation of transcription. *Mol. Microbiol.*, **15,** 649–660.
197. Ouhammouch, M., Adelman, K., Harvey, S. R., Orsini, G., and Brody, E. N. (1995) Bacteriophage T4 MotA and AsiA proteins suffice to direct *Escherichia coli* RNA polymerase to initiate transcription at T4 middle promoters. *Proc. Natl. Acad. Sci. USA*, **92,** 1451–1455.

198. Hinton, D. M., March-Amegadzie, R., Gerber, J. S., and Sharma, M. (1996) Characterization of pre-transcription complexes made at a bacteriophage T4 middle promoter: Involvement of the T4 MotA activator and the T4 AsiA protein, a sigma-70 binding protein, in the formation of the open complex. *J. Mol. Biol.*, **256,** 235–248.
199. Tinker, R. L., Williams, K. P., Kassavetis, G. A., and Geiduschek, E. P. (1994) Transcriptional activation by a DNA-tracking protein: Structural consequences of enhancement at the T4 late promoter. *Cell*, **77,** 225–237.
200. Tinker, R. L., Sanders, G. M., Severinov, K., Kassavetis, G. A., and Geiduschek, E. P. (1995) The COOH-terminal domain of the RNA polymerase α subunit in transcriptional enhancement and deactivation at the bacteriophage T4 late promoter. *J. Biol. Chem.*, **270,** 15899–15907.
201. Severinov, K., Fenyo, D., Severinova, E., Mustaev, A., Chait, B. T., Goldfarb, A., and Darst, S. A. (1994) The σ subunit conserved region 3 is part of '5′-face' of active center of *Escherichia coli* RNA polymerase. *J. Biol. Chem.*, **269,** 20826–20828.
202. Cannon, W. V., Chaney, M. K., Wang, X. Y., and Buck, M. (1997) Two domains within sigma-N (sigma-54) cooperate for DNA binding. *Proc. Natl. Acad. Sci. USA*, **94,** 5006–5011.

4 Repressors and activators

STEPHEN J. W. BUSBY

1. Introduction

The study of adaptive responses in bacteria led to the discovery of a panoply of gene regulatory proteins (1). These are mostly DNA-binding proteins and their role is to modulate promoter activity, thereby coupling the expression of specific genes to changes in the environment. Gene regulatory proteins are ubiquitous: they are essential for all processes of adaptation and differentiation in all living cells. At first sight this is a very complex topic, made incomprehensible by the sheer number of different factors and different modes of operation. The aim of this chapter is to tackle the central question of how gene regulatory proteins function, and to describe a small number of key examples from the microbial repertoire, that reveal principles applicable to most systems.

It is interesting to note that it was a genetic study of the *Escherichia coli* lactose operon that led to the discovery of the first example of this class of regulatory protein (2, 3). Monod and co-workers noted an abundant class of mutants in which β-galactosidase expression was constitutive rather than inducible. The observation that inducibility could be restored by introducing genetic material *in trans* led directly to the postulate that this material was coding for a repressor (and that the mutants had a defective repressor). This led to the suggestion that all regulation was due to repressors and that all genetic material was 'tied up' with repressors that 'lay awaiting' their cognate inducer. This view was challenged in the late sixties by experiments with the arabinose and maltose operons (4). Frequent mutations led to non-inducibility, but inducibility could be restored by introducing genetic material *in trans*. This led to the discovery of gene activator proteins: the mutations destroyed activator function but inducibility could be restored by genetic material encoding the activator.

Since the discovery of the genes encoding the *lac* repressor (*lacI*) and the arabinose (*araC*) and maltose (*malT*) activator proteins, dozens of genes encoding activators and regulators have been found and characterized. More than 5 per cent of *E. coli* genes code for such gene regulatory proteins (5): some are specific regulators acting at very few loci (e.g. the *lac* repressor), whilst others are global regulators controlling hundreds of genes in response to particular signals (e.g. the cyclic AMP receptor protein, CRP, which is triggered by the elevation of intracellular cAMP levels). Finally, some gene regulatory proteins appear to be 'bystanders', whose main function is to organize the *E. coli* chromosome, but have been 'co-opted' into playing a regulatory role at particular promoters (e.g. HNS and IHF proteins, see later).

2. Families of gene regulatory proteins

Gene manipulation techniques have simplified the characterization of activators and repressors, and hundreds of sequences are now available. Simple sequence analysis shows that most activators and repressors belong to a relatively small number of families, members of which share family traits. The main families of activators and repressors are listed in Table 1. It is important to realize that many transcription factors are made up of modules and that these modules can appear in a range of different proteins. For example, the LuxR-type DNA-binding module appears at the C-terminal of many response-regulators, as well as at the C-termini of diverse transcription factors (e.g. LuxR and MalT) and σ factors (e.g. σ^{70}).

In many cases, binding sites for activators and repressors have been identified by both genetics and biochemistry. The positions of base substitutions that interfere with activation (for activators) or repression (for repressors), together with the location of protected bases in footprinting experiments, have been sufficient to define factor-binding elements at many promoters. In most cases, activator-binding sites are located further upstream than repressor-binding sites (with respect to the transcript start site), although this is not a hard and fast rule (6). The majority of repressor-binding sites (operators) overlap the RNA polymerase (RNAP) binding site (defined from footprinting as bases between ~ -50 and $\sim +20$). At most activatable promoters served by RNAP containing σ^{70}, the activator binds between -30 and -100 (this applies to the most downstream activator at promoters regulated by multiple activators). For promoters served by RNAP containing σ^{54}, the activator often binds further upstream (6) (here I shall mainly concentrate on promoters served by σ^{70} but include some comments on σ^{54} later).

3. Simple activation

3.1 Models for activation

There are many promoters where a single transcription activator is sufficient to stimulate transcription initiation by RNA polymerase, and some of these can be studied in simple *in-vitro* systems consisting of promoter DNA, RNAP, and the activator. The best-understood cases are activation of the *E. coli lac* promoter by CRP, activation of the phage lambda P_{RM} promoter by the lambda repressor (cI), and activation of the *merT* promoter by MerR (18–20). These studies show that transcription activators accelerate the formation of open complexes that are similar to those forming at activator-independent promoters. This is a crucial point: activators do not create promoters, but do make preexisting poor promoters more efficient. Hence most activator-dependent promoters are as susceptible to mutations in the –10 and

Table 1 Major families of bacterial transcription factors

The Ara C family (7)

Examples:	*E. coli* AraC, MelR, RhaS, RhaR, SoxS Many homologues in other organisms
Domain structure:	N-terminal domain concerned with triggering by small ligand. C-terminal domain carries two helix–turn–helix motifs responsible for operator binding.
Main properties:	Transcription activators that overlap –35 region. Bind to ~18-bp sequence in absence and presence of ligand.

The LysR family (8)

Examples:	*E. coli* LysR, OxyR, MetR, CysB Many homologues in other organisms
Domain structure:	N-terminal domain carries helix–turn–helix motif responsible for operator binding. C-terminal domain concerned with triggering.
Main properties:	Co-inducer responsive transcription activators. Bind in absence and presence of ligand.

The CRP family (9)

Examples:	*E. coli* CRP and FNR Homologues in many other organisms
Domain structure:	C-terminal DNA-binding domain carries helix–turn–helix. N-terminal domain concerned with triggering.
Main properties:	Transcription activators. Binding to target is ligand dependent. Variety of promoter architectures.

The MerR family (10)

Examples:	*E. coli* SoxR and transposon-encoded MerR Homologues in other organisms
Domain structure:	No evidence for domains. N-terminal carries helix–turn–helix and C-terminal concerned with triggering.
Main properties:	Transcription activators. Binding to target is ligand independent.

The response-regulator family (11, 12)

Examples:	*E. coli* NarL, NarP, UhpA, OmpR, PhoB Many homologues all in other organisms
Domain structure:	N-terminal domain (the response domain) triggered by phosphorylation. C-terminal domain carries helix–turn–helix motif that binds DNA. There are two types of domain: the 'OmpR' module found in OmpR and PhoB etc., and the 'LuxR' module found in NarL, NarP, and UhpA etc. The 'LuxR' DNA-binding module is also found in some activators(e.g. LuxR and MalT)and in sigma factors (13).
Main properties:	Transcription activators that bind at a variety of positions in target promoters.

The σ^{54} bacterial enhancer-binding family (14, 15)

Examples:	*E. coli* FhlA, *Klebsiella pneumoniae* NifA and NtrC. Homologues in many organisms
Domain structure:	N-terminal domain responsible for triggering (some N-terminal domains arerelated to the response domains of the response-regulator family). C-terminal part carries helix–turn–helix motif that binds DNA and a segment that contacts $E\sigma^{54}$.
Main properties:	Transcription enhancer-like proteins that can bind well upstream of target promoters.

The Lac repressor family (16)

Examples:	*E. coli* LacI, GalR, PurR, CytR
Domain structure:	N-terminal carries helix–turn–helix motif that binds DNA. C-terminal carries segment responsible for triggering.
Main properties:	Transcription repressors. Proteins bind as dimer but some can form tetramer.

The MetJ repressor family (17)

Examples:	*E. coli* MetJ, phage P22 Arc and Mnt repressors
Domain structure:	Single domain.
Main properties:	Transcription repressors. Contact DNA via β strand.

–35 hexamer sequences as activator-independent promoters. This is vividly demonstrated at the *lac* promoter where CRP-dependent expression can be suppressed by substitutions in both the –10 or –35 hexamers and, conversely, the requirement for CRP can be simply short-circuited by improvement of the –10 hexamer (21). Moreover, CRP is essential to accelerate open-complex formation (it does this by simply improving the initial binding of RNAP to the promoter), but it is not needed for transcription initiation once the open complex has formed (22). Because of fierce competition for RNAP in the cell, small differences in promoter efficiency will result in big differences in expression. To be effective as a switch, any activator need only contribute a few kilocalories to an open complex (one or two hydrogen bonds). Stated simply, activator-independent promoters contain sufficient DNA sequence information in their –10, –35, and UP elements for RNAP docking. Activator-dependent promoters carry one or more defective elements and thus RNAP needs help from activator proteins. Activators can function to accelerate closed-complex formation, open-complex formation, or promoter clearance (23).

Two simple models have been advanced to explain how simple activators function. One model supposes that activators provide a direct contact point for RNA polymerase, whilst the other postulates that there is no direct activator–RNAP contact but that the activator alters the conformation of target DNA to make it more 'attractive' to RNAP. Although the primary mechanism of activation in most cases appears to involve direct contact, the case of MerR provides an important exception. It is instructive to consider the cases of the *lac*, lambda P_{RM}, and *merT* promoters separately.

3.2 The *E coli lac* promoter

Transcription activation at the *lac* promoter requires the binding of CRP to a site centred between base pairs –61 and –62 (position –61.5) upstream of the transcript start site (21). Two observations argue that the primary effect of CRP may be to alter the promoter conformation. First, CRP induces a bend in the DNA target greater than 90° upon binding (24), and, second, curved DNA sequences (introduced into the *lac* promoter by cloning) can supplant the requirement for CRP (25). However, the discovery that activation by CRP requires a seven amino acid surface-exposed β turn (residues 156–162: the CRP activating region) provides strong evidence of a crucial role for CRP–RNAP contacts. CRP carrying substitutions in this region can bind and bend *lac* promoter DNA normally and yet is unable to activate transcription or to interact cooperatively with RNAP (26). This shows that CRP-induced DNA bending alone is not sufficient for transcription activation: the activating effects of cloned curved sequences are likely to be by providing an independent path for transcription initiation (e.g. by providing contacts for αCTD).

Several lines of evidence show that the activating region is a contact site for αCTD and that this contact is essential for transcription activation (27, 28). First, photocross-linking indicates that the activating region of CRP and αCTD are in close proximity in the ternary complex between CRP, RNAP, and *lac* promoter DNA. Second, in

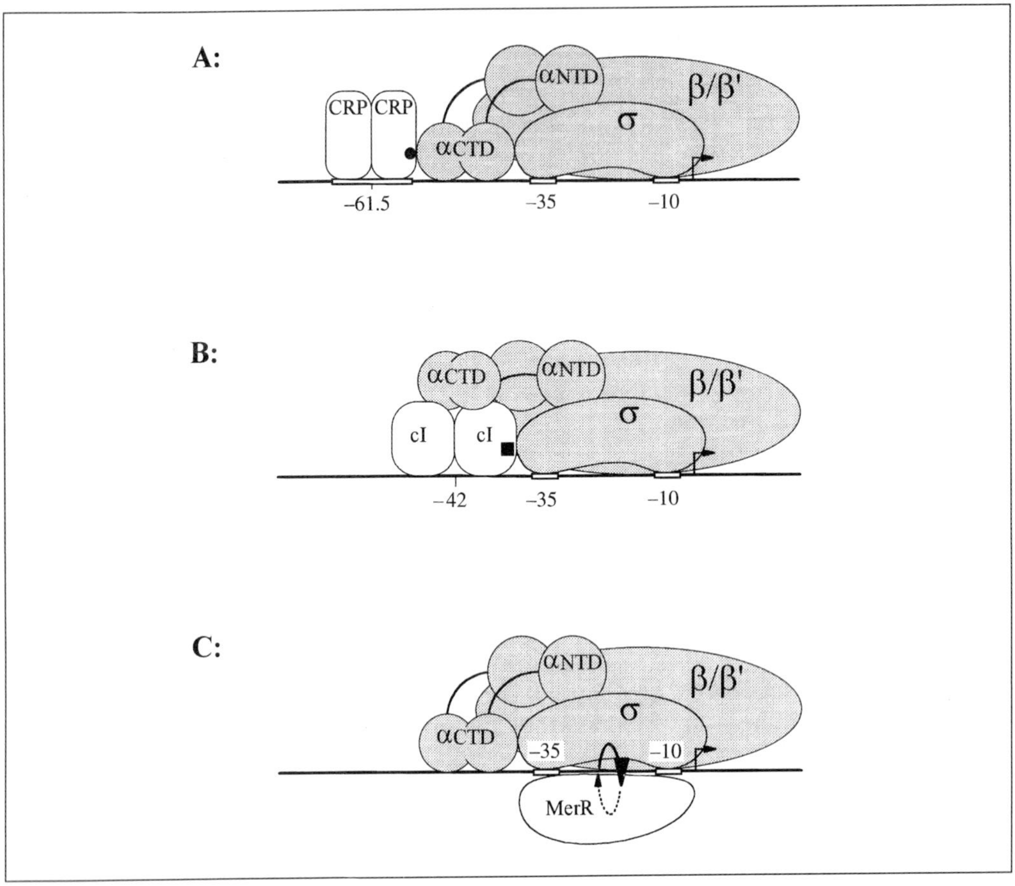

Fig. 1 Simple transcription activation by CRP, cI, and MerR. The figure shows the likely organisation of RNAP and activator subunits in transcriptionally competent complexes at different promoters. Although, for clarity, the DNA is drawn as a straight line, the path of the DNA is likely to be distorted with the bends facilitating the different contacts. A: CRP at the *lac* promoter—the activating region in the downstream subunit of the CRP dimer (filled circle) contacts αCTD, recruiting it to the promoter. B: cI at P_{RM}—the cI activating region (filled square) contacts region 4 of σ, most likely facilitating contacts with the –35 element. C: MerR at *pmerT*—MerR twists the promoter DNA so that the –10 and –35 elements become aligned correctly.

footprinting experiments, wild-type RNAP, but not RNAP derivatives containing C-terminally truncated α, protects the DNA segment between the DNA site for CRP and the –35 element. Third, CRP-dependent transcription is reduced by the removal of αCTD or by some single amino acid substitutions in αCTD (these identify the CRP contact site). Finally, interactions between purified α and CRP can be detected and these interactions are disrupted by substitutions in the CRP activating region. The simplest model is that CRP activates transcription by recruiting αCTD to the DNA immediately upstream of the *lac* –35 element, thereby tightening the binding of RNAP to the promoter (Fig. 1A). Stated simply, CRP compensates for defective promoter–RNAP interactions with protein–RNAP interactions (27, 28).

3.3 The phage lambda P_{RM} promoter

cI protein activates the lambda P_{RM} promoter by binding to a site centred at −42 and overlapping the −35 element. Again, strong evidence for protein–protein contacts playing a crucial role comes from the identification of single amino acid substitutions in cI that interfere with activation of P_{RM}: these substitutions identify a 10 amino acid activating region, rich in negatively charged side-chains, that is essential for transcription activation (29). Two lines of evidence show that the target for this activating region is region 4 of σ^{70}. First, the activating region in cI is located within the DNA-binding helix–turn–helix motif and is placed right next to σ^{70} in ternary cI:RNAP: promoter complexes (recall that bound cI overlaps the P_{RM} −35 hexamer which is the target for region 4 of σ^{70}). Second, a single amino acid substitution in region 4 of σ^{70} (R596H) is sufficient to suppress the effects of an activation-defective cI mutant (D38N) (30). Moreover this suppression is allele specific (i.e. cI-D38N functions with σ^{70}-R596H but not wild type σ^{70}) suggesting that D38 of cI and R596 of σ^{70} are in close proximity. The simplest model is that cI activates transcription by helping region 4 of σ^{70} bind to the −35 element, again tightening the binding of RNAP to the promoter (Fig. 1B).

3.4 αCTD and σ^{70} region 4 both carry contact sites for different activators

Transcription activation by a number of different activators (e.g. AraC, MalT, and PhoB) is affected by some single amino acid substitutions in σ^{70} region 4, and it is probable that this region carries contact sites for transcription activators that bind at or near the −35 region (as for cI) (31, 32). Similarly, with αCTD, substitutions interfere with the function of diverse activators (e.g. OmpR, OxyR, CysB, Ada) suggesting that αCTD carries contact sites for many activators besides CRP. Interestingly, different substitutions in αCTD have different effects with different activators, suggesting that αCTD-contacting activators do not all contact the same site (32–34). The same is true for substitutions in σ^{70} region 4. The simplest view of αCTD and σ^{70} region 4 is that they are DNA-binding modules that can be recruited to their targets by diverse contacts with diverse transcription activators: these contacts can be via any number of different surface exposed patches. In other words, activators need to make contact but do not need to 'push' a specific 'button' in RNAP.

Sequence analysis of different promoters (6) shows that the binding sites for different activators are found at diverse locations ranging from around −40 to upstream of −100, raising the question of how distally bound activators can make direct contact with RNAP. Of course, one possibility is that there is no direct contact, and that the effects of the activators are negotiated via conformational changes in the promoter DNA. However, there is little evidence for this and, in many cases where activators are distally bound, the intervening DNA can be bent such that direct contact is possible (often with the help of a supplementary factor: see later and

Chapters 6 and 10). In the case of activators that contact αCTD, diversity in binding site location is also made possible by the flexible linker between αCTD and αNTD, which permits αCTD to stretch to different locations (34, 35). This is particularly apparent at different CRP-dependent promoters triggered by the activating region–αCTD interaction: such promoters are found with the DNA site for CRP centred near –61, –71, –81 ,or –91 and all are dependent on the same CRP–RNAP interaction. The flexible linker between αCTD and the remainder of RNAP, together with bending of the intervening DNA, permits the establishment of the same CRP–αCTD–DNA interaction at promoters with different architectures (36, 37).

3.5 MerR and the *merT* promoter

The MerR protein directly activates transcription of microbial mercury resistance genes expressed from the *merT* promoter (20). Unusually, the DNA site for MerR is located immediately downstream of the –35 hexamer. Bound MerR covers the DNA between the –35 and –10 regions, but, surprisingly, there is no evidence for direct interactions between MerR and RNAP. The key to the action of MerR lies in the observation that the distance between the –10 and –35 elements at the *merT* promoter is greater than normal (19 base pairs instead of the optimal 17 base pairs) and that the MerR–Hg^{2+} complex causes a deformation in the *merT* promoter (38, 39). This deformation is a local underwinding of the spacer DNA by 33°, causing realignment of the –10 and –35 elements. Thus MerR activates transcription by provoking a change in the conformation of the target DNA that makes the *merT* promoter more attractive to RNA polymerase (Fig. 1C). Crucial evidence for this model comes from mutational studies: one or two base pair deletions that shorten the spacer between the –10 and –35 elements at the *merT* promoter result in a stronger promoter that is no longer dependent on MerR (40).

4. Control of transcription factor activity

Many different mechanisms can control the activity of transcription factors. For example, some bacterial factors are controlled by the binding of small ligands: in these cases, the ligand controls binding of the factor to specific DNA targets at promoters. Thus, the binding of CRP to the *lac* promoter is regulated by cyclic AMP: modulation of intracellular cAMP levels via adenyl cyclase activity couples CRP activity to the environment (41). When cAMP levels are low, CRP binds poorly and non-specifically to DNA, but when levels are higher the affinity of CRP for specific sites is increased by several orders of magnitude. In other cases, ligands trigger the removal of factors from specific target sites (as, for example, with members of the Lac repressor family).

As a general rule for repressors, small ligands act as co-repressors for biosynthetic genes (e.g. for MetJ at the *met* biosynthetic genes) and as inducers for catabolic genes (e.g. for LacI at the *lac* operon). In contrast, with activators, small ligands are essential for binding to promoters controlling catabolic genes (e.g. CRP at the *lac* promoter), but destabilize binding to promoters controlling biosynthetic genes (e.g. NagC at genes for *N*-acetyl glucosamine biosynthesis). Although this type of regulation is found frequently, there are also many cases where both the triggered and non-triggered transcription factor bind specifically to the target promoter. In these cases the trigger causes a reorganization or conformational change of the factor whilst it is anchored to the DNA. A good example is MerR which is triggered by Hg^{2+} ions (20). Although both unliganded MerR and the MerR–Hg^{2+} complex occupy the same site at the MerT promoter, unliganded bound MerR does not distort the *merT* promoter spacer: distortion is triggered by binding of Hg^{2+} ions to the preformed MerR–*merT* promoter complex. Similarly, the activity of members of the AraC family of transcription activators is triggered by an inducer ligand (e.g. AraC activity is triggered by arabinose) but, in each case, the protein binds to target promoters in both the presence and absence of inducer. In the case of the *araBAD* promoter, AraC protein in the absence of arabinose binds to two sites, *O2*, which is far upstream of the promoter, and *I2*, which is just upstream. Arabinose induces AraC to release *O2* and to occupy sites *I2* and *I1*, *I1* overlapping with the –35 region of the promoter (42). Transcription activation is due to occupation of site *I1* by AraC, likely because AraC at *I1* interacts directly with σ^{70} region 4 (probably in a manner similar to cI at the lambda P_{RM} promoter) (43). Thus, in this case, the ligand triggers a rearrangement of the occupancy of available sites. It is possible to rationalize this type of mechanism in terms of economy: for a factor that interacts at no more than a few promoters in the cell, it is worthwhile 'anchoring' the factor at target promoters rather than maintaining a pool of free-floating factor.

Apart from small ligand binding, two other mechanisms are used to control the activity of bacterial activators and repressors. First, the activity of a large number of factors (notably the response-regulator class) is controlled by covalent phosphorylation by membrane-bound kinases: this results in direct coupling of factor activity to events outside the cell (see Chapter 8). Second, the activity of many factors is controlled by their concentration in the cell, which, in turn, is controlled either by synthesis or turnover or both. The best examples of this are the phage lambda cI and cII proteins (44). cI levels in phage lambda lysogens are tightly autoregulated by the complex P_{RM} promoter, at which cI acts as both an activator (via contact with σ^{70} region 4) and as a repressor. The initial burst of cI expression is due to induction of a second promoter P_{RE} which is totally dependent on the transient appearance of cII protein (cII binds to a site overlapping the –35 region of P_{RE}). cII is subject to rapid degradation by proteolysis, notably by the host *hfl* protease. Phage lambda-encoded cIII protein acts to counter this proteolysis. Although the case of cI and cII is an extreme example, it serves to make the point that regulation of the level of a transcription factor can be just as effective as regulation by ligand. Indeed many operons involved in the specification of virulence encode at least one regulatory protein that positively autoregulates that operon. This sets up a situation in which triggering of the regulatory gene product positively induces its own synthesis, thus creating a very effective switch.

5. Complex activation

5.1 Types of complex activation

The regulation of most promoters is complex and expression is not simply dependent on one transcription activator. In many cases where two activators are involved, one activator is a 'global' regulator sensing a global metabolic signal and interacting at a large number of promoters, whilst the other is a 'specific' regulator, triggering specific responses to a specific inducer at a very small number of promoters (45). Examples of this are found at the *mal* and *mel* operons encoding genes for the catabolism of maltose and melibiose. Expression of these operons is co-dependent on CRP and the operon-specific regulators MalT and MelR. CRP is a global regulator that is triggered by elevations in cAMP levels (due to glucose starvation), whilst MalT and MelR are regulon-specific activators triggered by maltose and melibiose, respectively. Some other examples of complex activation are listed in Table 2.

Before considering the mechanisms responsible for co-dependence on two activators, it is important to understand that apparent co-dependence may not be due to both activators binding at the same promoter, but can be due to the synthesis of one activator being dependent on the activity of the other (Fig. 2A). For example, expression of the *melAB* operon (encoding an α-galactosidase and a melibiose transport protein) is strictly dependent on CRP and MelR, and yet the *melAB* promoter is activated by MelR alone (46). However, the *melR* promoter is totally dependent on CRP, thus explaining the dependence of *melAB* expression on two activators: in fact both the *melR* and *melAB* promoters are 'simple' in that they are both controlled by just one activator. Interestingly, binding of CRP to the *melR* promoter is only triggered at higher levels of cAMP. Thus, the melibiose-dependent trigger, MelR, is only produced in conditions of extreme glucose starvation: this appears to be a

Table 2 Some examples of *E. coli* promoters controlled by two activators

p*malE*–p*malK*	These divergent promoters are regulated by an array of three MalT monomers, three CRP dimers, and two MalT monomers (47).
p*araBAD*	Activated by AraC. Correct positioning of AraC requires CRP (42, 43).
p*melAB*	A simple promoter activated by MelR. MelR expression is strictly controlled by CRP (48).
p*nirB*	Promoter dependent on FNR binding around –41. Further stimulated by NarL or NarP binding around –69 (49).
p*narG*	Promoter dependent on FNR binding around –41. Activity co-dependent on NarL binding around –200 (50).
p*nrfA*	Promoter dependent on FNR binding around –41. Further stimulated by NarP binding around –69 but repressed by NarL (51).
p*ansB*	Promoter co-dependent on FNR and CRP, binding around –41 and –90, respectively (52).

The *malE* and *malK* promoters control genes responsible for maltose transport; the *araBAD* promoter controls genes responsible for arabinose catabolism; and the *melAB* promoter regulates genes necessary for melibiose catabolism and transport. The *nirB* and *nrfA* promoters regulate operons encoding NADH-dependent and formate-dependent nitrite reductases, respectively; the *narG* promoter regulates nitrate reductase expression; and the *ansB* promoter controls the asparaginase gene.

mechanism for avoiding the wasteful biosynthesis of a factor in conditions where the cell would not 'wish' to express the genes controlled by that factor (even if melibiose were available). In contrast, although synthesis of the maltose-specific activator, MalT, is also dependent on CRP, expression from several promoters of the maltose regulon requires simultaneous binding of both CRP and MalT for transcription initiation (e.g. the divergent *malK–malE* promoters).

5.2 Mechanisms of complex activation

5.2.1 A role for nucleoprotein structures

One of the first complex regulatory regions to be dissected was the *malK–malE* intracistronic regulatory region carrying the divergent *malK* and *malE* promoters. Both promoters are dependent on binding of both MalT and CRP: there are five DNA sites for MalT and three sites for CRP (47). An attractive suggestion (53) is that occupation of these sites leads to the formation of a nucleoprotein complex and it is this complex that triggers transcription activation (Fig. 2B(i)). Although this is probably too simple for the *malE–malK* case (see below), it is an attractive model and may well be applicable at a host of promoters (many of which are so complex that no serious mechanistic studies have been attempted). The distinguishing feature of this model is that it supposes that the different factors bind cooperatively to form a structure (reminiscent of the situation at complex eukaryotic promoters (54) where initiation complexes are formed by numerous heterologous interactions between the basal transcription apparatus and specific factors).

5.2.2 Repositioning mechanisms

Transcription activation at the *E. coli malK* promoter requires CRP binding to three sites, and MalT binding to two sites upstream of the bound CRP (sites 1′ and 2′) and three sites downstream (sites 3′, 4′, and 5′). In the absence of CRP, MalT occupies three distinct downstream sites, 3, 4, and 5 that overlap with sites 3′, 4′, and 5′. Crucially, MalT can only activate transcription when it occupies sites 3′, 4′, and 5′, likely due to direct interactions with σ^{70}. Thus the action of CRP (together with upstream-bound MalT) is to reposition downstream-bound MalT from an unproductive to a productive binding site (Fig. 2B(ii)). This observation suggests a clear molecular basis for co-activation: the second activator is required to 'nudge' the first (primary) activator from an abortive to a productive mode (55). Evidence to support this comes from two sources. First, mutations that improve MalT sites 3′, 4′, and 5′ decrease the CRP-dependence of the *malK* promoter. Second, there is no absolute requirement for CRP to trigger the repositioning, and activation can be triggered by integration host factor (IHF), a ubiquitous DNA-binding protein that can also bend target sequences (56). This is an interesting observation, as it shows how an apparently unrelated protein can be recruited into an activation mechanism, and it suggests that the crucial feature in repositioning might be the protein-induced bend.

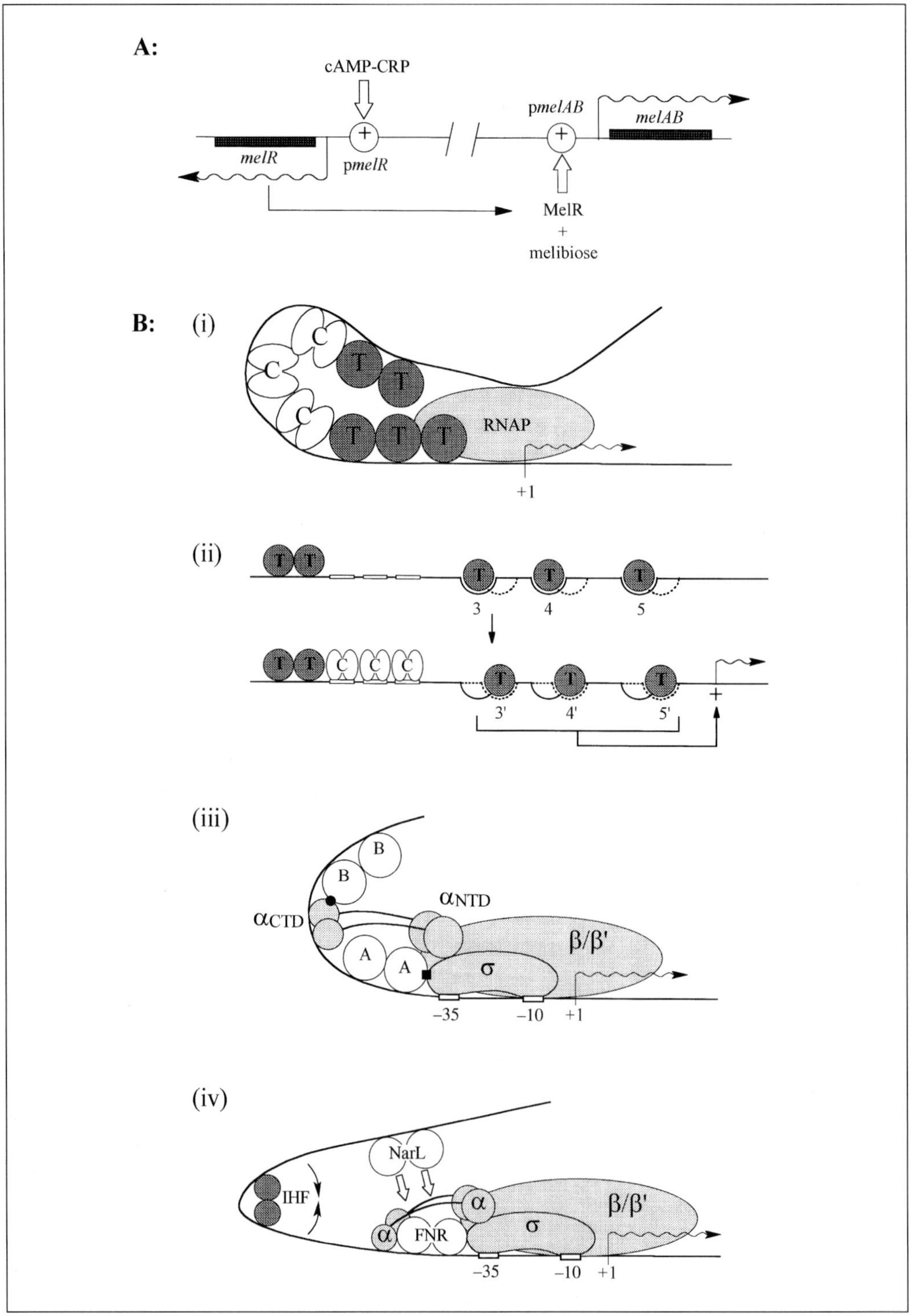
A:
cAMP-CRP
pmelAB
melAB
melR
pmelR
MelR
+
melibiose
B:
(i)
C
C
C
T
T
T
T
T
RNAP
+1
(ii)
T T
T
T
T
3
4
5
T T C C C
T
T
T
3'
4'
5'
+
(iii)
B
B
αCTD
αNTD
A
A
β/β'
σ
−35
−10
+1
(iv)
NarL
IHF
α
α
FNR
β/β'
σ
−35
−10
+1

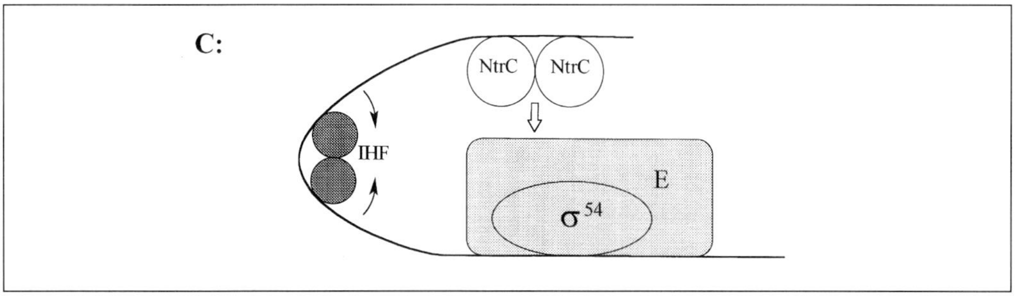

Fig. 2 Complex transcription activation. A: A case where synthesis of one activator is dependent on synthesis of another. The figure illustrates the divergent promoters at the *E. coli mel* operon regulatory region. Expression from the *melAB* promoter is completely dependent on melibiose-triggered MelR. Expression of MelR is dependent on the *melR* promoter which is completely dependent on cAMP–CRP. B: Cases where both activators are essential at the same promoter.(i) A nucleoprotein complex forms: three CRP (C) dimers and five MalT (T) monomers form a complex at the *E. coli malB* locus (containing the divergent *malE* and *malK* promoters). (ii) One activator repositions another: at the *E. coli malK* promoter, the binding of CRP, together with upstream-bound MalT, triggers the repositioning of downstream-bound MalT to a position where it is competent for transcription activation. (iii) Simultaneous touching: both activators A and B contact RNAP. Activator B makes contact via the flexible αCTD (filled circle), whilst activator A can contact σ. An example is the *E. coli ansB* promoter where A and B are FNR and CRP. (iv) Recruitment of bystanders: at the *E. coli narG* promoter, a DNA bending protein (IHF) co-activates by facilitating the interaction between a transcription activator (NarL) and RNAP. C: Activation at σ^{54}-dependent promoters. The figure shows a typical σ^{54}-dependent promoter where IHF promotes contact between an activator (e.g. NtrC) and RNAP. Although the primary role of IHF is to induce a bend that facilitates activator–RNAP contacts, a secondary role may be to suppress promiscuous activation by other transcription factors (87).

5.2.3 Simultaneous touching mechanisms

An alternative mechanism for complex activation by two activators is to suppose that both activators make direct contact with RNAP. This mechanism is found in many cases where one of the activators overlaps the –35 region and makes contacts with σ^{70}. For example, expression from the *E. coli ansB* promoter (that controls the asparaginase gene) is dependent on both FNR and CRP, binding around positions –41 and –91, respectively (52). Bound FNR is dimeric and overlaps the –35 region, most likely making direct contact with region 4 of σ^{70} via an activating region on the downstream subunit. Co-activation by CRP requires the CRP activating region to make contact with αCTD (57). The simplest model (Fig. 2B(iii)) is to suppose that the FNR dimer displaces αCTD from its preferred location at the promoter, and αCTD thereby becomes available as a 'ligand' for a second activator. A similar result has been found with a synthetic derivative of the phage lambda P_{RM} promoter, which is dependent on cI protein bound at a site overlapping the –35 region and contacting region 4 of σ^{70} (Fig. 1B). This promoter can be super-activated by CRP if a DNA site for CRP is introduced upstream: super-activation is dependent on the CRP activating region showing that CRP–αCTD interactions are essential (58). Simultaneous touching models have also been invoked to explain synergy at a number of both naturally occurring and synthetic promoters where optimal expression is dependent on the binding of two CRP or FNR dimers (usually with one located around –41 and the other around –91) (58–60). In all cases the upstream-bound dimer appears to contact αCTD.

Although simultaneous touching provides an attractive model to account for co-dependent activation, it is necessary to explain why the binding of one activator is insufficient to trigger transcription initiation (this is not a problem with the re-positioning model for co-dependent activation). This is especially crucial at FNR-dependent promoters such as the *E. coli ansB* promoter which is totally inactive in the absence of upstream-bound CRP (57). The simplest explanation is to suppose that the αCTD, which is displaced when FNR makes contact with region 4 of σ^{70}, is somehow inhibitory to transcription activation, and that this inhibition is overcome by contact with the upstream activator. Thus, the upstream activator behaves as an anti-inhibitor as well as an activator. This may be an important factor in explaining observed differences between the 10–20 naturally occurring FNR-dependent promoters that have been characterized to date (61). In almost every case the DNA site for FNR is located around –40: in some cases (e.g. the *frd* promoter controlling fumarate reductase expression) FNR alone is sufficient for substantial promoter activity, whereas in other cases (e.g. the *narG* promoter controlling nitrate reductase expression) a co-activator bound upstream, NarL or NarP is required. The simplest explanation is to suppose that the degree to which αCTD is inhibitory (and thus the promoter is dependent on a second activator) differs according to the promoter context.

5.2.4 Recruitment of 'bystanders'

All bacteria contain a number of abundant small proteins that appear to play a role in maintaining DNA structure (62). Amongst these are the histone-like proteins HU and IHF, and HNS and Fis. HU and HNS were originally identified as 'non-specific' DNA binding proteins present in sufficiently large amounts to be considered as proteins that could structure the bacterial chromosome. IHF and Fis were discovered as host functions essential for DNA rearrangements involving phage lambda and Mu (63, 64). Although the activity of these proteins is not triggered by any specific signal, they have been recruited at a number of promoters where they play specific roles in activation and repression mechanisms.

Fis and IHF can both behave as 'simple' activators, binding upstream of promoter elements in a number of cases: it is not clear whether activation is via direct contact with RNAP or via the induction of a conformation change in the promoter DNA (63, 64). IHF plays an important role at some 'complex' promoters. For example, the *E. coli narG* promoter is co-activated by FNR and NarL which bind around –41 and –200, respectively. IHF binding to target sites located between bound FNR and NarL is essential for activation (65). The simplest model is that IHF induces a bend that brings the upstream-bound NarL near to the FNR–RNAP complex, facilitating contacts with the distally bound activator (Fig. 2B(iv)).

5.2.5 Activation at σ^{54}-dependent promoters

Open complex formation by RNAP containing σ^{70}-related sigmas proceeds via the formation of a closed complex in which the promoter duplex is not unwound.

Promoters that are positively regulated can be limited at the level of both closed complex formation and isomerization from the closed to open complex (in which the duplex is unwound, see Chapter 3). At some promoters, open complexes can form in the absence of an activator. In contrast, RNAP carrying σ^{54} ($E\sigma^{54}$) can form only closed complexes at target promoters and cannot proceed to the open complex: closed complexes containing $E\sigma^{54}$ are totally dependent on upstream-bound activators for progression to the open complex (14, 66, 67). These activators, which belong to the NtrC family (or the σ^{54} bacterial enhancer-binding protein family), bind 100–200 bp upstream from the promoter elements, but remain functional when positioned further upstream. Activation requires ATP hydrolysis and a contact between the bound activator and bound $E\sigma^{54}$. This necessitates looping of the intervening DNA. In some cases looping is facilitated by IHF which, as at the σ^{70}-dependent *narG* promoter, acts by bending DNA to facilitate contacts between two distally bound proteins (Fig. 2C).

6. Simple repression

6.1 The *lac* repressor

Simple repression occurs when a bound repressor covers essential promoter elements and thus blocks access by RNAP (Fig. 3A). The *lac* repressor is usually taken as the paradigm for 'simple' repression although, in fact, the *lac* repressor is not so simple because of the existence of multiple binding sites (see below). The *lac* repressor is a tetramer of identical subunits and belongs to a large family of repressors (16, 68). Family members have a DNA-binding helix–turn–helix near the N-terminal, and recognition of cognate operators is modulated by small ligand binding. The principal operator (*O1*) at the *lac* promoter is a pseudosymmetric sequence that overlaps the transcription start site and accommodates two repressor subunits. Initially it was supposed that RNAP would be simply sterically blocked by repressor. However, some *in vitro* studies have suggested that ternary repressor:RNAP:promoter complexes can form, but the RNAP fails to make the usual downstream contacts, prompting a model in which RNAP is waiting at the promoter ready to initiate transcription upon arrival of the inducer and departure of the repressor (69; but see 70 for discussion).

A crucial factor in determining the efficiency of repression of any promoter is the lifetime of the repressor–promoter complex. Although the half-time of the *lac* repressor–operator complex is a few minutes, the promoter can be occupied by RNAP during the brief time that the promoter is free (for simplicity, most experiments have been performed with CRP-independent mutants of the *lac* promoter). Clearly there will be competition between repressor and RNAP, and it is this competition that will determine the efficiency of repression. In fact the competition is complex, depending on the exact juxtaposition of the RNAP- and repressor-binding sites and the kinetic characteristics of the promoter. As a general rule, repression becomes less efficient as the operator is distanced from the transcript start, implying that repressors are more efficient at interfering with transcription initiation than with transcript elongation.

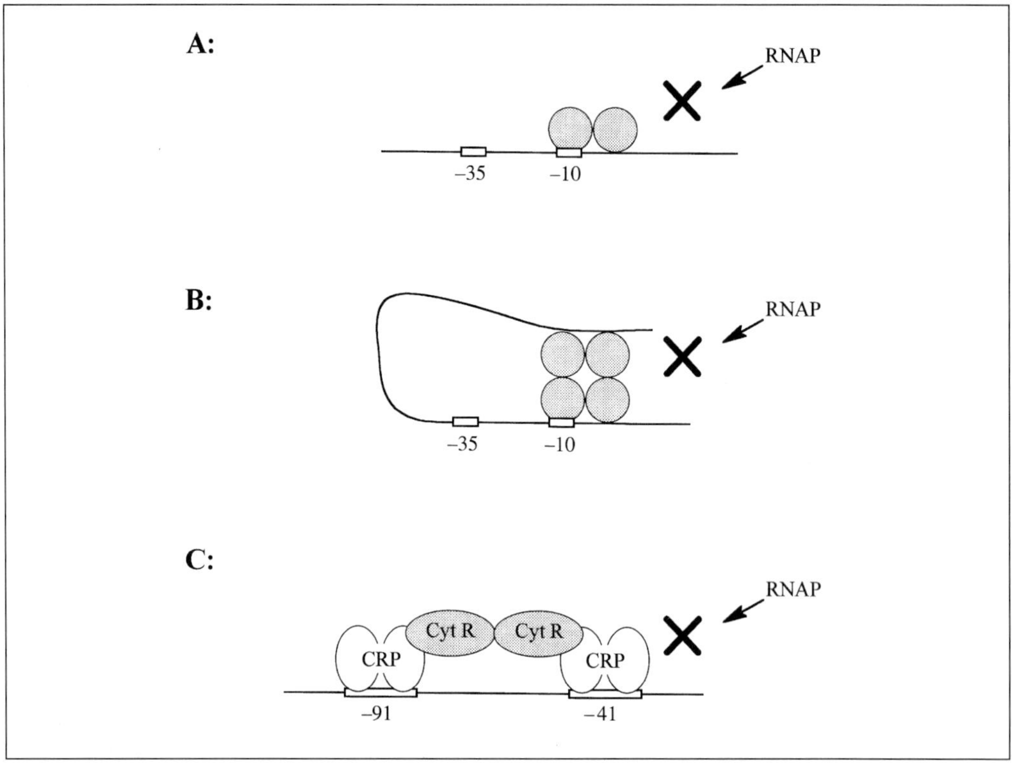

Fig. 3 Mechanisms of repression. A: Simple repression—bound repressor directly overlaps a promoter element and occludes RNAP. B: Interaction between distant repressors—distally bound repressors interact via a loop that enhances repression. C: Repressors that act via activators—a repressor (CytR) binds to an activator (CRP). In the example, CytR recognizes tandem-bound CRP

Promoters that are limited at the stage of the initial binding of RNAP (K_b) are more sensitive to repression than promoters that are limited at the stage of isomerization to the open complex (k_f) or promoter clearance. This is most simply understood in terms of RNAP–repressor competition: RNAP will be a better competitor if the promoter is limited at a step after the initial binding of RNAP.

6.2 Multiple repressor binding sites

At most repressible promoters, the DNA site for the repressor is located near to the transcript start, and in many cases two or more copies of the operator sequence are found. For example, promoters repressed by the MetJ or TrpR repressors (and many others) can contain two, three, four, or five copies of the operator sequence (71, 72). In these cases, cooperative interactions between tandemly bound repressors help to exclude RNAP. In some instances, operators are duplicated but the second copy of the operator is remote from the transcript start. In these cases, repression due to repressor binding at the remote site requires DNA looping and repressor aggregation

(Fig. 3B). In the case of the *lac* operon, supplementary operator sites *O2* and *O3* are located 400 bp downstream and 93 bp upstream from *O1*: whilst *O1* alone is sufficient to ensure a good level of repression, optimal repression is contingent on the presence of both *O2* and *O3* (68, 70, 73). The simplest model to explain this is to suppose that tetrameric repressor simultaneously binds to *O1* and *O2* or *O3* and that this involves looping of the intervening DNA. In some cases binding of multiple repressors to distant sites is essential for repression and not just an optional extra: for example, repression by the GalR repressor and the AraC protein (in the absence of arabinose it acts as a repressor) requires binding to distant sites via looping of the intervening sequences (74, 75). An interesting case appears at the divergent *nagE* and *nagB* promoters. These promoters are repressed by NagC binding at two sites separated by ~100 bp that overlap the two promoters: binding of NagC induces loop formation. Under inducing conditions (i.e. when the operators are not occupied by NagC) both promoters are activated by CRP, which binds between the two NagC operators. Under non-inducing conditions (i.e. when the operators are occupied by NagC) CRP can also bind and appears to stabilize binding of NagC. The simplest model is to suppose that the CRP-induced bend stabilizes loop formation (76).

6.3 Repressors come in many forms

Most repressors are proteins 'dedicated' to repressing a small number of promoters and induced by a specific trigger. However, there are a number of cases where regulatory proteins fulfil a dual role, for example the phage lambda cI repressor represses the lambda P_L and P_R promoters whilst activating the promoter P_{RM} for its own synthesis (44). Similarly AraC, MerR, TyrR, and many more proteins can act as both repressors and activators (4). Clearly, if repression is simply due to interference with the access of RNAP to a target promoter, then any activator also has the potential to behave as a repressor. It is instructive to consider CRP and FNR, which are both activators, provided that they are correctly positioned with respect to promoter elements at a transcription start. Incorrect positioning of either protein will result in cAMP- or anaerobically induced repression of overlapping promoters, and thus CRP and FNR can behave as both activators and repressors. This explains the observation that the synthesis of a small number of proteins is derepressed in *crp* and *fnr* mutants (41).

Finally, Fis, IHF, and HNS have been recruited as repressors at a number of promoters. For Fis and IHF, repression is via binding to specific sites at target promoters, in some cases by acting as a simple repressor, the sites overlapping the –10 or –35 elements (63, 64). In other cases, IHF sites overlap sites for an essential activator, and thus the activator is displaced by IHF (63). Repression by HNS is more complex as HNS appears to recognize a DNA structure rather than a short operator sequence. Repression by HNS is due to ~100-bp blocks of sequence that are located either upstream or downstream from the target promoter. The simplest explanation is that these blocks act as nucleation sites for HNS which then silences neighbouring regions of the chromosome (77). Induction of promoters repressed by HNS can be triggered by changes in DNA structure (most likely supercoiling) which are linked to environmental factors such as osmolarity or temperature. It is also possible that the HNS 'blockade' can be lifted by competition with transcription factors. For example, CRP-dependent activation of the *pap* operon may be partially due to displacement of HNS (induction is CRP-independent in a mutant *hns* background) (78).

7. Complex repression and anti-activation

It is usually assumed that the crucial feature of repressors is their ability to bind DNA: thus occupation, whether by one or multiply bound repressor molecules, is assumed to be both necessary and sufficient to shut out RNAP. However, there are a number of cases where the situation is not so simple, and repression is 'active' involving heterologous interactions between the repressor and other factors. A good example is the promoter of the *crp* gene, which is autoregulated by CRP which binds to a single site downstream of the *crp* transcript start at around +40 (79). In this case, repression is not due to bound CRP blocking RNAP elongation from the *crp* promoter, but rather due to the bound CRP activating a cryptic antisense promoter. Bound CRP recruits RNAP to this promoter, and it is the bound RNAP that blocks the *crp* promoter (note that repression is not due to antisense RNA made from this promoter). It is also likely that some repressors function actively by making a contact with RNAP that then 'jams' the RNAP in an unproductive mode (70) (this is likely to be important for those repressors that can also function as activators).

Perhaps the best example of repression involving protein–protein interactions is the CytR repressor, which represses the expression of genes scattered at ~10 different locations on the chromosome, involved in nucleoside catabolism and transport (80). Expression of all of these genes is dependent on CRP, which binds at tandem sites located around –40 and –93 at each promoter. In each case, activation is due to the CRP bound at –40, though in some cases the upstream-bound CRP acts synergically (as described above in Section 4.5.2.3). Although CytR is a member of the LacI family, alone it can bind only poorly to target promoters: for many years this was a puzzle as it was unclear how it could repress transcription. The key observation is that CytR only binds to target promoters when CRP is bound: rather than recognizing a base sequence, CytR recognizes the array of tandemly bound CRP dimers separated by ~53 bp (81, 82). Single amino acid substitutions in CRP that prevent CytR-dependent repression whilst not interfering with transcription activation have been isolated (83). These identify a contact site for CytR in CRP, a surface exposed region that is distinct from the activating region and the DNA-binding motif. CytR-dependent repression thus involves repressor binding to an activator: the repressor acts as an anti-activator (Fig. 3C).

8. Perspectives

For any regulatory region we need to know what are the factors that interact and why have they been selected. In most cases, we now know which factors are operating (though there are still areas of ignorance, see, for example, Reference 84) but the question 'why?' is more difficult. However, it is a safe bet to suppose that the interplay of activators and repressors at any target promoter ensures that the bacterium makes the right product in the right quantity at the right time. Thus, many promoters are under the dual control of both a global and a specific regulator which, often, are a repressor and an activator (as at the *lac* promoter) or two activators (as at the *mal* promoters). Thus expression is tied to a dual signal: both the global signal (cAMP in the case of the *lac* and *mal* promoters) and the specific signal (lactose or maltose) need to be present. Interestingly, there are very few cases where regulation is effected by two repressors acting in tandem (the best examples are promoters repressed by both CytR and DeoR). Perhaps this reflects a preference for activators over repressors: repressors are costly, since they need to be maintained during periods of silence, whilst activators need only appear when the conditions are right. In most cases, we can now see the logic in why things are organized the way they are (or at least we can now suggest explanations to satisfy our curiosity, see Chapter 12). Furthermore, it is also clear that the rules and principles outlined in this chapter, though mostly taken from the *E. coli* world, do operate in all other microorganisms, although some microorganisms use different combinations of the same players to bring about the same end-effect. A good example is catabolite repression: whilst in *E. coli* this is dominated by cyclic AMP and its receptor, Gram-positive organisms lack cyclic AMP and use repressors (members of the *lac* repressor family) to regulate gene expression in response to a surfeit of glucose (85) (although nothing is simple, and *E. coli* also contains Cra protein (86), a *lac* repressor homologue also involved in catabolite repression/activation).

Finally, the question of the mechanisms of action of activators and repressors continue to pose problems, for three main reasons. First, we still lack structural information for any bacterial RNAP, second, we have but a sketchy understanding of the kinetics of the initiation process, and, third, we do not really understand the importance of the bacterial folded chromosome structure in regulation. However, some of these gaps in our knowledge are now being filled: new structural and kinetic data will be the starting point for understanding the making and breaking of the complex macromolecular structures that regulate gene expression.

References

1. Yanofsky, C. (1992) Transcriptional regulation: Elegance in design and discovery. In *Transcriptional regulation* (ed. S. McKnight and K. Yamamoto), p. 3. Cold Spring Harbor Press, Cold Spring Harbor, NY.
2. Cohn, M. (1980) In memoriam. In *The operon* (ed. J. Miller and W. Reznikoff), p. 1. Cold Spring Harbor Press, Cold Spring Harbor, NY.
3. Beckwith, J. (1980) *lac*: The genetic system. In *The operon* (ed. J. Miller and W. Reznikoff), p. 11. Cold Spring Harbor Press, Cold Spring Harbor, NY,
4. Raibaud, O. and Schwartz, M. (1984) Positive control of transcription initiation in bacteria. *Ann. Rev. Genet.*, **18**, 173.

5. Blattner, F., Plunkett, G., Bloch, C., Perna, N., Burland, V., Riley, M., Collado-Vides, J., Glasner, J., Rode, C., Mayhew, G., Gregor, J., Davis, N., Kirkpatrick, H., Goeden, M., Rose, D., Mau, B., and Shao, Y. (1997) The complete genome sequence of *Escherichia coli* K-12. *Science*, **277**, 1453.
6. Gralla, J. and Collado-Vides, J. (1996) Organisation and function of transcriptional regulatory elements. In Escherichia coli *and* Salmonella (ed. F. Neidhardt), p. 1232. ASM Press, Washington DC.
7. Gallegos, M-T., Schleif, R., Bairoch, A., Hofmann, K., and Ramos, J. (1997) AraC/ XylS family of transcriptional regulators. *Microbiology and Molecular Biology Reviews*, **61**, 393.
8. Schell, M. (1993) Molecular biology of the LysR family of transcriptional regulators. *Ann. Rev. Microbiol.*, **47**, 597.
9. Spiro, S. (1994) The FNR family of transcriptional regulators. *Antonie van Leewenhoek*, **66**, 23.
10. Summers, A. (1992) Untwist and shout- a heavy metal-responsive transcription regulator. *J. Bacteriol.*, **174**, 3097.
11. Gross, R., Arico, B., and Rappuoli, R. (1989) Families of bacterial signal-transducing proteins. *Mol. Microbiol.*, **3**, 1661.
12. Parkinson, J. and Kofoid, E. (1992) Communication modules in bacterial signalling proteins. *Ann. Rev. Genet.*, **26**, 71.
13. Henikoff, S., Wallace, J., and Brown, J. (1990) Finding protein similarities with nucleotide sequence databases. *Methods Enzymol.*, **183**, 111.
14. North, A., Klose, K., Stedman, K., and Kustu, S. (1993) Prokaryotic enhancer-binding proteins reflect eukaryotic-like modularity: the puzzle of nitrogen regulatory protein C. *J. Bacteriol.*, **175**, 4627.
15. Morett, E. and Segovia, L. (1993) The σ^{54} bacterial enhancer-binding protein family: mechanism of action and phylogenetic relationship of their functional domains. *J. Bacteriol.*, **175**, 6067.
16. Schumacher, K., Choi, K., Zalkin, H., and Brennan, R. (1994) Crystal-structure of LacI member, PurR, bound to DNA-minor groove binding by alpha helices. *Science*, **266**, 763.
17. Phillips, S. (1991) Specific β-sheet interactions. *Current Opinion in Structural Biology*, **1**, 89.
18. Malan, T., Kolb, A., Buc, H., and McClure, W. (1984) Mechanism of CRP–cAMP activation of *lac* operon transcription: activation of the P1 promoter. *J. Mol. Biol.*, **180**, 881.
19. Hawley, D. and McClure, W. (1983) The effect of a lambda repressor mutation on the activation of transcription initiation from the lambda *PRM* promoter. *Cell*, **32**, 327
20. O' Halloran, T., Frantz, B., Shin, M., Ralston, D., and Wright, J. (1989) The MerR heavy metal receptor mediates positive activation in a topologically novel transcription complex. *Cell*, **56**, 119.
21. Reznikoff, W. and Abelson, J. (1980) The *lac* promoter. In *The operon* (ed. J. Miller and W. Reznikoff), p. 221. Cold Spring Harbor Press, Cold Spring Harbor, NY.
22. Tagami, H. and Aiba, H. (1995) Role of CRP in transcription activation at the *Escherichia coli lac* promoter: CRP is dispensable after the formation of open complex. *Nucleic Acids Res.*, **23**, 599.
23. McClure, W. (1985) Mechanism and control of transcription initiation in prokaryotes. *Ann. Rev. Biochem.*, **54**, 171.
24. Schultz, S., Shields, S., and Steitz, T. (1991) Crystal structure of a CAP–DNA complex: the DNA is bent by 90°. *Science*, **253**, 1001.
25. Bracco, L., Kotlarz, D., Kolb, A., Diekmann, S., and Buc, H. (1989) Synthetic curved DNA sequences can act as transcriptional activators in *Escherichia coli. EMBO J.*, **8**, 4289.

26. Zhou, Y., Zhang, X., and Ebright, R. (1993) Identification of the activating region of CAP: isolation and characterization of mutants of CAP specifically defective in transcription activation. *Proc. Natl. Acad. Sci. USA*, **90**, 6081.
27. Ebright, R. (1993) Transcription activation at Class I CAP-dependent promoters. *Mol. Microbiol.*, **8**, 797.
28. Busby, S. and Ebright, R. (1994) Promoter structure, promoter recognition, and transcription activation in procaryotes. *Cell*, **79**, 743.
29. Hochschild, A., Irwin, N., and Ptashne, M. (1983) Repressor structure and the mechanism of positive control. *Cell*, **32**, 319.
30. Li, M., Moyle, H., and Susskind, M. (1994) Target of the transcriptional activation function of phage lambda cI protein. *Science*, **263**, 75.
31. Ishihama, A. (1993) Protein–protein communication within the transcription apparatus. *J. Bacteriol.*, **175**, 2483.
32. Rhodius, V. and Busby, S. (1998) Positive activation of gene expression. *Current Opinion in Microbiology*, **1**, 152–159.
33. Russo, F. and Silhavy, T. (1992) Alpha: the Cinderella subunit of RNA polymerase. *J. Biol Chem.*, **267**, 14515.
34. Ebright, R. and Busby, S. (1995) *Escherichia coli* RNA polymerase α subunit: structure and function. *Current Opinion in Genetics and Development*, **5**, 197.
35. Blatter, E., Ross, W., Tang, H., Gourse, R., and Ebright, R. (1994) Domain organisation of RNA polymerase α subunit: C-terminal 85 amino acids constitute a domain capable of dimerization and DNA binding. *Cell*, **78**, 889.
36. Zhou, Y., Merkel, T., and Ebright, R. (1994) Characterization of the activating region of *Escherichia coli* catabolite gene activator protein (CAP). II. Role at Class I and Class II CAP-dependent promoters. *J. Mol Biol.*, **243**, 603.
37. Zhou, Y., Pendergrast, S., Bell, A., Williams, R., Busby, S., and Ebright, R. (1994) The functional subunit of a dimeric transcription activator protein depends on promoter architecture. *EMBO J.*, **13**, 4545.
38. Frantz, B. and O'Halloran, T. (1990) DNA distortion accompanies transcriptional activation by the metal-responsive gene-regulatory protein MerR. *Biochemistry*, **29**, 4747.
39. Ansari, A., Chael, M., and O'Halloran, T. (1992) Allosteric underwinding of DNA is a critical step in positive control of transcription by Hg-MerR. *Nature*, **355**, 87.
40. Parkhill, J. and Brown, N. (1990) Site-specific insertion and deletion mutants in the mer promoter operator region of Tn501; the nineteen base pair spacer is essential for normal induction of the promoter by MerR. *Nucleic Acids Res.*, **17**, 5157.
41. Kolb, A., Busby, S., Buc, H., Garges, S., and Adhya, S. (1993) Transcriptional regulation by cAMP and its receptor protein. *Ann. Rev. Biochem.*, **62**, 749.
42. Lobell, R. and Schleif, R. (1990) DNA looping and unlooping by AraC protein. *Science*, **250**, 528.
43. Schleif, R. (1996) Two positively regulated systems, *ara* and *mal*. In Escherichia coli *and* Salmonella (ed. F. Neidhardt), p. 1300. ASM Press, Washington, DC.
44. Ptashne, M. (1986) *A genetic switch: Gene control and phage lambda*. Blackwell Scientific Publications, Oxford & Cell Press, Boston.
45. Gottesman, S. (1984) Bacterial regulation: global regulatory networks. *Ann. Rev. Genet.*, **18**, 415.
46. Webster, C., Gaston, K., and Busby, S. (1988) Transcription from the *Escherichia coli melR* promoter is dependent on the cyclic AMP receptor protein. *Gene*, **68**, 297.

47. Raibaud, O., Vidal-Ingigliardi, D., and Richet, E. (1989) A complex nucleoprotein structure involved in activation of two divergent *Escherichia coli* promoters. *J. Mol. Biol.*, **205**, 471.
48. Webster, C., Gardner, L., and Busby, S. (1989) The *Escherichia coli melR* gene encodes a DNA-binding protein with affinity for specific sequences located in the melibiose-operon regulatory region. *Gene*, **83**, 207.
49. Tyson, K., Bell, A., Cole, J., and Busby, S. (1993) Definition of nitrite and nitrate response elements at the anaerobically inducible *Escherichia coli nirB* promoter. *Mol. Microbiol.*, **7**, 151.
50. Dong, X-R., Li, S., and de Moss, J. (1992) Upstream elements required for NarL-mediated activation of transcription from the narGHJI promoter of *Escherichia coli*. *J. Biol. Chem.*, **267**, 14122.
51. Darwin, A., Tyson, K., Busby, S., and Stewart, V. (1997) Differential regulation by the homologous response regulators NarL and NarP of *Escherichia coli* K12 depends on DNA binding site arrangement. *Mol. Microbiol.*, **25** 583.
52. Jennings, M. and Beacham, I. (1993) Co-dependent positive regulation of the *ansB* promoter of *Escherichia coli* by CRP and the FNR protein: a molecular analysis. *Mol. Microbiol.*, **9**, 155.
53. Raibaud, O. (1989) Nucleoprotein structures at positively regulated bacterial promoters: homology with replication origins and some hypotheses on the quaternary structure of the activator proteins in these complexes. *Mol. Microbiol.*, **3**, 455.
54. Latchman, D. (1995) *Gene regulation: A eucaryotic perspective* (2nd edn). Chapman and Hall, London.
55. Richet, E., Vidal-Ingigliardi, D., and Raibaud, O. (1991) A new mechanism for coactivation of transcription: repositioning of an activator triggered by the binding of a second activator. *Cell*, **66**, 1185.
56. Richet, E., and Sogaard-Andersen, L. (1994) CRP induces the repositioning of MalT at the *Escherichia coli malKp* promoter primarily through DNA bending. *EMBO J.*, **13**, 4558.
57. Scott, S., Busby, S., and Beacham, I. (1995) Transcriptional coactivation at the *ansB* promoters: involvement of the activating regions of CRP and FNR when bound in tandem. *Mol. Microbiol.*, **18**, 521.
58. Joung, J., Koepp, D., and Hochschild, A. (1994) Synergistic activation of transcription by bacteriophage lambda cI protein and cyclic AMP receptor protein. *Science*, **265**, 1863.
59. Joung, J., Le, L., and Hochschild, A. (1993) Synergistic activation of transcription by *Escherichia coli* cyclic AMP receptor protein. *Proc. Natl. Acad. Sci.*, **90**, 3083.
60. Busby, S., West, D., Lawes, M., Webster, C., Ishihama, A., and Kolb, A. (1994) Transcription activation by the *Escherichia coli* cyclic AMP receptor protein. Receptors bound in tandem can interact synergistically. *J. Mol. Biol.*, **241**, 341.
61. Guest, J., Green, J., Irvine, A., and Spiro, S. (1996) The FNR modulon and FNR regulated gene expression. In *Regulation of Gene Expression in* Escherichia coli (ed. E. Lin and A. Lynch), Ch 16, p. 317. R. G. Landes Biomedical Publishers, Austin, Texas.
62. Drlica, K. and Rouviere-Yaniv, J. (1987) Histone-like proteins of bacteria. *Microbiological Reviews*, **51**, 301.
63. Goosen, N. and van de Putte, P. (1995) The regulation of transcription initiation by integration host factor. *Mol. Microbiol.*, **16**, 1.
64. Finkel, S. and Johnson, R. (1992) The Fis protein: it's not just for DNA inversion anymore. *Mol. Microbiol.*, **6**, 3257.

65. Schroder, I., Darie, S., and Gunsalus, R. (1993) Activation of the *Escherichia coli* nitrate reductase (*narGHJI*) operon by NarL and FNR requires integration host factor. *J. Biol. Chem.*, **268**, 771.
66. Weiss, D., Klose, K., Hoover, T., North, A., Porter, S., Wedel, A., and Kustu, S. (1992) Procaryotic transcriptional enhancers. In *Transcriptional regulation* (ed. S. McKnight and K. Yamamoto), p. 667. Cold Spring Harbor Press, Cold Spring Harbor, NY.
67. Merrick, M. (1993) In a class of its own- the RNA polymerase sigma factor σ^{54} (σ^{N}). *Mol. Microbiol.*, **10**, 903.
68. Gralla, J. (1992) *lac* repressor. In *Transcriptional regulation* (ed. S. McKnight and K. Yamamoto), p. 629. Cold Spring Harbor Press, Cold Spring Harbor, NY.
69. Lee, J. and Goldfarb, A. (1991) *Lac* repressor acts by modifying the initial transcribing complex so that it cannot leave the promoter. *Cell*, **66**, 793.
70. Muller-Hill, B. (1998) Some repressors of bacterial transcription. *Current Opinion in Microbiology*, **1**, 145.
71. Sigler, P. (1992) The molecular mechanism of trp repression. In *Transcriptional regulation* (ed. S. McKnight and K. Yamamoto), p. 475. Cold Spring Harbor Press, Cold Spring Harbor, NY.
72. Davidson, B. and Saint Girons, I. (1989) The *Escherichia coli* regulatory protein MetJ binds to a tandemly repeated 8 bp palindrome. *Mol. Microbiol.*, **3**, 1639.
73. Chakerian, A. and Matthews, K. (1992) Effect of *lac* repressor oligomerization on regulatory outcome. *Mol. Microbiol.*, **6**, 963.
74. Schleif, R. (1992) DNA looping. *Ann. Rev. Biochem.*, **61**, 199.
75. Adhya, S. (1989) Multipartite genetic control elements: communication by DNA loops. *Ann. Rev. Genet.*, **23**, 227.
76. Plumbridge, J. and Kolb, A. (1998) DNA bending and expression of the divergent nagE-B operons. *Nucl. Acids Res.*, **26**, 1254
77. Higgins, C., Hinton, J., Hulton, C., Owen-Hughes, T., Pavitt, G., and Seirafi, A. (1990) Protein H1: a role for chromatin structure in the regulation of bacterial gene expression and virulence? *Mol. Microbiol.*, **4**, 2007.
78. Goransson, M., Sonden, B., Nilsson, P., Dagberg, B., Forsman, K., Emanuelsson, K., and Uhlin, B. (1990) Transcriptional silencing and thermoregulation of gene expression in *Escherichia coli*. *Nature*, **344**, 682.
79. Hanamura, A. and Aiba, H. (1991) Molecular mechanism of negative autoregulation of *Escherichia coli crp* gene. *Nucl. Acids Res.*, **16**, 4413.
80. Valentin-Hansen, P., Sogaard-Andersen, L., and Pedersen, H. (1996) A flexible partnership: the CytR anti-activator and the cAMP–CRP activator protein, comrades in transcription control. *Mol. Microbiol.*, **20**, 461.
81. Sogaard-Andersen, L., Pedersen, H., Holst, B., and Valentin-Hansen, P. (1991) A novel function of the cAMP-CRP complex in *Escherichia coli*: cAMP–CRP functions as an adaptor for the CytR repressor in the *deo* operon. *Mol. Microbiol.*, **5**, 969.
82. Sogaard-Andersen, L. and Valentin-Hansen, P. (1993) Protein–protein interactions in gene regulation: the cAMP–CRP complex sets the specificity of a second DNA-binding protein, the CytR repressor. *Cell*, **75**, 557.
83. Sogaard-Andersen, L., Mironov, A., Pedersen, H., Sukhodelets, V., and Valentin-Hansen, P. (1991) Single amino acid substitutions in the cAMP receptor protein specifically abolish regulation by the CytR repressor in *Escherichia coli*. *Proc. Natl. Acad. Sci.*, **88**, 4921.
84. Green, J. and Guest, J. (1994) Regulation of transcription at the ndh promoter of *Escherichia coli* by FNR and novel factors. *Mol. Microbiol.*, **12**, 433.

85. Hueck, C. and Hillen, W. (1995) Catabolite repression in *Bacillus subtilis*: a global regulatory mechanism for the Gram-positive bacteria? *Mol. Microbiol.*, **15**, 395.
86. Saier, M. and Ramseier, T. (1996) The catabolite repressor/activator (Cra) protein of enteric bacteria. *J. Bacteriol.*, **178**, 3411.
87. Perez-Martin, J. and De Lorenzo, V. (1995) Integration host factor suppresses promiscuous activation of the sigma 54-dependent *Pu* promoter of *Pseudomonas putida*. *Proc. Natl. Acad. Sci.*, **92**, 7277.

5 | Post-transcriptional control

ZHIPING GU and PAUL S. LOVETT

1. Introduction

The regulation of gene expression can potentially occur at several stages in the step-wise transfer of information from a gene to its protein product. While transcription initiation appears to be pivotal in the control of expression of many genes, modern methodologies for gene and RNA manipulation have identified various forms of post-transcriptional control in bacteria and eukaryotes. Post-transcriptional control formally refers to all regulation that occurs after transcription. However, this chapter will focus on forms of control that specifically affect translation.

In bacteria, the most widely studied examples of post-transcriptional control are those affecting mRNA stability, translation initiation, and programmed frame-shifting. A feature common among these forms of regulation is that mRNA is the regulatory target. Recent studies in bacteria and eukaryotes have revealed additional examples of translational control in which the regulatory target is the ribosome or rRNA. Examination of these systems has offered a plausible explanation for several complex regulatory phenotypes that may influence translation rates and translational pausing.

Distinguishing post-transcriptional regulation from transcriptional regulation is ideally based on demonstrating that an observed change in the level of the protein product of a gene is independent of changes in transcription. While this criterion can, with varying degrees of success, be applied to certain post-transcriptionally regulated genes (e.g. 1), a commonly used *in vivo* approach is initially simpler and occasionally provides a more valid assessment of the form of regulation; e.g. since mRNA stability may be enhanced during translation (see later), induction of a gene that is translationally regulated could cause an *apparent* enhancement of the level of the corresponding mRNA. The 'first approximation' approach to analyze patterns of gene regulation in bacteria typically involves *in vivo* fusions between a reporter gene, such as *lacZ*, and signals for transcription or translation of the gene in question (2). For genes that are regulated at both the transcriptional and post-transcriptional levels other methods are clearly required to delineate the contributions of each to the overall regulation.

Outlined below are seven mechanistically distinct categories known to result in translational control in bacteria.

2. mRNA stability as a regulatory mechanism

mRNA species that exist intact in cells for a long time should provide a more stable template for protein synthesis than short lived mRNAs. If these two hypothetical classes of mRNAs are translated at comparable rates, the products of long-lived mRNAs will exceed those of the short-lived transcripts. This reasoning is supported by compelling observations which indicate a positive correlation between mRNA stability and the level of gene expression (3–5).

2.1 The exo-ribonucleases

RNA decay in cells is mediated by ribonucleases (6, 7). Eight exo-ribonucleases have been identified in *E. coli* and each digests RNA in a 3′ to 5′ direction (8). Two such enzymes, RNaseII and polynucleotide phosphorylase (PNPase) (9, 10), appear critical to mRNA turnover (11). Indeed these two enzyme activities may serve overlapping functions within the cell. Mutations that inactivate either of the genes individually, *rrb* and *pnp*, respectively, have little effect on the average half-life of bulk mRNA but simultaneous inactivation of both genes results in a lethal phenotype and an increase in the half-life of bulk mRNA (11).

Stem-loop structures commonly seen at the 3′ end of transcripts impede the rate of exo-ribonuclease digestion *in vitro* (3). However, the time that a stem-loop structure impedes decay *in vitro* is shorter than the time the stem-loop retards decay *in vivo* (12). This led to the discovery of an eight-protein complex, the Exo-ribonuclease Impeding Factor (EIF). The complex has been shown to impede the processive activity of PNPase at a stem-loop *in vitro*, and binding of the complex to the stem-loop is mediated by PNPase (13).

2.2 The endo-ribonucleases

The rate limiting step in mRNA decay appears to be the endo-nucleolytic removal of stabilization elements (stem-loop structures) which impede the processive degradation by exo-ribonucleases (3). The two most well-understood endo-ribonucleases are RNaseE and RNaseIII (14–19). An endo-ribonuclease that exposes a stem-loop free 3′ end will increase the susceptibility of upstream sequences to exo-nucleolytic digestion. In contrast, endo-ribonuclease exposure of a 5′ end lacking secondary structure would not normally be thought to enhance decay given the known specificity of exo-ribonuclease digestion in bacteria. However, since endo-ribonuclease cleavage that exposes an unstructured 5′ end will frequently separate downstream sequences from sites of translation initiation, it is conceivable that it is the loss of ribosomes from the mRNA which enhances its decay rate (20 and see later). Additionally, the recent identification of a protein complex that includes both an exo-ribonuclease (PNPase) and a 5′ endo-ribonuclease (RNaseE) suggests that both activities may act in a coordinated manner (21, 22). For example, polyA tails on the 3′ ends of bacterial transcripts are now known to increase the rate of decay in *E. coli* (23). 3′ polyA sequences could facilitate transcript folding, bringing 5′ and 3′ ends into close proximity, possibly allowing processing by the endo- and exo-ribonuclease complex (24).

2.3 Control of mRNA stability

2.3.1 The lambda *int*/*sib* system

Removal of stem-loop structures enhances the *in vivo* susceptibility of upstream sequences to exo-ribonuclease digestion, as demonstrated using l*int* mRNA as a model. RNaseIII is an endo-ribonuclease that cleaves at specific double stranded structures in RNA (15). Transcripts that contain the λ*int* coding sequence also contain a downstream *sib* site that can form an RNaseIII susceptible stem-loop structure (25). RNaseIII cleavage at *sib* increases the susceptibility of the upstream *int* region to degradation by the exo-ribonuclease PNPase. An inactivating mutation in *rnc*, the gene for RNaseIII, and a mutation that eliminates *sib* as a site of RNaseIII cleavage dramatically increase the stability of *int* mRNA (25).

2.3.2 Induction-dependent decay of the *puf* operon

The *puf* operon of *Rhodobacter capsulatus* provides a direct demonstration that an inducer can alter mRNA decay rates and thereby affect gene expression (26, 27). Under low oxygen tension, the photosynthetic bacterium *R. capsulatus* forms an intracytoplasmic membrane (ICM) that is anchored to a newly assembled photosynthetic apparatus which consists of the reaction center (RC) and the light harvesting complexes B800–850 (LHII) and B876 (LHI) (28). The *puf* operon encodes LHI proteins (*pufB* and *pufA*), the RC complex (*pufM* and *pufL*), and two additional proteins (the products of *pufQ* and *pufX*). The proposed gene order is *pufQBALMX*. Northern blots reveal two transcript species of 0.5 kb and 2.7 kb (26). The smaller encompasses *pufBA* sequences and is a relatively stable degradation product of the larger transcript which spans *pufBALMX*. Because of its lower decay rate, the concentration of *pufBA* transcripts is ninefold higher than *pufBALMX* transcripts. This approximates the 12-fold ratio of the LHI proteins:the RC complex proteins during oxygen limitation (29). These observations suggest that regions of a single polycistronic transcript are subject to differential rates of decay under specific environmental conditions.

Several structures within *puf* operon mRNA could facilitate differential decay of portions of the polycistronic transcript. First, a stabilizer sequence (a stem-loop) is found at the 3′ end of *pufA*. Deletion of the stabilizer decreases the stability of the *pufBA* transcript (30). Second, a destabilizer sequence, of about 30 nts, is found in the *pufL* coding sequence and seems to include an RNaseE sensitive site (31). The destabilizer is responsible for the accelerated degradation of *pufBALMX* transcripts. Indeed, insertion of the destabilizer into the *pufBA* transcript increases its rate of decay (32). Third, an oxygen sensing element is believed to overlap the destabilizer, suggesting that oxygen limitation may govern the initiating cleavage by RNaseE (33).

2.3.3 5′ stabilizer sequences

Early studies of mRNA decay in bacteria suggested chemical dissolution occurred in 5′ to 3′ direction (34). Since 5′ to 3′ exo-ribonucleases have not been detected in bacteria in spite of extensive searches (5′ to 3′ exo-ribonucleases do exist in eukaryotes; 35 and references therein), observations made in bacterial cells could not be correlated with the known enzyme activities. However, in recent years numerous reports have established that the stability of many mRNA species is most certainly determined by sequences at the 5′ end, again arguing for a processive effect initiating at the 5′ terminus.

ompA mRNA is long lived, with a 17 min half-life (36). Fusions of 5′ segments of *ompA* transcripts to the short-lived *bla* mRNA (3 min half-life) increases the half-life of the fusion mRNA. Emory *et al.* (37) demonstrated that a stem-loop structure at the extreme 5′ end of *ompA* transcripts was responsible for the long-lived character of the mRNA. This structure, hp1, apparently in conjunction with a 30 nt sequence surrounding the *ompA* ribosome binding site seem to be the prime determinants of stability. The studies by Emory *et al.* suggest a model for processive decay in which an endo-ribonuclease binds to single-stranded RNA at the 5′ end; if the 5′ end is not single stranded the presumptive nuclease (perhaps RNaseE; 38) cannot gain a foothold and the message is resistant to decay by this process. Once the nuclease is bound to the 5′ end of the mRNA, it moves 3′ on the mRNA making endo-nucleolytic cuts at every susceptible site. The resulting fragments are digested by 3′ to 5′ exo-ribonucleases. This model is appealing since proteins (or ribosomes) that bind to the 5′ end of mRNA may block either the binding of the endo-ribonuclease or its processive movement down the message.

ermC is an erythromycin resistance gene that is induced by erythromycin. Regulation is by translation attenuation (see later). In this form of regulation, the inducing antibiotic stalls a ribosome in the leader region of *ermC* transcripts. During induction, the half-life of *ermC* transcripts increases more than 10-fold. Several observations are consistent with the notion that the ribosome stalled in the leader might act to impede the processive 5′ to 3′ endo-ribonuclease, as suggested by the studies of Emory *et al.* (37). For instance, gene fusions demonstrate that the 5′ end of *ermC* transcripts is responsible for induction-dependent stability (39, 40). Moreover, induced-stability does not require translation of the *ermC* coding sequence (41). Another gene regulated by translation attenuation, *catA112*, is induced by chloramphenicol (42). The half-life of *catA112* mRNA increases 15-fold during induction, suggesting that the ribosome stalled in the *cat* leader may also act as a blockade to a candidate 5′ to 3′ endo-ribonuclease (43). Thus, for both *erm* and *cat*, induction both enables downstream translation (see later) and stabilizes the transcript.

2.4 Translation effects on mRNA stability

Translation is commonly thought to stabilize mRNA (44). It is currently believed that translation protection from decay is due to shielding of endonuclease vulnerable sites by the physical presence of the ribosome, or the act of translation abolishes a region(s) of secondary structure that may be needed for an initiating endo-ribonucleolytic cleavage (20, 44, 45). However, it has been estimated that a given site on actively translated *lacZ* mRNA is not protected by ribosomes about 76 per cent of the time (46). Thus, direct shielding by ribosomes would seem of less consequence than the changes in mRNA folding that likely result from active translation.

3. Translational repression

Repression of translation initially appears comparable to the well-studied systems of transcriptional repression. However, close inspection of several systems demonstrates fundamental similarities among translational repressors that are only infrequently seen in transcription repression. Foremost is the fact that many proteins which mediate translational repression have a second function that is unrelated to their role as repressors. Secondly, translational repressors frequently regulate their own synthesis, a feature termed autogenous regulation. Translational repression was first observed in gene systems of phage T4 and subsequently in the control of ribosomal protein expression, and more recently in the control of tryptophan biosynthesis in *B. subtilis*.

3.1 Translational repression of ribosomal protein synthesis

Bacterial ribosomes consist of more than 50 proteins and three species of rRNA, the 16S, 23S, and 5S rRNA molecules. Since ribosomes constitute about half the mass of a bacterial cell, ribosome synthesis utilizes a large proportion of the available biosynthetic capability. Coordinated synthesis of r-proteins and rRNA is also critical, and this coordination is largely a result of translational repression. Genes for the r-proteins are arranged in several operons. R-protein mediated autogenous regulation has been identified for about half of the operons (47). In multigene r-protein operons, the protein product of only one of the genes is the autoregulator. In all known examples of r-protein regulation, an autoregulatory protein only influences the translation of those messages transcribed from its own operon. Moreover, aside from two known exceptions (47), the *non-regulatory* role of autoregulatory r-proteins is to bind to *rRNA* at an early step in ribosome assembly.

Autogenous translational repression results from the binding of the regulatory protein to a region located at the 5′ end of its own transcript. The binding typically blocks translation initiation, but in one case may decrease mRNA stability (47). Since rRNA binding is the non-regulatory function of most autoregulatory r-proteins, a competition should exist between protein binding to rRNA versus protein binding to the leader region of mRNA. Furthermore, one would assume that a preference would exist for rRNA binding. Remarkably, in those instances where this comparison has been made, the affinity of an autoregulatory protein for naked rRNA and naked mRNA is very similar (48, 49), suggesting that preferential binding to rRNA must depend on other features such as cooperativity between r-proteins on rRNA or an rRNA conformation that is not seen in *in vitro* experiments.

3.1.1 Repression of translation of the alpha operon

Among the many r-protein operons under autogenous control (47), the regulation of the alpha operon provides a useful overview. The alpha operon of *E. coli* contains five genes. The fourth gene encodes the alpha subunit of RNA polymerase and the remainder encode r-proteins. The third gene in the operon encodes r-protein S4, the translational autoregulator of the operon. S4 binds to the leader region of the alpha operon (48, 50, 51). The binding site extends beyond the ribosome binding site for the operon and contains two pseudoknots (51, 52). Unlike the mRNA binding sites for several other autoregulatory r-proteins, the S4 binding site in mRNA shows no sequence nor obvious structural similarity to the proposed binding site for S4 in 16S rRNA (53–55). Mutations in the leader pseudoknot structure have been identified that prevent translational repression by S4 but do not block S4 binding (56). These data and kinetic studies summarized in References 57 and 58 suggest that translational repression may not result from a direct competition for the operon ribosome binding site between S4 protein and a ribosome. Rather, S4 binding to its mRNA target appears to induce a conformational change in the mRNA that prevents translation initiation.

Translational repression of r-protein mRNA has been most extensively studied in *E. coli*. However, the pattern appears present in other bacteria, such as *Bacillus subtilis* (59, 60), arguing for the generality of this mechanism in the control of r-protein operons.

3.2 Translational repression in bacteriophage T4

Three extensively studied examples of translational repression come from the T4 phage system (61). Each regulates the translation of mRNAs encoding DNA replication proteins. Indeed, two of the translational repressors, gp32 and gp43, are themselves proteins essential for DNA replication. In contrast, the RegA translational repressor has no known function other than repression, and RegA is not essential for growth of wild-type T4. Here we describe the mechanism through which gp32 is believed to mediate translational repression.

3.2.1 T4 gene 32 protein is a DNA binding protein and a translational repressor

Gene 32 protein (gp32) binds to single-stranded DNA and is essential for replication, repair, and recombination of T4 DNA (62, 63). Synthesis of gp32 is related to the number of single-stranded targets available in a cell (64–66), and many years ago it was demonstrated that excess gp32 can block the *in vitro* translation of gp32 transcripts at protein concentrations, 1 to 5 μM, that do not retard the translation of other mRNAs (67, 68). At protein levels in excess of 10 μM, gp32 non-specifically represses a wide range of mRNAs.

The binding affinity of gp32 protein for single-stranded DNA is about three- to four-fold greater than for gp32 mRNA and about 25-fold greater than for general mRNA (66). The basis for the selective inhibition of gp32 mRNA translation resides in a unique 'binding trigger' found in gene 32 transcript leaders (69). The binding trigger is a pseudoknot located at the 5′ end of the gene 32 transcripts. The pseudoknot is adjacent to the region of RNA to which gp32 can bind non-specifically when added at very high protein concentrations. Mutations that delete the pseudoknot structure reduce the specific (high efficiency) gp32 binding to the mRNA *in vitro* and abolish autorepresson *in vivo* (70). Gp32 contains a zinc binding motif which may be essential to its specific interaction with the pseudoknot (71).

3.3 Translational repression of non-ribosomal bacterial genes

Aside from repression of translation of r-protein operons, translational repression in bacterial systems is uncommon. Recently described examples detected in *B. subtilis* and *E. coli* are therefore noteworthy (72–74).

3.3.1 Tryptophan regulation in *B. subtilis*

TRAP is a protein encoded by the *B. subtilis mtrB* gene (75,76). In the presence of tryptophan, TRAP negatively regulates transcription of the *B. subtilis trp* operon (77–80). A complex of TRAP and tryptophan binds to an attenuation structure immediately upstream of the first gene in the operon, *trpE*. This causes termination of transcription prior to *trpE*. TRAP also mediates translational repression of two genes involved in tryptophan biosynthesis, *trpE* and *trpG* (72); and here we focus on *trpG* regulation.

Tryptophan negatively regulates expression of *trpG*, a tryptophan biosynthetic gene that is present in an operon, the folic acid operon, distinct from that for the other *trp* genes (81,82). The *trpG* product is essential for both folic acid and tryptophan biosynthesis and while *trpG* is tryptophan regulated, the other genes of the folic acid operon are not. Recent studies indicate that tryptophan allows TRAP to bind to a region of folate operon mRNA that includes the *trpG* ribosome binding site, where the protein may prevent ribosome loading (72, 73). Two potential binding sites have been proposed (72, 73), although current evidence strongly argues that the *trpG* mRNA binding site is probably a series of nine G/UAG triplets, each separated by two or more nts (73). TRAP consists of 11 subunits (83), and it has been suggested that each subunit binds to a separate G/UAG triplet in mRNA (73). Since this model allows only nine protein subunits to be occupied during binding to *trpG* mRNA, binding to *trpG* mRNA may be less efficient than the binding to the *trp* attenuator. Hence, translational repression of *trpG* may be incomplete. This could be the mechanism that provides a low level of *trpG* product for folic acid biosynthesis during tryptophan repression.

3.3.2 Threonyl-tRNA synthetase in *E. coli*

Two mechanisms have been found that regulate tRNA synthetase expression in bacteria. In *B. subtilis*, and perhaps other Gram-positive bacteria, regulation of tRNA synthetases is mediated by a novel form of transcription attenuation (84–86). By contrast, the regulation of threonyl-tRNA synthetase in *E. coli* is through translational repression.

E. coli threonyl-tRNA synthetase (ThrS) represses its own translation by binding to the leader region of its mRNA and blocking ribosome loading at the ribosome binding site (74). The synthetase binds to the leader due to a structural similarity between mRNA leader sequences and the anti-codon arm of threonyl-tRNA. Four domains in the leader were analyzed for their contribution to the observed binding site for the synthetase, and domain 4 appears to be critical for binding although its role is unclear. Domain 1 contains the ribosome binding site and it is inferred that synthetase binding to domain 2 interferes with ribosome loading in domain 1 through steric hindrance. The sequence of domain 3 does not appear to be critical for the regulation, and it is suspected that domain 3 provides appropriate spacing.

4. Antisense RNA in the control of translation

Antisense RNAs are short (50 to 400 nt) transcripts that bind to sense regions of DNA (or RNA) and thereby alter downstream replication, transcription, or translation. The biochemical basis for antisense control of gene expression is currently a very important area of research, since appropriate application of this technology in eukaryotic cells may provide the means to control unregulated (and pathological) gene expression (87). Here we focus on two examples of antisense regulation of translation which individually contribute to osmoregulation and plasmid transfer in *E. coli*.

4.1 Osmoregulation of OmpF and OmpC

The *ompF* and *ompC* genes map at distant sites on the *E. coli* chromosome (88, 89) and each encodes a membrane protein that acts as a pore for the passage of small, hydrophilic molecules (90). As the osmolarity of growth medium is increased, the expression of *ompC* increases, but expression of *ompF* decreases (89, 91). This regulation is governed by *envZ* and *ompR*, two genes at the *ompB* locus (92, 93). *envZ* encodes a membrane protein thought to sense environmental changes in osmolarity (94–98). As osmolarity increases, EnvZ, which autophosphorylates at His-243, transfers a phosphate to Asp residues in the OmpR protein (96, 99, 100); at low osmolarity, EnvZ can remove phosphates from OmpR. Thus, EnvZ and OmpR represent the fundamental elements of a classical two-component regulatory system (101). Phosphorylated OmpR is a transcriptional activator of *ompC* and a repressor of *ompF*.

Immediately upstream of *ompC* is a divergently transcribed gene *micF*, whose expression is also enhanced by OmpR in response to an increase in osmolarity (102, 103). The RNA product of *micF* is highly complementary to the 5′ end of *ompF* transcripts, spanning approximately 70 nts that include the *ompF* ribosome binding site and initiation codon. Moreover, when *micF* is cloned on a high copy plasmid, expression of *ompF* is nearly eliminated (102). These results suggested that *ompF* is regulated at two levels, repression of transcription by phosphorylated OmpR and repression of translation by *micF* RNA.

Seemingly contradictory observations have been made regarding the importance of the post-transcriptional control of *ompF* by *micF* RNA versus the repression of *ompF* transcription by OmpR. Recent studies indicate that the two forms of regulation are indeed complementary (104). At low to moderate levels of osmolarity (represented as 0 to 6 per cent sucrose), the 60 per cent reduction in *ompF* expression seen in wild-type cells is not observed in cells deleted for *micF*. However, at higher osmolarity the reduction of *ompF* expression is the same in the deletant and in the wild-type parent. Thus it seems likely that in this system, post-transcriptional control mediates a response to subtle environmental changes, while transcriptional control abolishes *ompF* expression at excessive levels of osmolarity.

4.2 FinP control of plasmid transfer

Members of the IncF1 plasmid incompatibility group are self-transmissible plasmids that contain most of the genes essential for transfer in a single large operon (105). Transcription of this operon depends on the protein product of an upstream gene *traJ*. Expression of *traJ* is inhibited by two fertility inhibition genes *finP* and *finO* (106). The *finP* gene overlaps with the 5′ end of *traJ*, and is oriented opposite to *traJ*. Transcription of *finP* produces a 78-nt untranslated RNA molecule (FinP), that is complementary to the 5′ untranslated region of *traJ* mRNA and can form a stable duplex with *traJ* mRNA both *in vitro* and *in vivo* (107–109). Since the duplexed region includes the *traJ* ribosome binding site, FinP may block translation of *traJ* mRNA by preventing ribosome loading. This mechanism of action of an antisense RNA has been seen in additional examples of antisense-mediated translational control (110).

Genetic analysis of the interaction of FinP with *traJ* mRNA has provided evidence that the initial contact between the two RNAs is through an interaction of complementary loops (109, 111), generating a so-called 'kissing complex' (112–114) which can lead to complete duplex formation between two RNAs. However, the *finP–traJ* system is unusual since repression *in vivo* also requires the protein product of the plasmid *finO* gene (115). A comparison of the lengths of *traJ* transcripts *in vivo* during the expression of FinP revealed that in the absence of the *finO* product full-length transcripts were observed (116). In the presence of the *finO* product, truncated *traJ* transcripts were detected (116). The sequence of the site cleaved suggested that cleavage might be mediated by an RNase III like activity. At present it is speculated that the *finO* protein product may stabilize FinP, allowing it to form a stable duplex with *traJ* mRNA. The duplex both occludes the *traJ* ribosome binding site and leads to the formation of an RNAase III cleavage site within the complex (116).

5. Programmed frameshifting in the post-transcriptional control of release factor 2

Changes in reading frame can produce different proteins from the same coding sequence. Several examples of this principle for maximizing the utility of a single nt sequence have been described in bacteria (117, 118). However, the response of a change in reading frame to the intracellular levels of a particular protein is unusual, and demonstrates another level of translation at which cells can control gene expression.

Release factor 2 (RF2) is essential for complete termination of translation at UGA codons, which includes release of both the nascent peptide and ribosome from the mRNA. The gene for the RF2 protein contains an inframe UGA codon early in its coding sequence (119). The nt sequence of the RF2 gene demonstrated that complete translation of RF2 requires a +1 change of reading frame at the UGA codon. When the intracellular levels of RF2 are low, a ribosome encountering the UGA codon will idle at that codon, awaiting RF2 for peptide and ribosome release from the mRNA (120–122). The duration of the stalling allows the ribosome to select and resume translation in a new reading frame. However, the change in reading frame is clearly not random, but is dictated by a sequence in the mRNA immediately upstream from the UGA codon. This sequence resembles a ribosome binding site and mutations that diminish its complementarity with 16S rRNA diminish or abolish programmed frameshifting *in vivo* (123).

Studies in which the UGA codon has been replaced by sense codons have demonstrated that sense codons whose decoding slows translation, at varying efficiencies, can promote the change of frame needed for complete RF2 translation (124, 125). Such changes in reading frame are independent of the level of RF2 (126).

Collectively, the data indicate three important features for frameshifting in the RF2 system:

1. The codon at the frameshift site must slow or pause translation.
2. The ribosome binding site immediately 5′ to the shift site presumably stabilizes the ribosome and may facilitate the shift to the +1 reading frame.
3. The P site tRNA at the sense codon which immediately precedes the UGA nonsense codon must be capable of rephasing to the +1 frame.

In the wild-type RF2, recognition of the +1 frame depends on wobble of the anticodon for the P site tRNA.

6. Regulation by translation attenuation

Genes that specify inducible resistance to chloramphenicol (the *cat* genes) or erythromycin (the *erm* genes) in Gram-positive bacteria and inducible chloramphenicol resistance in Gram-negatives (the *cmlA* gene, which encodes a membrane protein) appear to be regulated by a common mechanism termed translation attenuation (42, 127–131). In this form of translational control, the mRNA for the resistance determinant contains a domain of secondary structure that sequesters the ribosome binding site. Immediately upstream is a short open reading frame, the leader. It was originally speculated that an inducing antibiotic could cause ribosome stalling in the leader, which would lead to destabilization of the secondary structure. This would free the normally sequestered ribosome binding site. Hence, antibiotic induction of gene expression would result from the activation of translation initiation of the resistance determinant. All experimental evidence to date supports the general outlines of this model.

For both *cat* and *erm* genes it has been demonstrated that induction requires translation of the leader (132–134). This is consistent with the model since a ribosome should be capable of translating to the vicinity of the secondary structure for drug-mediated stalling to provoke destabilization of the RNA secondary structure. However, virtually all *cat* and *erm* genes are induced uniquely by the antibiotic to which each confers resistance, chloramphenicol and erythromycin, respectively. Since both chloramphenicol and erythromycin inhibit ribosome elongation, albeit by different mechanisms, stalling in the leader cannot be due to non-specific inhibition of ribosome elongation. This interpretation is also supported by a unique finding. The gene *catA86* (formerly *cat-86*) is induced by chloramphenicol and less efficiently by the nucleoside antibiotic amicetin (135). Specific missense mutations in the leader abolish induction by amicetin but have no effect on induction by chloramphenicol (136), arguing for an active role of the leader peptide in drug induction.

Missense mutations in both the *erm* and *cat* leaders can prevent induction of the corresponding gene by the relevant inducer (134, 137). In the case of *catA86*, the inhibitory effect of the mutations is not complemented *in trans* by a wild-type copy of the leader (129). Thus, the leader peptide exerts its influence *in cis*.

6.1 Active role for the regulatory leader

Induction of *cat* depends on the site specific stalling of a ribosome with its aminoacyl site at leader codon 6 (133). In the case of *ermC*, induction appears to require ribosome stalling at leader codon 9 (138). Thus, the leader encoded peptide upstream from the stall site contains the biologically relevant information for selecting the site of stalling and probably for responding to a particular antibiotic. Recent studies of the leader peptides for the *catA86* and *cmlA* genes demonstrate the mechanism through which leader peptides select the stall site.

The first five codons of the *catA86* leader specify a peptide, MVKTD, that is an *in-vitro* inhibitor of peptidyl transferase, the catalytic activity of the ribosome that forms peptide bonds during translation (139, 140). Shortening the peptide at the N- or C-terminus eliminates inhibition. Inhibition is not simply due to the charge of the peptide since the reverse sequence of the peptide, DTKVM, is not inhibitory (139). Missense mutations in the leader that diminish or block the *in vivo* induction of *catA86* by chloramphenicol correspond to amino acid substitutions in the peptide that appropriately diminish or abolish its activity as a peptidyl transferase inhibitor *in vitro* (141).

The inhibitor five-residue peptide has recently been shown to be an RNA binding peptide whose target is the peptidyl transferase center of large subunit rRNA from bacteria, yeast, and archebacteria (142–144). *In vitro* footprinting studies, using T1 and cobra venom nucleases, have demonstrated that the peptide alters the conformation of domains IV and V of 23S and 26S rRNA (143). It has been proposed that the nascent five-residue peptide selects the site of ribosome stalling in the *catA86* leader by pausing translation through an alteration of the conformation of 23S rRNA. This observation is highly significant since 23S rRNA is the (or a) target for antibiotics that induce genes regulated by translation attenuation such as chloramphenicol, erythromycin, and amicetin (145–147), and studies by Noller *et al.* have strongly implicated 23S rRNA as a critical component of peptidyl transferase (148).

The first eight codons of the *cmlA* leader encode an eight-residue peptide, MSTSKNAD, that behaves identically to the *cat* five-residue peptide as an *in-vitro* inhibitor of peptidyl transferase (142, 143). The conformational changes in rRNA brought about by the *cat* five-residue and the *cmlA* eight-residue peptides are indistinguishable.

An as yet unresolved question in translation attenuation regulation is the role of the leader peptide in determining which antibiotic will serve as inducer. *catA86* may also serve as a model for such studies since this gene is induced by amicetin as well as by chloramphenicol, and missense mutations in the leader have been identified that eliminate the genes response to amicetin without affecting inducibility by chloramphenicol (136).

7. *cis* effects of the nascent peptide on translation

A universal feature of translation is that the nascent peptide product is maintained within the ribosome until about 20 to 40 amino acids have been joined through the formation of peptide bonds. It is at that point the peptide begins to emerge from the protection afforded by the ribosome. Recent studies, noted above, indicate that a nascent peptide can interact with components of its translating ribosome and modify the function of only that ribosome.

7.1 Autoinduction of *catA86*

The site of stalling in the *catA86* leader that will induce gene expression was identified by amino acid starvation experiments (133, 149). Starvation that delivers the aminoacyl site of the ribosome to leader codon 6 induces the gene in the absence of any antibiotic. Starvation that stalls the ribosome at earlier or later codons of the leader, respectively, fails to induce the gene or has only a minimal activating effect. When leader codon 6 was replaced with any of the three translation stop codons, an alternative way to pause translation at the induction site, the *cat* gene was constitutively expressed (150). The level of constitutive expression, termed autoinduction, was about 2/3 that observed during complete chloramphenicol induction of the wild-type gene. Autoinduction was found to depend on the amino acid sequence of the leader encoded peptide. Missense mutations within the first five leader codons that are known to block induction by chloramphenicol also block autoinduction (150). Therefore, autoinduction could plausibly be explained if the nascent five-residue peptide were an inhibitor of translation termination, in addition to its known role as an inhibitor of peptidyl transferase. When this was directly tested, the five-residue peptide was found to be an effective inhibitor of translation termination *in vitro* (151). This observation is very important since it demonstrates that a nascent peptide can communicate with its translating ribosome *in vivo* and bring about selective alterations in only that ribosome.

The *catA86* five-residue peptide is known to alter the conformation of specific domains of 23S rRNA (143). Thus, it is conceivable that rRNA may be involved in translation termination as well as peptidyl transferase, an idea already suggested by previous work (148, 152). Geballe and Morris (153) independently suggested a mechanism of action similar to that proposed to explain autoinduction of *catA86*, in an attempt to explain the role of u(pstream)ORFs in the reduction of translation of downstream coding sequences for several eukaryotic genes (154–157). If ribosome release at the termination codon for a uORF is prevented by interference of the uORF's peptide product with ribosome release, the stalled ribosome would block ribosome scanning to the downstream ORF. To our knowledge, the possible role of these eukaryotic uORF peptides as rRNA binding peptides has not been examined.

7.2 Ribosome hopping in T4 topoisomerase expression

Within the mRNA coding sequence for a subunit of bacteriophage T4 topoisomerase there are 50 contiguous nts that are not translated into amino acids (158). Since other T4 genes involved in DNA replication contain self-splicing introns, the occurrence of the insert in the topoisomerase transcript might appear to be another such example. However, the topoisomerase insert is not removed from the mRNA. Rather, the evidence from *in vivo* and *in vitro* translation of the insert-containing transcript demonstrates the translating ribosome skips over the insert, joining amino acids encoded by nts upstream from the insert to amino acids encoded by sequences immediately 3′ to the insert. This novel translational 'skipping' has been termed ribosome hopping or translational bypass (159).

In a series of elegant experiments (159), it has been established that there are three critical elements for ribosome hopping in the T4 system: a region of secondary structure at the 5′ end of the insert; identity, or near identity, of the 'takeoff' and 'landing' codons; and the amino acid sequence of the nascent peptide immediately upstream from the insert. It has been supposed that the role of the nascent peptide might be to displace P site tRNA in a manner analogous to the peptide antibiotic edeine (160). Although appealing, as yet there is no direct evidence to support this idea.

A specific mutation, *hop-1*, has recently been identified in the gene for ribosomal protein L9 which partially substitutes for the 5′ secondary structure in ribosome hopping (160). This mutation is extremely important since it implies an active role of the ribosome in this unique form of translational control. Further dissection of the roles of both the nascent peptide and the ribosome is needed to provide an understanding of ribosome hopping at the biochemical level.

The biological relevance of the insert in the topoisomerase gene is unclear, since deletion of the insert does not interfere with expression of the protein product. However, the occurrence of self-splicing introns in functionally related T4 genes suggests a possible ancestral origin for the topoisomerase insert.

7.3 Rhodanese release from the ribosome is influenced by its N-terminal sequence

The gene for the eukaryotic protein rhodanese can be efficiently expressed in a coupled *in vitro* transcription/translation system derived from *E. coli* components (161). Despite the efficiency of expression, most of the protein remains bound to the ribosome as a peptidyl tRNA–ribosome complex, and its release requires the presence of several *E. coli* chaperones. It therefore appears that proper folding of rhodanese is critical to its release from the ribosome after completion of translation.

A peptide corresponding to the N-terminal region of rhodanese is known to interfere with the refolding of a denatured form of the enzyme, and the addition of an N-terminal 17-residue peptide blocked the chaperone-dependent release of the nascent wild-type protein from ribosomes (162). It is conceivable that during translation, portions of the protein (perhaps the N-terminal region) can interact with the ribosome and prevent release at the rhodanese translation termination codon. Correct folding of the protein, facilitated by the chaperones, may pull the interfering portion of the protein from the ribosome allowing normal termination to occur.

8. mRNA recoding

The 'classical' rules for transcription of DNA into RNA and for decoding mRNA into amino acid sequences allow the sequence of a protein to be inferred from the sequence of nts in the corresponding gene. Variations to these rules exist. The impact that introns and RNA editing have on the flow of information from DNA to protein are two of several well-studied examples. In recent years, several examples have been described in which a sequence of nts in mRNA is interpreted by the cell in a manner which differs from that predicted by the 'classical' rules. These variations, termed recoding (163), do not involve changes in the sequence of the RNA. Recoding causes the protein that is specified by a sequence of mRNA nts to differ (slightly or substantially) from that which would be predicted by direct interpretation of the genetic code. Detailed study of recoding examples can inform us of the elaborate side reactions that are needed to bypass the normal decoding process. The most broadly studied example of recoding is that which allows the translation termination codon UGA to specify the amino acid selenocysteine.

8.1 Incorporation of selenocysteine at UGA

The amino acid selenocysteine is extremely rare in nature and has been identified in about 10 proteins (164). In proteins which contain selenocysteine, the amino acid is typically present as a single residue in the active site for catalysis. Substituting cysteine for the selenocysteine residue normally seen in *E. coli* formate dehydrogenase decreases its catalytic activity 380-fold (165). This demonstrates that selenocysteine is very important for certain critical catalytic activities.

Selenocysteine is inserted into proteins by decoding UGA, normally a translation termination codon, as a sense codon (166–168). This unusual decoding event requires highly specialized cellular components. First, a specialized tRNA molecule, $tRNA^{sec}$ which contains an anti-codon complementary to UGA, is charged with serine (169). Serine is then modified to selenocysteine in a two-step process. Second, a critical mRNA secondary structure domain, termed SECIS, is present immediately 3′ to the UGA codon that is to be the site of selenocysteine decoding (170). The primary sequence of SECIS appears to differ among the various selenocysteine-incorporating mRNAs, arguing that the function of the structure requires a specific folding pattern. Third, a protein designated as SELB is critical for the recoding. SELB, the product of the *E. coli selB* gene, is probably derived from elongation factor Tu, since the two proteins share extensive amino acid homology over 240 N-terminal residues (171). Like EF-Tu, each SELB molecule binds one molecule of GTP.

The SELB binding site is a region of SECIS that is distal to the mRNA backbone. When the UGA codon is in the A site of a translating ribosome, the loop region of SECIS directs bound SELB to a location between the ribosomal subunits. The evidence suggests that this attraction allows SELB to function as a specialized elongation factor solely used for incorporation of selenocysteine at UGA (172, 173).

9. Conclusions

Ribosome loading on mRNA at the ribosome binding site sequence is critical for translation initiation and half of the post-translational control mechanisms discussed influence translation at this level (174–177). Indeed, control of the availability of the ribosome binding site also effects the growth rate dependent regulation of *gnd* expression, a gene encoding an enzyme of central carbon metabolism (180). Blocking the function of a ribosome binding site is achieved by protein or RNA binding to the site, or in its vicinity. In the case of regulation by antisense RNA, translation attenuation, and certain forms of protein-mediated repression of translation, the RNA duplex or the bound protein is in direct competition with ribosomes for binding to the ribosome binding site. In other cases, the bound regulatory molecule appears to alter the conformation of the ribosome binding site region, apparently decreasing its ability to be recognized by ribosomes.

The role of mRNA stability in controlling gene expression falls into two categories. First, a cell's need for the product of a particular gene presumably selects for the evolution of a net level of expression. The appropriate expression level can be met by varying initiation of transcription and/or translation. Modifying the stability of the mRNA by appropriate positioning of stabilizing sequences within the message would also enhance expression levels by providing a longer lived template for protein synthesis. Second, the *puf* operon provides a valuable illustration of how specific sequences in mRNA can serve as responding elements for the initiation of selective degradation of mRNA under specific environmental conditions.

The concept of recoding provides cells with a means to circumvent or modify the normal decoding process to achieve a specific need. The example of selenocysteine decoding of UGA illustrates the utility of recoding, although other examples exist (163). Selenocysteine at the active site of formate dehydrogenase dramatically increases catalytic activity over that observed when the identical position is occupied by cysteine. However, the rare occurrence of selenocysteine in proteins suggests that it is unlikely to be a critical amino acid at most other sites. Otherwise, not only would the amino acid be more common, but it is logical to assume a codon for selenocysteine would have evolved.

The *cis* effects of nascent peptides on translating ribosomes is a new concept in translational control and appears to be most well characterized as a contributor to translation attenuation regulation of inducible chloramphenicol resistance genes. In this role, the nascent peptide seems to select the site of ribosome stalling in the regulatory leader by interfering with peptide bond formation. The *cat* and *cmlA* leader peptides have recently been shown to be binding peptides for large subunit rRNA. Since both leader peptides modify the conformation of only the rRNA domains at the peptidyl transferase center, it is believed that this modification is the basis for peptidyl transferase inhibition. The *cat* leader peptide is also an inhibitor of translation termination (151). Thus, it is conceivable that the ribosomal activities for peptide bond formation and translation termination require participation of rRNA.

Recoding of UGA as a selenocysteine codon demonstrates a form of translational control whose goal could not be achieved by any obvious alteration in transcription. In other forms of translational control, such as antisense regulation by *micF* and certain examples of r-protein repression of translation, translational control appears to be complementary to transcriptional control. In yet other examples of translational control, such as translation attenuation, it is not clear why control of translation was chosen over control of transcription. However, for the investigator each system of post-transcriptional control can offer valuable insights into the complexities of translation.

References

1. Duvall, E. J. and Lovett, P. S. (1986) Chloramphenicol induces translation of the mRNA for a chloramphenicol resistance gene in *Bacillus subtilis*. *Proc. Natl. Acad. Sci. USA*, **83**, 3939–3943.
2. Silhavy, T. J. and Beckwith, J. R. (1985) Uses of *lac* fusions for the study of biological problems. *Microbiol. Rev.*, **49**, 398–418.
3. Higgins, C. F., Peltz, S. W., and Jacobson, A. (1992) Turnover of mRNA in prokaryotes and lower eukaryotes. *Curr. Opin. Genet. Develop.*, **2**, 739–767.
4. Peterson, C. (1992) Control of functional mRNA stability in bacteria: multiple mechanisms of nucleolytic and non-nucleolytic inactivation. *Mol. Microbiol.*, **6**, 277–282.
5. Arraiano, C. M. (1993) Post-transcriptional control of gene expression: bacterial mRNA degradation. *World J. Microb. Biotech.*, **9**, 421–432.

6. Belasco, J. G. and Higgins, C. F. (1988) Mechanisms of mRNA decay in bacteria: a perspective. *Gene*, **72**, 15–23.
7. Deutscher, M. P. and Zhang, J. (1990) Ribonucleases: diversity and regulation. In *Posttranscriptional control of gene expression* (ed. J. McCarthy and M. Tuite), NATO ASI Series H: Cell biology Vol. 49, pp. 1–12. NATO Scientific Affairs Division.
8. Deutscher, M. P. (1993) Promiscuous exoribnuclease of *Escherichia coli*, *J. Bacteriol.*, **176**, 4577–4583.
9. Spahr, P. F. (1964) Purification and properties of ribnuclease II from *Escherichia coli*. *J. Biol. Chem.*, **239**, 3716–3726.
10. Reiner, A. M. (1969) Isolation and mapping of polynucleotide phosphorylase mutants of *Escherichia coli*. *J. Bacteriol.*, **94**, 1431–1436.
11. Donovan, W. P. and Kushner, S. R. (1986) Polynucleotide phosphorylase and ribonuclease II are required for cell viability and mRNA turnover in *Escherichia coli*. *Proc. Natl. Acad. Sci. USA*, **83**, 120–124.
12. McLaren, R. S., Newbury, S., Dance, C. S. C., Causton, H. C., and Higgins, C. F. (1991) mRNA degradation by processive 3′-5′ exo-ribonuclease *in vitro* and the implications for prokaryotic mRNA decay *in vivo*. *J. Mol. Biol.*, **221**, 81–95.
13. Causton, H., Py, B., McLaren, R. S., and Higgins, C. F. (1994) mRNA degradation in *Escherichia coli*: a novel factor which impedes the exoribonucleolytic activity of PNPase at stem-loop structures. *Mol. Microbiol.*, **14**, 731–761.
14. Robertson, H. D., Webster, R. E., and Zinder, N. D. (1968) Purification and properties of ribonuclease III from *Escherichia coli*. *J. Biol. Chem.*, **243**, 82–91.
15. Robertson, H. D. (1982) *Escherichia coli* ribonuclease III cleavage sites. *Cell*, **30**, 669–672.
16. Misra, T. K. and Apirion, D. (1979) RNase E, an RNA processing enzyme from *E. coli*. *J. Biol. Chem.*, **254**, 11154–11159.
17. Ehretsmann, C. P., Carpousis, A. J., and Krisch, H. M. (1992) Specificity of *Escherichia coli* endo-ribonuclease RNase E: *in vivo* and *in vitro* analysis of mutants in a bacteriophage T4 mRNA processing site. *Genes and Develop.*, **6**, 149–159.
18. McDowall, K. J., Kaberdin, V. R., Wu, S. W., Cohen, S. N., and Lin Chao, S. (1995). Site-specific RNase E cleavage of oligonucleotides and inhibition by stem-loops. *Nature*, **376**, 287–290.
19. Babitzke, P., Granger, L., Olszewski, L., and Kushner, S. R. (1993) Analysis of mRNA decay and rRNA processing in *Escherichia coli* multiple mutants carrying a deletion in RNase III. *J. Bacteriol.*, **176**, 229–239.
20. Cole, J. R. and Nomura, M. (1986) Changes in the half-life of ribosomal protein messenger RNA caused by translational repression. *J. Mol. Biol.*, **188**, 383–392.
21. Carpousis, A. J., Van Houwe, G., Ehretsmann, C., and Krisch, H. M. (1994) Copurification of *E. coli* RNAase E and PNPase: evidence for a specific association between two enzymes important in RNA processing and degradation. *Cell*, **76**, 889–900.
22. Py, B., Causton, H., Mudd, E. A., and Higgins, C. F. (1994) A protein complex mediating mRNA degradation in *Escherichia coli*. *Mol. Microbiol.*, **14**, 717–729.
23. O'Hara, E. B., Chekanova, J. A., Ingle, C. A., Kushner, Z. R., Peters, E., and Kushner, S. R. (1995) Polyadenylation helps regulate mRNA decay in *Escherichia coli*. *Proc. Natl. Acad. Sci. USA*, **92**, 1807–1811.
24. Cohen, S. N. (1995) Surprises at the 3′ end of prokaryotic RNA. *Cell*, **80**, 829–832.
25. Guarneros, G., Montanez, C., Hernandez, T., and Court, D. (1982) Posttranscriptional control of bacteriophage l *int* gene expression from a site distal to the gene. *Proc. Natl. Acad. Sci. USA*, **79**, 6238–6244.

26. Belasco, J. G., Beatty, J. T., Adams, C. W., von Gabain, A., and Cohen, S. N. (1985) Differential expression of photosynthesis genes in *R. capsulata* results from segmental differences in stability within the polycistronic *rxcA* transcript. *Cell*, **40**, 171–181.
27. Klug, G. and Cohen, S. N. (1990) Rate-limiting endonucleolytic cleavage of the 2.7 kb *puf* mRNA of *R. capsulata* is influenced by oxygen. In, *Molecular biology of membrane-bound complexes in photosynthetic bacteria* (ed. G. Drews), pp. 110–122. Plenum, London.
28. Drews, G. and Oelze, J. (1981) Organization and differentiation of membranes of phototrophic bacteria. *Adv. Microbial Physiol.*, **22**, 1–97.
29. Drews, G. (1985) Structure and functional organization of light harvesting complexes and photochemical reaction centers in membranes of phototropic bacteria. *Microbiol. Rev.*, **49**, 59–70.
30. Chen, C. Y., Beatty, J. T., Cohen, S. N., and Belasco, J. G. (1988) An intercistronic stem-loop structure functions as an mRNA decay terminator necessary but insufficient for *puf* mRNA stability. *Cell*, **52**, 609–619.
31. Klug, G., Jock, S., and Rothfuchs, R. (1992) The rate of decay of *Rhodobacter capsulatus*-specific *puf* mRNA segments is differentially affected by RNase E activity in *Escherichia coli*. *Gene*, **121**, 96–102.
32. Chen, C. Y. and Belasco, J. G. (1988) Degradation of *pufLMX* mRNA in *Rhodobacter capsulatus* is initiated by nonrandom endonucleolytic cleavage. *J. Bacteriol.*, **172**, 4578–4586.
33. Klug, G. (1991) Endonucleolytic degradation of *puf* mRNA in *Rhodobacter capsulatus* is influenced by oxygen. *Proc. Natl. Acad. Sci. USA*, **88**, 176–180.
34. Schlessinger, E., Jacobs, K. A., Gupta, R. S., Kano, Y., and Imamoto, F. (1977) Decay of individual *Escherichia coli trp* messenger RNA molecules is sequentially ordered. *J. Mol. Biol.*, **110**, 421–439.
35. Stevens, A. and Poole, T. L. (1995) 5′-exonuclease-2 of *Saccharomyces cerevisiae*. *J. Biol. Chem.*, **270**, 16063–16069.
36. von Gabain, A., Belasco, J. G., Schottel, J. L., Chang, A. C. Y., and Cohen, S. N. (1983) Decay of mRNA in *Escherichia coli*: Investigation of the fate of specific segments of transcripts. *Proc. Natl. Acad. Sci. USA*, **80**, 653–657.
37. Emory, S. A. and Belasco, J. G. (1992) A 5′-terminal stem-loop can stabilize mRNA in *E. coli*. *Genes Devel.*, **6**, 135–148.
38. Bouvet, P. and Belasco, J. G., (1992) Control of RNase E-mediated RNA degradation by 5′-terminal base pairing in *E. coli*. *Nature (London)*, **360**, 488–491.
39. Sandler, P. and Weisblum, B. (1989) Erythromycin-induced ribosome stall in the *ermA* leader: A barricade to 5′-to-3′ nucleolytic cleavage of the *ermA* transcript. *J. Bacteriol.*, **171**, 6680–6688.
40. Bechhofer, D. H. and Zen, K. (1989) Mechanism of erythromycin-induced *ermC* mRNA stability in *Bacillus subtilis*. *J. Bacteriol.*, **171**, 5803–5811.
41. Bechhofer, D. H. and Dubnau, D. (1987) Induced mRNA stability in *Bacillus subtilis*. *Proc. Natl. Acad. Sci. USA*, **84**, 498–502.
42. Bruckner, R. and Matzura, H. (1985) Regulation of the inducible chloramphenicol acetylatransferase gene of the *Staphylococcus aureus* plasmid pUB112. *EMBO J.*, **4**, 2295–2300.
43. Dreher, J. and Matzura, J. (1991) Chloramphenicol-induced stabilization of *cat* messenger RNA in *Bacillus subtilis*. *Mol. Microbiol.*, **5**, 3025–3034.
44. Kennell, D. (1986) The instability of mRNA in bacteria. In *Maximizing gene expression* (ed. W. S. Reznikoff and L. Gold), pp. 101–142. Butterworths, Stoneham, Massachusetts.

45. Nisson, G., Belasco, J. G., Cohen, S. N., and von Gabain, A. (1987) Effect of premature termination of translation on mRNA stability depends on the site of ribosome release. *Proc. Natl. Acad. Sci. USA*, **84**, 4890–4894.
46. Pedersen, C. (1992) Control of functional mRNA stability in bacteria: multiple mechanisms of nucleolytic and non-nucleolytic inactivation. *Mol. Microbiol.*, **6**, 277–282.
47. Zengel, J. M. and Lindahl, L. (1994) Diverse mechanisms for regulating ribosomal protein synthesis in *Escherichia coli*, *Prog. Nucleic Acid Res. and Mol. Biol.*, **47**, 331- 370.
48. Deckman, I. C. and Draper, D. E. (1985) Specific interaction between ribosomal protein S4 and the alpha operon messenger RNA. *Biochemistry*, **24**, 7860–7865.
49. Gregory, R. J., Cahill, P. B., Thurlow, D. L., and Zimmermann, R. A. (1988) Interaction of *Escherichia coli* ribosomal protein S8 with its binding sites in ribosomal RNA and messenger RNA. *J. Mol. Biol.*, **204**, 295–307.
50. Deckman, I. C., Draper, D. E., and Thomas, M. S. (1987) S-4 alpha mRNA translation repression complex. I. Thermodynamics of formation. *J. Mol. Biol.*, **196**, 313–322.
51. Deckman, I. C. and Draper, D. E. (1987) S4-alpha mRNA translation repression complex. II. Secondary structures of the RNA regulatory site in the presence and absence of S4. *J. Mol. Biol.*, **196**, 323–330.
52. Tang, C. and Draper, D. E. (1989) Unusual mRNA pseudoknot structure is recognized by a protein translational repressor. *Cell*, **57**, 531–536.
53. Vartikar, J. V. and Draper, D. E. (1989) S4–16S ribosomal RNA complex. Binding constant measurements and specific recognition of a 460-nucleotide region. *J. Mol. Biol.*, **209**, 221–234.
54. Stern, S., Wilson, R. C., and Noller, H. F. (1986) Localization of the binding site for protein S4 on 16S ribosomal RNA by chemical and enzymatic probing and primer extension. *J. Mol. Biol.*, **192**, 101–110.
55. Sapag, A, Vartikar, J. V., and Draper, D. E. (1990) Dissection of the 16S rRNA binding site for ribosomal protein S4. *Biochim. Biophys. Acta*, **1050**, 34–37.
56. Tang, C. K. and Draper, D. E. (1990) Evidence for allosteric coupling between the ribosome and repressor binding sites of a translationally regulated mRNA. *Biochemistry*, **29**, 4434–4439.
57. Spedding, G., Gluick, T. C., and Draper, D. E. (1993) Ribosome initiation complex formation with the pseudoknotted alpha operon messenger RNA. *J. Mol. Biol.*, **229**, 609–622.
58. Spedding, G. and Draper, D. E. (1993) Allosteric mechanism for translational repression in the *Escherichia coli* alpha operon. *Proc. Natl. Acad. Sci. USA*, **90**, 4399–4403.
59. Grundy, F. J. and Henkin, T. M. (1991) The *rpsD* gene, encoding ribosomal protein S4, is autogenously regulated in *Bacillus subitlis*. *J. Bacteriol.*, **173**, 4595–4602.
60. Grundy, F. J. and Henkin, T. M. (1992) Characterization of the *Bacillus subtilis rpsD* regulatory target site. *J. Bacteriol.*, **176**, 6763–6770.
61. Miller, E. S., Karam, J. D., and Spicer, E. (1994) Control of translation initiation: mRNA structure and protein repressors. In *Molecular biology of bacteriophage T4* (ed. J. D. Karam), pp. 193–205. American Society for Microbiology, Washington, DC.
62. Karpel, R. L. (1990) T4 bacteriophage gene 32 protein, In *The biology of nonspecific DNA protein interactions* (ed. A. Revzin), pp. 103–130. CRC Press, Boca Raton, Fla.
63. Shamoo, Y., Spicer, E. K., Coleman, J. E., and Konigsberg, W. H. (1994) Correlating structure to function in proteins: T4 GP32 as a prototype. In *Molecular biology of bacteriophage T4* (ed. J. D. Karam), pp. 301–304. American Society for Microbiology, Washington, DC.
64. Krisch, H. M., Bolle, A., and Epstein, R. H. (1976) Regulation of the synthesis of bacteriophage T4 gene 32 protein. *J. Mol. Biol.*, **88**, 89–104.

65. Gold, L., O'Farrell, P., and Russel, M. (1976) Regulation of gene 32 expression during bacteriophage T4 infection of *Escherichia coli*. *J. Biol. Chem.*, **251**, 7251–7262.
66. Lemaire, G., Gold, L., and Yarus, M. (1978). Autogenous translational repression of bacteriophage T4 gene 32 expression *in vitro*. *J. Mol. Biol.*, **126**, 73–90.
67. Russell, M., Gold, L., Morrissett, H., and O'Farrell, P. (1976) Translational autogenous regulation of gene 32 expression during bacteriophage T4 infection. *J. Biol. Chem.*, **251**, 7263–7270.
68. Krisch, H. M., Van Houwe, G., Belin, D., Gibbs, W., and Epstein, R. H. (1977) Regulation of the expression of bacteriophage T4 genes 32 and 43. *Virology*, **78**, 87–98.
69. McPheeters, D. S., Stormo, G. D., and Gold, L. (1988) Autogenous regulatory site on the bacteriophage T4 gene 32 messenger RNA. *J. Mol. Biol.*, **201**, 517–535.
70. Shamoo, Y., Tam, A., Konigsberg, W. H., and Williams, K. R. (1993) Translational repression by the bacteriophage T4 gene 32 protein involves specific recognition of an RNA pseudoknot structure. *J. Mol. Biol.*, **232**, 89–104.
71. Shamoo, Y., Webster, K. R., Williams, K. R., and Konigsberg, W. H. (1991) A retroviruslike zinc domain is essential for translational repression of bacteriophage T4 gene 32. *J. Biol. Chem.*, **266**, 7967–7970.
72. Babitzke, P., Stults, J. T., Shire, S. J., and Yanofsky, C. (1994) TRAP, the *trp* RNA-binding attenuation protein of *Bacillus subtilis*, is a multisubunit complex that appears to recognize G/UAG repeats in the *trpEDCFBA* and *trpG* transcripts. *J. Biol. Chem.*, **269**, 16597–16604.
73. Yang, M., De Saizieu, A., Van Loon, A. P. G. M., and Gollnick, P. (1995) Translation of *trpG* in *Bacillus subtilis* is regulated by the *trp* RNA-binding attenuation protein (TRAP). *J. Bacteriol.*, **177**, 4272–4278.
74. Brunel, C., Romby, P., Moine, H., Caillet, J., Grunberg-Manago, M., Springer, M., Ehresmann, B., and Ehresmann, C. (1993) Translation regulation of the *Escherichia coli* threonyl-tRNA synthetase gene: Structural and functional importance of the *thrS* operator domains. *Biochimie*, **75**, 1167–1179.
75. Gollnick, P. (1994) Regulation of the *B. subtilis trp* operon by an RNA binding protein. *Mol. Microbiol.*, **11**, 991–997.
76. Gollnick, P., Ishino, S., Kuroda, M. I., Henner, D., and Yanofsky, C. (1990) The *mtr* locus is a two gene operon required for transcription attenuation in the *trp* operon of *Bacillus subtilis*. *Proc. Natl. Acad. Sci. USA*, **87**, 8726–8730.
77. Babitzke, P. and Yanofsky, C. (1993) Reconstitution of *Bacillus subtilis trp* attenuation *in vitro* with TRAP, the *trp* RNA-binding attenuation protein. *Proc. Natl. Acad. Sci. USA*, **90**, 133–137.
78. Kuroda, M. I., Shimotsu, H., Henner, D. J., and Yanofsky, C. (1986) Regulatory elements common to the *Bacillus pumilus* and *Bacillus subtilis trp* operons. *J. Bacteriol.*, **167**, 792–798.
79. Otridge, J. and Gollnick, P. (1993) MtrB from *Bacillus subtilis* binds specifically to *trp* leader RNA in a tryptophan dependent manner. *Proc. Natl. Acad. Sci. USA*, **90**, 128–132.
80. Shimotsu, H., Kuroda, M. I., Yanofsky, C., and Henner, D. J. (1986) Novel form of transcription attenuation regulates expression of the *Bacillus subtilis* tryptophan operon. *J. Bacteriol.*, **166**, 461–471.
81. Kane, J. F. (1977) Regulation of a common amidotransferase subunit. *J. Bacteriol.*, **132**, 419–425.
82. Slock, J., Stahly, E. P., Han, C.-Y., Six, E. W., and Crawford, I. P. (1990) An apparent *Bacillus subtilis* folic acid biosynthetic operon containing *pab*, an amphibolic *trpG* gene, a third gene required for synthesis of *para*-aminobenzoic acid, and the dihydropteroate synthase gene. *J. Bacteriol.*, **172**, 7211–7226.

83. Antson, A. A., Otridge, J., Brzozowski, A. M., Dodson, E. J., Dodson, G. G., Wilson, K. S., Smith, T. M., Yang, M., Kurecki, T., and Gollnick, P. (1995) The structure of the *trp* RNA-binding attenuation protein. *Nature (London)*, **376**, 693–700.
84. Grundy, F. and Henkin, T. M. (1993) tRNA as a positive regulator of transcription attenuation in *B. subtilis*. *Cell*, **74**, 475–482.
85. Henkin, T. M. (1994) tRNA-directed transcription antitermination. *Mol. Microbiol.*, **13**, 381–387.
86. Condon, C., Grunberg-Manago, M., and Putzer, H. (1996) Aminoacyl-tRNA synthetase regulation in *Bacillus subtilis*. *Biochimie*, **78**, 381–389.
87. Neckers, L., Whitesell,, P., Rosolen, A., and Geselowitz, D. A. (1992) Antisense inhibition of oncogene expression. *Crit. Rev. Oncog.*, **3**, 176–231.
88. Sato, T. and Yura, T. (1979) Chromosomal location and expression of the structural gene for outer membrane protein 1a of *Escherichia coli* K-12 and of the homologous gene of *Salmonella typhimurium*. *J. Bacteriol.*, **139**, 468–477.
89. Van Alphen, L., Lugtenberg, B., Van Boxtel, R., Hack, A. M., Verhoef, C., and Hauekes, L. (1979) *meoA* is the structural gene for the outer membrane protein C of *Escherichia coli* K-12. *Mol. Gen. Genet.*, **169**, 147–155.
90. Inouye, I. (ed.) (1979) Bacterial outer membranes: Biogenesis and function, pp. 1–12. Wiley, New York.
91. Lugtenberg, B., Peters, R., Bernheimer, H., and Berendsen, W. (1976) Influence of cultural conditions and mutations on the composition of the outer membrane proteins of *Escherichia coli*. *Mol. Gen. Genet.*, **147**, 251–262.
92. Hall, M. N. and Silhavy, T. J. (1981) The *ompB* locus and the regulation of the major outer membrane porin proteins of *Escherichia coli*. *J. Mol. Biol.*, **146**, 23–43.
93. Hall, M. N. and Silhavy, T. J. (1981) Genetic analysis of the *ompB* locus in *Escherichia coli* K-12. *J. Mol. Biol.*, **151**, 1–15.
94. Aiba, H. and Mizuno, T. (1990) Phosphorylation of a bacterial activator protein, OmpR, by a protein kinase, EnvZ, stimulates the transcription of the *ompF* and *ompC* genes in *Escherichia coli*. *FEBS Lett.*, **261**, 19–22.
95. Aiba, H., Nakasai, F., Mizushima, S., and Mizuno, T. (1989) Evidence for the physiological importance of phosphotransfer between the two regulatory components EnvZ and OmpR, in osmoregulation in *Escherichia coli*. *J. Biol. Chem.*, **264**, 14090–14094.
96. Forst, S., Delgado, J., and Inouye, M. (1989) Phosphorylation of OmpR by the osmosensor EnvZ, a protein involved in osmoregulation of OmpF and OmpC in *Escherichia coli*. *Proc. Natl. Acad. Sci. USA*, **86**, 6052–6056.
97. Igo, M. M., Ninfa, A. J., and Silhavy, T. J. (1989) A bacterial environmental sensor that functions as a protein kinase and stimulates transcriptional activation. *Genes Dev.*, **3**, 598–605.
98. Forst, S., Comeau, D., Norioko, S., and Inouye, I. (1987) Localization and membrane topology of EnvZ, a protein involved in osmoregulation of OmpF and OmpC in *Escherichia coli*. *J. Biol. Chem.*, **262**, 16433–16438.
99. Slauch, J. M. and Silhavy, T. J. (1989) Genetic analysis of the switch that controls porin gene expression in *Escherichia coli* K-12. *J. Mol. Biol.*, **210**, 281–292.
100. Tokishita, S., Yamada, H., Aiba, H., and Mizuno, T. (1990) Trans-membrane signal transduction and osmoregulation in *Escherichia coli*. II. The osmotic sensor, EnvZ, located in the isolated cytoplasmic membrane displays its phosphorylation and dephosphorylation abilities as to the activator protein, OmpR. *J. Biochem.*, **108**, 488–493.
101. Ronson, C., Nixon, B. T., and Ausubel, R. M. (1987) Conserved domains in bacterial regulatory proteins that respond to environmental stimuli. *Cell*, **49**, 579–581.

102. Mizuno, T., Chou, M.-Y., and Inouye, M. (1984) A unique mechanism regulating gene expression: translational inhibition by a complementary RNA transcript (micRNA). *Proc. Natl. Acad. Sci. USA*, **81**, 1966–1970.
103. Coyer, J., Andersen, J., Forst, S. A., Inouye, M., and Delihas, N. (1990) *micF* RNA in *ompB* mutants of *Escherichia coli*: different pathways regulate *micF* RNA levels in response to osmolarity and temperature change. *J. Bacteriol.*, **172**, 4143–4150.
104. Ramani, N., Hedeshian, M., and Freundlich, M. (1994) *micF* antisense RNA has a major role in osmoregulation of OmpF in *Escherichia coli*. *J. Bacteriol.*, **176**, 5005–5010.
105. Ippen-Ihler, K. and Minkley, E. G. (1986) The conjugation system of F, the fertility system of *Escherichia coli*. *Annu. Rev. Genet.*, **20**, 593–624.
106. Willets, N. S. and Skurray, R. (1987) Structure and function of the F factor and mechanism of conjugation. In Escherichia coli *and* Salmonella typhimurium: *Cellular and molecular biology* (ed. F. C. Neidhardt), pp. 1110–1133. American Society for Microbiology, Washington, DC.
107. Millineaux, P. and Willetts, N. S. (1985) Promoters in the transfer region of F. In *Plasmids in bacteria* (ed. D. R. Helinski, S. N. Cohen, D. B. Clewell, D. A. Jackson, and A. Hollander), pp. 605–614. Plenum, New York.
108. Dempsey, W. (1987) Transcript analysis of the plasmid R100 *traJ* and *finP* genes. *Mol. Gen. Genet.*, **209**, 533–544.
109. Finlay, B. B., Frost, L. S., Paranchych, W., and Willetts, N. S. (1986) Nucleotide sequences of five IncF plasmid *finP* alleles. *J. Bacteriol.*, **167**, 764–767.
110. Simons, R. W. and Kleckner, N. (1988) Biological regulation by antisense RNA in prokaryotes. *Annu. Rev. Genet.*, **22**, 567–600.
111. Koraimann, G., Koraimann, C., Koronakis, V., Schlager, S., and Hogenauer, G. (1991) Repression and derepression of conjugation of plasmid R1 by wild-type and mutated *finP* antisense RNA. *Mol. Microbiol.*, **5**, 77–87.
112. Tomizawa, J. (1984) Control of ColE1 plasmid replication: the process of binding of RNA 1 to the primer transcript. *Cell*, **38**, 861–870.
113. Persson, C., Wagner, E. G. H., and Nordstrom, K. (1988) Control of replication of plasmid R1: kinetics of *in vitro* interaction between the antisense RNA, CopA, and its target, CopT. *EMBO J.*, **9**, 3767–3776.
114. Persson, C., Wagner, E. G. H., and Nordstrom, K. (1990) Control of replication of plasmid R1: formation of an initial transient complex is rate limiting for antisense RNA-target RNA pairing. *EMBO J.*, **9**, 3777–3785.
115. van Biesen, T. and Frost, L. S. (1992) Differential levels of fertility inhibition among F-like conjugative plasmids are related to the cellular concentration of *finO* mRNA. *Mol. Microbiol.*, **6**, 771–780.
116. van Biesen, T., Soderbom, F., Gerhart, E., Wagner, J., and Frost, L. S. (1993) Structural and functional analyses of the FinP antisense RNA regulatory system of the F conjugative plasmid. *Mol. Microbiol.*, **10**, 35–43.
117. Weiss, R. B., Dunn, D. M., Atkins, J. F., and Gesteland, R. F. (1987) Slippery runs, shifty stops, backward steps, and forward hops: P2, P1, +1, +2, +5, and +6 ribosomal frameshifting. *Cold Spring Harbor Symp. Quant. Biol.*, **52**, 687–693.
118. Tsuchihashi, Z. and Kornberg, A. (1990) Translational frameshifting generates the gamma subunit of DNA polymerase III holozyme. *Proc. Natl. Acad. Sci. USA*, **87**, 2516–2520.
119. Craigen, W. J., Cook, R. G., Tate, W. P., and Caskey, C. T. (1985) Bacterial peptide chain release factors: Conserved primary structure and possible frameshift regulation of release factor 2. *Proc. Natl. Acad. Sci. USA*, **82**, 3616–3620.

120. Craigen, W. J. and Caskey, C. T. (1986) Expression of peptide chain release factor 2 requires high-efficiency frameshift. *Nature*, **322**, 273–276.
121. Cameron Donly, B., Edgar, C. D., Adamski, F. M., and Tate, W. P. (1990) Frameshift autoregulation in the gene for *Escherichia coli* release factor 2: partly functional mutants result in frameshift enhancement. *Nucleic Acids Res.*, **18**, 6517–6522.
122. Adamski, F. M., Cameron Donly, B., and Tate, W. P. (1993) Competition between frameshifting, termination and suppression at the frameshift site in the *Escherichia coli* release factor-2 mRNA. *Nucleic Acids Res.*, **21**, 5076–5078.
123. Weiss, R. B., Dunn, D. M., Dahlberg, A. E., Atkins, J. F., and Gesteland, R. F. (1988) Reading frame switch caused by base-pair formation between the 3′ end of 16S rRNA and the mRNA during elongation of protein synthesis in *Escherichia coli*. *EMBO J.*, **7**, 1503–1507.
124. Curran, J. F. and Yarus, M. (1988) Use of tRNA suppressors to probe regulation of *Escherichia coli* release factor 2. *J. Mol. Biol.*, **203**, 76–83.
125. Curran, J. F. and Yarus, M. (1989) Rates of aminoacyl-tRNA selection at 29 sense codons *in vivo*. *J. Mol. Biol.*, **209**, 65–77.
126. Sipley, J. and Goldman, E. (1993) Increased ribosomal accuracy increases a programmed translational frameshift in *Escherichia coli*. *Proc. Natl. Acad. Sci. USA*, **90**, 2315–2319.
127. Dubnau, D. (1984) Translational attenuation: the regulation of bacterial resistance to the macrolide-lincosamide-streptogramin B antibiotics. *Crit. Rev. Biochem.*, **16**, 103–132.
128. Weisblum, B. (1983) Inducible resistance to macrolides, lincosamides, and streptogramin B type antibiotics: the resistance phenotype, its biological diversity and structural elements that regulate expression, In *Gene function in prokaryotes* (ed. J. Beckwith, J. Davis, and J. A. Gallant), pp. 91–121. Cold Spring Harbor Laboratory, Cold Spring Harbor , NY.
129. Lovett, P. S. (1990) Translational attenuation as the regulator of inducible *cat* genes. *J. Bacteriol.*, **172**, 1–6.
130. Stokes, J. W. and Hall, R. M. (1991) Sequence analysis of the inducible chloramphenicol resistance determinant in the *Tn*1696 integron suggests regulation by translational attenuation. *Plasmid*, **26**, 10–19.
131. Dorman, C. J. and Foster, T. J. (1985) Posttranscriptional regulation of the inducible nonenzymatic chloramphenicol resistance determinant of IncP plasmid R26. *J. Bacteriol.*, **161**, 147–152.
132. Dubnau, D. (1985) Induction of *ermC* requires translation of the leader peptide. *EMBO J.*, 4. 533–537.
133. Alexieva, Z., Duvall, E. J., Ambulos, N. P., Jr., Kim, U. J., and Lovett, P. S. (1988) Chloramphenicol induction of *cat-86* requires ribosome stalling at a specific site in the regulatory leader. *Proc. Natl. Acad. Sci. USA*, **85**, 3057–3061.
134. Dick, T. and Matzura, H. (1988) Positioning ribosomes on leader mRNA for translational activation of the message of an inducible *Staphylococcus aureus cat* gene. *Mol. Gen. Genet.*, **21**, 108–111.
135. Duvall, E. J., Mongkolsuk, S., Kim, U. J., Lovett, P. S., Henkin, T. M., and Chambliss, G. H. (1985) Induction of the chloramphenicol acetyltransferase gene *cat-86* through the action of the ribosomal antibiotic amicetin: involvement of a *Bacillus subtilis* ribosomal component in *cat* induction. *J. Bacteriol.*, **161**, 665–672.
136. Kim, U. J., Ambulos, N. P., Jr, Duvall, E. J., Lorton, M. A., and Lovett, P. S. (1988) Site in the *cat-86* leader critical to induction by amicetin. *J. Bacteriol.*, **170**, 2933–2938.
137. Mayfield, M. and Weisblum, B. (1989) *ermC* leader peptide. Amino acid sequence critical for induction by translational attenuation. *J. Mol. Biol.*, **206**, 69–79.

138. Kadam, S. K. (1989) Induction of *ermC* methylase in the absence of macrolide antibiotics and by pseudomonic acid A. *J. Bacteriol.*, **171**, 4518–4520.
139. Gu, Z., Rogers, E. J., and Lovett, P. S. (1993) Peptidyl transferase inhibition by the nascent leader peptide of an inducible *cat* gene. *J. Bacteriol.*, **176**, 5309–5313.
140. Monro, R. E. and Marcker, K. A. (1967) Ribosome catalysed reaction of puromycin with a formylmethionine-containing oligonucleotide. *J. Mol. Biol.*, **25**, 347–350.
141. Gu, Z., Harrod, R., Rogers, E. J., and Lovett, P. S. (1994) Properties of a pentapeptide inhibitor of peptidyltransferase that is essential for *cat* gene regulation by translation attenuation. *J. Bacteriol.*, **176**, 6238–6244.
142. Gu, Z., Harrod, R., Rogers, E. J., and Lovett, P. S. (1994) Anti-peptidyl transferase leader peptides of attenuation regulated chloramphenicol resistance genes. *Proc. Natl. Acad. Sci. USA*, **91**, 5612–5616.
143. Harrod, R. and Lovett, P. S. (1995) Peptide inhibitors of peptidyl transferase alter the conformation of domains IV and V of large subunit rRNA: A model for nascent peptide control of translation. *Proc. Natl. Acad. Sci. USA*, **92**, 8650–8654.
144. Lovett, P. S. (1994) Nascent peptide regulation of translation. *J. Bacteriol.*, **176**, 6415–6417.
145. Moazed, D. and Noller, H. F. (1987) Chloramphenicol, erythromycin, carbomycin and vernamycin B protect overlapping sites in the peptidyl transferase region of 23S ribosomal RNA. *Biochemie*, **69**, 879–884.
146. Leviev, I. G., Rodriguez-Fonseca, C., Phan, H., Garrett, R. A., Heilek, G., Noller, H. F., and Mankin, A. S. (1994) A conserved secondary structural motif in 23S rRNA defines the site of interaction of amicetin, a universal inhibitor of peptide bond formation. *EMBO J.*, **13**, 1682–1686.
147. Gu, Z. and Lovett, P. S. (1995) A gratuitous inducer of *cat-86*, amicetin, inhibits bacterial peptidyl transferase. *J. Bacteriol.*, **177**, 3616–3618.
148. Noller, H. F., Hoffarth, V., and Zimniak, L. (1992) Unusual resistance of peptidyl transferase to protein extraction procedures. *Science*, **256**, 1416–1419.
149. Duvall, E. J., Ambulos, N. P., Jr, and Lovett, P. S. (1987) Drug-free induction of a chloramphenicol acetyltransferase gene in *Bacillus subtilis* by stalling ribosomes in a regulatory leader. *J. Bacteriol.*, **169**, 4235–4241.
150. Rogers, E. J. and Lovett, P. S. (1994) The *cis*-effect of a nascent peptide on its translating ribosome: influence of the *cat-86* leader pentapeptide on translation termination at leader codon 6. *Mol. Microbiol.*, **12**, 181–186.
151. Moffat, J. G., Tate, W. P., and Lovett, P. S. (1994) The leader peptides of attenuation-regulated chloramphenicol resistance genes inhibit translational termination. *J. Bacteriol.*, **176**, 7115–7117.
152. Tate, W. P., Adamski, F. M., Brown, C. M., Dalphin, M. E., Gray, J. P., Horsfield, J. A., McCaughan, K. K., Moffat, J. G., Powell, R. J., Timms, K. M., and Trotman, C. N. A. (1993) Translational stop signals: Evolution, decoding for protein synthesis and recoding for alternative events. In *The translational apparatus. Structure, function, regulation, evolution* (ed. K. H. Nierhaus, F. Franceschi, A. R. Subramanian, V. A. Erdmann, and B. Wittmann-Liebold), pp.253–262. Plenum Press, New York.
153. Geballe, A. P. and Morris, D. R. (1994) Initiation codons within 5′-leaders of mRNAs as regulators of translation. *TIBS*, **19**, 159–164.
154. Hill, J. R. and Morris, D. R. (1992) Cell-specific translation of *S*-adenosylmethionine decarboxylase mRNA. Regulation by the 5′ transcript leader. *J. Biol. Chem.*, **267**, 21886–21893.
155. Degnin, C. R., Schleiss, M. R., Cao, J., and Geballe, A. P. (1993) Translational inhibition

mediated by a short upstream open reading frame in the human cytomegalovirus gpUL4 (gp48) transcript. *J. Virol.*, **67**, 5514–5521.

156. Delbecq, P., Werner, M., Feller, A., Filipkowski, R. K., Messenguy, F., and Pierard, A. (1994) A segment of mRNA encoding the leader peptide of the *CPA1* gene confers repression by arginine on a heterologous yeast gene transcript. *Mol. Cell. Biol.*, **14**, 2378–2390.
157. Parola, A. L. and Kobilka, B. K. (1994) The peptide product of a 5′ leader cistron in the b_2 adrenergic receptor mRNA inhibits receptor synthesis. *J. Biol. Chem.*, **269**, 4497–4505.
158. Huang, W. M., Ao, S., Casjens, S., Orlandi, R., Zeikus, R., Weiss, R., Winge, D., and Fang, M. (1988) A persistent untranslated sequence within bacteriophage T4 topoisomerase gene 60. *Science*, **239**, 1005–1012.
159. Weiss, R. B., Huang, W. M., and Dunn, D. M. (1990) A nascent peptide is required for ribosomal bypass of the coding gap in bacteriophage T4 gene 60. *Cell*, **62**, 117–126.
160. Herbst, K. L., Nichols, L. M., Gesteland, R. F., and Weiss, R. B. (1994) A mutation in ribosomal protein L9 affects ribosomal hopping during translation of gene *60* from bacteriophage T4. *Proc. Natl. Acad. Sci. USA*, **91**, 12525–12529.
161. Kudlicki, W., Odom, O. W., Kramer, G., and Hardesty, B. (1994) Activation and release of enzymatically inactive, full-length rhodanese that is bound to ribosomes as peptidyl-tRNA. *J. Biol. Chem.*, **269**, 16549–16553.
162. Kudlicki, W., Odom, O. W., Kramer, G., Hardesty, B., Merrill, G. A., and Horowitz, P. M. (1995) The importance of the N-terminal segment for DNAJ-mediated folding of rhodanese while bound to ribosomes as peptidyl-tRNA. *J. Biol. Chem.*, **270**, 10650–10657.
163. Gesteland, R., Weiss, R., and Atkins, J. (1992) Recoding: reprogrammed genetic decoding. *Science*, **257**, 1640–1641.
164. Bock, A., Forchhammer, K., Heider, J., and Baron, C. (1991) Selenoprotein synthesis: an expansion of the genetic code. *TIBS*, **16**, 463–467.
165. Axley, M. J., Bock, A., and Stadtman, T. C. (1991) Catalytic properties of an *Escherichia coli* formate dehydrogenase mutant in which sulfur replaces selenium. *Proc. Natl. Acad. Sci. USA*, **88**, 8450–8454.
166. Zinoni, F., Birkmann, A., Leinfelder, W., and Bock, A. (1987) Cotranslational insertion of selenocysteine into formate dehydrogenase from *Escherichia coli* directed by a UGA codon. *Proc. Natl. Acad. Sci. USA*, **84**, 3156–3160.
167. Forchhammer, K., Leinfelder, W., and Bock, A. (1989) Identification of a novel translation factor necessary for the incorporation of selenocysteine into protein. *Nature*, **342**, 453–456.
168. Berg, B. L., Li, J., Heider, J., and Stewart, V. (1991) Nitrate-inducible formate dehydrogenase in *Escherichia coli* K-12. I. Nucleotide sequence of the *fdnGHI* operon and evidence that opal (UGA) encodes selenocysteine. *J. Biol. Chem.*, **266**, 22380–22385.
169. Leinfelder, W., Zehelein, E., Mandrand-Berthelot, M.-A., and Bock, A. (1988) Gene for a novel tRNA species that accepts L-serine and cotranslationally inserts selenocysteine. *Nature*, **331**, 723–725.
170. Heider, J., Baron, C., and Bock, A. (1992) Coding from a distance: dissection of the mRNA determinants required for the incorporation of selenocysteine into protein. *EMBO J.*, **11**, 3769–3766.
171. Forchhammer, K., Rucknagel, K.-P., and Bock, A. (1990) Purification and biochemical characterization of SELB, a translation factor involved in selenoprotein synthesis. *J. Biol. Chem.*, **265**, 9346–9350.
172. Baron, C., Heider, J., and Bock, A. (1993) Interaction of translation factor SELB with the

formate dehydrogenase H selenopolypeptide mRNA. *Proc. Natl. Acad. Sci. USA*, **90**, 4181–4185.

173. Ringquist, S., Schneider, D., Gibson, T., Baron, C., Bock, A., and Gold, L. (1994) Recognition of the mRNA selenocysteine insertion sequence by the specialized translational elongation factor SELB. *Genes Dev.*, **8**, 376–385.
174. de Smit, M. H. and van Duin, J. (1990) Secondary structure of the ribosome binding site determines translational efficiency: A quantitative analysis. *Proc. Natl. Acad. Sci. USA*, **87**, 7668–7672.
175. de Smit, M. H. and van Duin, J. (1990) Control of prokaryotic translational initiation by mRNA secondary structure. *Prog. Nucleic Acids Res.*, **38**, 1–35.
176. Schultz, V. P. and Resnikoff, W. S. (1991) Translation initiation of IS*50*R readthrough transcripts. *J. Mol. Biol.*, **221**, 65–80.
177. Gold, L. (1988) Posttranscriptional regulatory mechanisms in *Escherichia coli*. *Annu. Rev. Biochem.*, 57, 199–233.
178. Carter-Muenchau, P. and Wolf, R. E., Jr (1989) Growth rate dependent regulation of 6-phosphogluconate dehydrogenase level mediated by an anti-Shine–Dalgarno sequence located within the *Escherichia coli gnd* structural gene. *Proc. Natl. Acad. Sci. USA*, **86**, 1138–1142.

6 Prokaryotic DNA topology and gene expression

KARL DRLICA*, ERDEN-DALAI WU, CHANG-RUNG CHEN, JIAN-YING WANG, XILIN ZHAO, CHEN XU, LIN QIU, MUHAMMAD MALIK, SAMUEL KAYMAN, and S. MARVIN FRIEDMAN

1. Introduction

Systems that regulate initiation of transcription are built on the binding of specific proteins to their cognate DNA sites. Complexity is generated by interactions of the primary DNA recognition proteins with themselves and with other proteins, sometimes through formation of distinct structures that either block or facilitate transcription. All of these interactions occur on a chromosome that undergoes structural changes mediated by other proteins (topoisomerases and DNA bending proteins) that alter the interactions between specific regulatory proteins and DNA. The aim of this chapter is to provide a framework for understanding how the dynamics of chromosome topology influences gene expression.

The DNA of the bacterial chromosome is a long, circular molecule that is negatively supercoiled. Intuitively, this means that the DNA is torsionally strained due to a deficiency of duplex turns relative to a linear DNA of the same length. This deficiency is called linking deficit, a term sometimes used to signify negative supercoiling when measured on extracted DNA. The intracellular level of supercoiling is controlled by enzymes (DNA topoisomerases) whose function can be perturbed by mutations and by antibiotics; consequently, supercoiling has become one of the best understood topological features of the chromosome. Two key observations have emerged: (1) negative supercoiling facilitates both the DNA looping and the strand separation important for many activities of DNA, including initiation of transcription, and (2) supercoiling changes in response to environmental alterations. Thus, supercoiling may contribute to the ability of bacteria to vary gene expression as the environment changes.

Bacterial DNA topoisomerases operate by a strand-passage mechanism (1). If DNA strand passage is *intra*molecular, supercoiling is altered. A number of genetic, biochemical, and physiological experiments have identified DNA gyrase (DNA

*To whom correspondence should be addressed

topoisomerase II) (2) and DNA topoisomerase I (omega protein) (3) as the major enzymes controlling supercoiling in bacterial cells. *Inter*molecular strand passage links (catenates) or unlinks (decatenates) circular DNA molecules. If replication does not completely separate the daughter chromosomes during duplication, then interlinking will prevent proper cell division. Topoisomerase IV (4, 5) appears to be the principal bacterial decatenase (6).

The activities of small DNA bending proteins represent another type of protein-mediated manipulation of DNA structure. The best studied are HU (7), IHF (8), FIS (9, 10), and H-NS (11). For HU, IHF, and FIS, the bending property is used to bring distant sequence elements together, and this activity has been exploited by phages, plasmids, and transposons to facilitate the formation of site-specific protein–DNA interactions involved in recombination and initiation of replication. Indeed, most of our understanding of these three proteins comes from examination of their use by parasitic DNA elements. Subsequent studies uncovered additional involvement of these proteins in initiation of transcription. For H-NS, the relevance of its ability to bend DNA is not yet clear. It generally acts as a silencer of gene expression, apparently by binding to curved regions of DNA located in or near promoters.

The following discussion focuses first on DNA topoisomerases and then on DNA supercoiling, both as a global, environment-sensitive aspect of DNA structure and as a local phenomenon influenced by transcription and replication. Then several ways are considered in which promoters may respond to changes in supercoiling, some of which involve DNA bending proteins. The chapter concludes with brief descriptions of HU, IHF, FIS, and H-NS.

2. DNA topoisomerases

2.1 Gyrase and topoisomerase I

Early biochemical characterization of gyrase and topoisomerase I revealed that these enzymes have substrate preferences that could allow them to produce a set level of supercoiling: gyrase, the supercoiling enzyme, is much more active on a relaxed substrate, and topoisomerase I, the relaxing enzyme, is more active on highly supercoiled DNA. The physiological importance of the opposing activities of gyrase and topoisomerase I was established when defects in *topA*, the gene encoding topoisomerase I, were found to be compensated by gyrase mutations that lowered supercoiling (12, 13). Further evidence for supercoiling control came from the observation that lowering supercoiling by inhibiting gyrase (14) causes an increase in expression of gyrase (15) and a decrease in *topA* expression (16). Moreover, raising supercoiling above normal levels (17) elicited an increase in *topA* expression (18). Thus, there seems to be homeostatic control of supercoiling and expression of gyrase and topoisomerase I, although elevated supercoiling levels do not block gyrase expression (19). The existence of mechanisms for control of supercoiling was also suggested by the finding that topoisomerase activity can compensate for effects of growth temperature and intercalating dyes on DNA helical pitch (20, 21).

The early observation that gyrase also relaxes DNA (2) was often overlooked, since topoisomerase I is so potent. However, in the mid-1980s anti-gyrase agents were found to relax DNA at the same rate in the presence or absence of topoisomerase I (22). Thus, it seemed unlikely that supercoiling levels were set by the competing action of these two topoisomerases. Instead, gyrase controls supercoiling by both introducing *and* removing supercoils (23, 24); topoisomerase I acts as a safety valve to prevent accumulation of excess supercoiling (23). This general idea is developed in Section 6.3.

2.2 Topoisomerase III

Topoisomerase III (25) relaxes negative supercoils and decatenates DNA *in vitro* (26, 27). Sequence comparisons have shown that *topB*, the gene encoding topoisomerase III, is identical to *mutR* (28), a gene defined by mutations that result in a high frequency of chromosomal deletions. Cells deficient in topoisomerase III grow vigorously; consequently, it has been difficult to assign a function to the enzyme. Since there is little evidence that topoisomerase III influences DNA structure in a way that affects gene expression, this enzyme is not considered further in this chapter.

2.3 Topoisomerase IV

Topoisomerase IV is a homologue of gyrase that closely resembles eukaryotic topoisomerase II. It exhibits two activities, relaxation of supercoils and decatenation of interlinked circles (4). The relaxing activity can compensate for the absence of topoisomerase I, if the genes encoding topoisomerase IV are overexpressed from a plasmid (4, 29) and probably if the chromosomal region encoding topoisomerase IV is amplified (30). Thus the relaxing activity of topoisomerase IV is probably quite weak, and there is no indication that this enzyme normally plays a role in modulating supercoiling.

A temperature-sensitive topoisomerase IV mutation causes intracellular accumulation of plasmid catenanes not observed with a comparable gyrase mutant (6, 31). Thus topoisomerase IV is probably the major intracellular decatenating activity. Gyrase, which shows poor decatenation activity relative to topoisomerase IV *in vitro* (32, 33), may serve as a back-up, since overproduction of gyrase does suppress a deficiency of topoisomerase IV (4). The accumulation of doublet chromosomes by a temperature-sensitive gyrase mutant (34) probably reflects a negative effect of supercoil relaxation on topoisomerase IV-mediated decatenation rather than a direct effect of the loss of gyrase activity.

3. Cellular energetics, environment, and the control of supercoiling

The structure of purified DNA is sensitive to changes in environment. For example, helical pitch, the number of base pairs per turn, increases when the concentration of certain salts increases, and pitch decreases when temperature increases. If the ends of

DNA are fixed, as would be the case in circular DNA, then changes in pitch change DNA supercoiling. Thus we expect certain environmental factors to have an intrinsic effect on DNA supercoiling. Sometimes the topoisomerases correct for the environmental effects and maintain intracellular superhelical tension at a constant value (20, 21). In other cases, which are discussed below, topoisomerase action alters superhelical tension in a way that appears to be influenced by cellular energetics, factors that are themselves subject to environmental influence.

Early studies with purified gyrase showed that the enzyme requires ATP to introduce supercoils and that gyrase relaxes DNA in the absence of ATP (2, 35). However, it was not until the late 1980s that it became clear that the ratio of [ATP] to [ADP] determines the level of supercoiling established by gyrase (36). High ratios of [ATP] to [ADP] produce high levels of supercoiling and *vice versa*, regardless of initial levels. This concept, plus the observation that gyrase might be a major source of relaxation *in vivo* (22), focused attention on gyrase as the principal regulator of supercoiling. At about the same time, several environmental alterations were found to change supercoiling, as reflected in DNA extracted from cells (reviewed in Reference 23). Entry into stationary phase (37) and growth at low pH (38) both reduce supercoiling, while growth at high osmolarity (39–41) or low oxygen tension (42, 43) increase supercoiling. These observations are of central importance, since the ability to respond to environmental change is required for supercoiling to have a controlling influence over processes such as transcription.

An experimental connection among environmental effects, supercoiling, and cellular energetics emerged from the observation that a shift to anaerobic conditions caused both supercoiling and the ratio of [ATP] to [ADP] to transiently drop in the same time frame (43). The connection was extended when addition of NaCl to cultures transiently increased both supercoiling and [ATP]/[ADP] (41). These two correlations between supercoiling and [ATP]/[ADP], as well as correlations derived from steady-state experiments, support the hypothesis that cellular energetics is an important factor in determining the level of supercoiling, acting through its influence on DNA gyrase (23, 24).

Recently, the energetics hypothesis has been solidified by manipulation of the cellular H^+-ATPase concentration to vary the ratio of [ATP] to [ADP] over a broad range (44). At low values of [ATP]/[ADP] plasmid supercoiling is very sensitive to cellular energetics, but in the range normally observed in growing cells (ratios of 1 to 10), supercoiling is not as sensitive as expected from *in-vitro* comparisons (36). Intracellular factors, which probably include topoisomerase I, may prevent excessive supercoils from accumulating, thereby reducing the sensitivity of supercoiling to [ATP]/[ADP]. Thus environmental conditions can have two types of effect on DNA topology. One is the direct effect on helical pitch, which is sometimes corrected to maintain constant intracellular superhelical tension. The other is an indirect effect mediated by DNA gyrase in response to changes in the ratio of [ATP] to [ADP]. In both cases we see gyrase as the regulator of supercoiling and topoisomerase I as a safety valve. In the next section a third type of environmental effect is described in which *both* gyrase and topoisomerase I respond by rapidly relaxing DNA.

4. DNA relaxation and exposure to high temperature

The effect of temperature on DNA supercoiling has been studied most extensively with shift experiments, usually through sudden increases in temperature (heat shock). An early event following heat shock is induction of a specialized sigma factor (σ^{32}), the *rpoH* gene product (45). σ^{32} controls a regulon that includes the GroEL and DnaK proteins, chaperones that prevent misfolding of other proteins (46). Since DNA supercoiling relaxes quickly during heat shock, it has been suggested that supercoiling is a crucial part of the heat shock response (47). Indeed, DnaK and GroEL are induced when relaxation is generated by means other than heat shock, such as by treatment with ethanol, novobiocin, or by overexpression of CcdB (LetD; a toxic, gyrase-inhibiting product of the F plasmid) (47, 48). Heat-induced relaxation occurs in an *rpoH* (σ^{32}) mutant (47), suggesting that relaxation is more likely to be a part of the induction of the response than a result of it.

Heat shock mediated DNA relaxation is transient, and after about 10 min supercoiling recovers to preshock levels (47). If recovery of supercoiling is prevented by a temperature-sensitive mutation of *gyrA*, expression of both DnaK and GroEL remains elevated. Thus recovery of supercoils may be associated with decay of the response.

Gyrase and topoisomerase I are each sufficient to produce heat-induced relaxation: relaxation is blocked by oxolinic acid, an inhibitor of gyrase, but only when topoisomerase I is absent (49). What causes gyrase to relax DNA is unknown. We speculate that a drop in ATP/ADP occurs shortly after heat shock, but that has not been demonstrated. There is, however, a correlation between a post-heat shock increase in [ATP]/[ADP] and supercoiling (50). Particularly puzzling is why heat shock induced relaxation is not always seen (50). This observation raises questions about whether relaxation is required for induction of heat shock genes, since a causal relationship has not been demonstrated.

More details are available for the participation of topoisomerase I in relaxation during heat shock. Four promoters drive the *topA* gene, and one, P1, is a typical σ^{32}-dependent heat shock promoter (51, 52). This promoter contributes to maintenance of topoisomerase I at a constant level during heat shock, and in its absence heat shock lowers topoisomerase I protein to 25 per cent of wild-type levels (52). In such cases cell survival is 10 per cent of wild-type levels following a 52 °C heat treatment, and lower levels of the DnaK and GroEL proteins are synthesized following a shift to 42 °C (52). It has been suggested the topoisomerase I effect may be related to R-loop formation (52). R-loops are short regions of denatured DNA that are stabilized by formation of mRNA–DNA hybrids. As pointed out in the following section, high levels of negative supercoils accumulate behind active transcription complexes, which could facilitate R-loop formation. The accumulation of R-loops might be high enough in cells lacking topoisomerase I that they would interfere with expression of the heat shock genes. Still undefined is how topoisomerase I facilitates heat-induced relaxation when its level remains constant (53).

Supercoiling, as measured with isolated plasmid DNA, also varies with growth temperature. In the case of *E. coli*, lowering the temperature from 37 °C to 17 °C relaxes plasmid DNA, while raising it from 17 °C to 37 °C increases supercoiling (20). A similar phenomenon occurs with *Yersinia enterocolitica* (54). These topological changes, which roughly counter temperature-induced alterations in DNA twist and thereby DNA supercoiling (55, 56), support the idea that under some conditions gyrase and topoisomerase I maintain a set level of supercoiling (20). However, this response to temperature is not universal. With *Bacillus subtilis*, lowering the temperature causes an increase, rather than a decrease in plasmid supercoiling (57). It may be that the ecological niche occupied by a microbe strongly influences how supercoiling responds to temperature. Fewer supercoils at higher growth temperatures has also been reported for *Shigella flexneri* (58), *Salmonella typhimurium* (58), and the archaebacterium *Haloferax volcanii* (59). These observations now need to be fit with the *E. coli/B. subtilis* models by standardizing growth conditions.

Another relationship between temperature and supercoiling is seen with hyperthermophiles, organisms that grow at temperatures near the melting point of DNA (for reviews see References 60 and 61). While extensive duplex denaturation is probably prevented by polyamines (62) and DNA-binding proteins (63–65), the hyperthermophiles also contain an enzyme called reverse gyrase (66, 67). When purified, this enzyme is able to introduce positive supercoils into DNA. Since positive supercoils tend to disfavor duplex denaturation, reverse gyrase activity could be an additional DNA stabilizing factor. Negative supercoils, on the other hand, would aggravate the denaturation problem.

Two forms of reverse gyrase have been found in archaebacteria. Members of one category, recovered from *Sulfolobus shibatae* and *Desulfurococcus amylolyticus*, are single-subunit enzymes of molecular weight approximately 130 kD. These proteins have ATPase and topoisomerase activity residing in separate domains (68) that exhibit homologies to helicases and to type I topoisomerases (69). The connection with helicases is mechanistically interesting because movement of a helicase-like activity through DNA would generate negative supercoils behind the enzyme and positive ones ahead of it. If the enzyme's relaxing activity preferentially removed the negative supercoils, the net effect would be introduction of positive supercoils. The other form, recovered from *Methanopyrus kandleri*, has two kinds of subunit with molecular weights of 50 kD and 150 kD (70). In this case the ATPase and helicase activities are in different subunits.

Measurement of linking deficit with small circular DNA extracted from hyperthermophiles indicates that *in vivo* DNA may be relaxed at high temperature (59), providing that the topological effects of intracellular proteins are ignored. Initially this observation suggested a fundamental difference between hyperthermophiles and organisms such as *E. coli*, since linking deficit in the latter increased when temperature increased (20). However, we subsequently found that a gyrase mutant that grows at a normally lethal temperature has partially relaxed DNA (71). This led us to propose that even under exponential growth conditions at least partial relaxation is probably a general phenomenon at elevated temperatures (71).

A plausible evolutionary scenario emerges if we assume that ancient bacteria lived at high temperature and that under those conditions their DNA was relaxed. As mesophilic bacteria evolved, gyrase and negative supercoiling might have been acquired to facilitate DNA strand separation. At the same time, promoters would have evolved to exploit the structural changes in DNA. While many of the genes required for growth at high temperature would have evolved to function in supercoiled DNA, some may have retained a promoter preference for the relaxed DNA characteristic of hyperthermophiles. The heat shock genes might fall into this category. It will now be interesting to learn whether genes specialized for growth at high temperature have a common promoter structure that shows optimal activity when the template DNA is relaxed.

5. Effects of transcription on supercoiling

When examining the effect of *topA* mutations on plasmid DNA supercoiling, we noticed that the absence of topoisomerase I could lead to hypernegative supercoiling (12, 72). The level of supercoiling was so high that it exceeded that generated *in vitro* by gyrase (73). We were surprised to find that not every plasmid exhibited hypernegative supercoiling when extracted from *topA* mutants: derivatives of pBR322 lacking the *tetA* region had normal levels of supercoiling (72). This indicated that there must be nucleotide sequences in pBR322 responsible for determining supercoiling levels. Deletion and insertion analysis then identified the source of hypernegative supercoiling as transcription of *tetA*.

Several years earlier hyper*positive* supercoiling of pBR322 had been observed when cells were treated with novobiocin, an inhibitor of gyrase (74). This observation, plus the *topA*-transcription conclusion described above, led Liu and Wang to argue that hypernegative and hyperpositive supercoiling are related, that they both arise from transcription complexes constraining the plasmid into two topological domains (75). Positive supercoils, which would ordinarily be relaxed by gyrase, are expected to arise ahead of transcription complexes; negative supercoils, which would be relaxed by topoisomerase I, are expected to arise behind transcription complexes. An imbalance between topoisomerase I and gyrase would lead to either hypernegative or hyperpositive supercoiling, depending on which enzyme activity was deficient. The dependence of hyperpositive supercoiling on transcription was soon confirmed (76), and gyrase was shown to be localized downstream from actively transcribing promoters using a quinolone inhibitor of gyrase to detect sites of gyrase–DNA interaction (77).

An important aspect of the Liu–Wang translocation model is restriction of DNA rotation by the moving transcription complex. Restricted rotation could be achieved in part by the coupling of transcription and translation, since ribosome-attached mRNA is expected to produce a drag that would retard rotation. The *tetA* gene of pBR322 is particularly well suited for revealing translocation-dependent changes in supercoiling, since the nascent Tet protein is membrane-bound, a feature that would virtually block rotation of the transcription complex. As predicted, deletion of the translation initiation signal for *tet* eliminated hypernegative supercoiling in *topA* mutants (78), as did inhibition of RNA or protein synthesis (79). Moreover, hypernegative supercoiling could be observed with a number of other genes whose products were thought to be membrane-bound (79). Translation, however, is not absolutely required for observing transcription-dependent supercoiling *in vitro* (80). Indeed, other DNA helix-tracking processes, such as movement of either the UvrAB complex or the simian virus 40 large tumor antigen along DNA, can also generate domains of supercoiling, as inferred from the ability of bacterial topoisomerase I to selectively detect and remove negative supercoils arising from the translocation (81, 82).

Direct evidence that transcription-dependent hypernegative supercoiling exists inside living cells has been obtained by measuring the formation of Z-DNA in plasmids. In this assay, increases in negative supercoiling are expected to reduce the length of alternating GC tracts required to generate Z-form DNA, which is then detected by the reactivity of B–Z junctions to osmium tetroxide. When alternating GC tracts of various lengths are placed 50 bp upstream from an actively transcribed gene, much shorter tracts are able to adopt Z-form structures than when inserted downstream from the gene (83, 84). Since hypernegative supercoiling is not detected when the Z-forming structures are moved to a position 800 bp upstream from the induced promoter, the wave of negative supercoiling decays over short distances. These intracellular studies, when combined with results from the genetic and *in-vitro* approaches described above, led to general acceptance of the idea that transcription complexes affect local supercoiling levels by their movement through DNA.

Negative supercoils following transcription complexes may be stabilized by formation of R-loops. These short regions of denatured DNA contain mRNA–DNA hybrids that can be removed by an enzyme called RNase H. In the absence of topoisomerase I, the negative supercoiling would not be quickly relieved, and that would favor formation of extensive R-loops. If extensive R-looping adversely affects transcription, we would begin to understand why a mutant deficient in both topoisomerase I and RNase H is not viable and why overexpression of RNase H can suppress a *topA* defect (85). Gyrase must also be involved, since reduced gyrase activity can suppress defects associated with the absence of RNase H (85).

Since R-loops would stabilize denatured regions and effectively relax intracellular superhelical tension, R-looping would indirectly lead to an increase in linking deficit as gyrase responds to the relaxation. Thus R-looping would tend to amplify the effect of transcription on supercoiling as measured after extraction and removal of RNA–DNA hybrids. As expected, topoisomerase I, RNase A, or RNase H eliminate hypernegative supercoiling generated *in vitro* by the combination of transcription and gyrase activity (86).

The effects of translocation-induced supercoiling on gene expression have been studied most extensively with the *leu-500* promoter mutation of *Salmonella typhimurium*, a mutation that is suppressed by mutation of *topA* (87, 88). Since *topA* mutations raise supercoiling (12) and since suppression is eliminated by treatment of cells with coumermycin (89), a drug that lowers supercoiling (14), it seemed clear that suppression is due to an increase in supercoiling. However suppression of the *leu-500* promoter is not observed when the promoter is on a plasmid (90), unless a second, active promoter directs negative supercoils at the *leu-500* promoter (91, 92). The effectiveness of the supercoil-generating promoter depends on its strength and on the distance between it and *leu500* (in a case in which the *lac* promoter was used to provide negative supercoils in the absence of topoisomerase I, the effective distance between the promoters was less than 250 bp (93)). In *topA*$^+$ cells, translocation-induced supercoiling is likely to be restricted to a much shorter range, although an active promoter placed 154 bp upstream from the supercoil-sensitive *virB* promoter of *Shigella flexneri* does overcome the repressive action of H-NS binding at *virB* (94). From these data we conclude that translocation-induced supercoiling can affect promoter activity in the range of hundreds of nucleotides.

In summary, negative supercoiling is tightly controlled by gyrase and topoisomerase I, with the latter probably acting primarily as a safety valve. Changes in cellular energetics alter the balance between the relaxing and supercoiling activities of gyrase, enabling supercoiling to respond to environment. Temperature has complex effects, altering supercoiling directly by changing DNA helical pitch and indirectly through changes in topoisomerase activities. Shifts to high temperature enlist both gyrase and topoisomerase I to relax DNA. These global changes in supercoiling are modified locally by movement of replication forks and transcription complexes, and both gyrase and topoisomerase I act to correct local perturbations. Since formation of R-loops is likely to interfere with gene expression, it is important that topoisomerase I relaxes negative supercoils to minimize R-loop formation. R-loop formation might be a principal reason why high levels of negative supercoiling are detrimental to cell growth.

It is important to point out that translocation-induced supercoiling is not the major generator of intracellular supercoiling (76, 80): supercoils are introduced into superinfecting bacteriophage lambda DNA (35), which undergoes no transcription; treatment of wild-type cells with rifampicin has no effect on plasmid linking deficit (43, 44); and topoisomerases are highly efficient at correcting supercoiling perturbations *in vivo* (95). Nevertheless, local regions of negative supercoiling are likely to exist immediately behind transcription complexes, and these could, in principle, stimulate promoter activity for the very gene that is being transcribed. Indeed, local supercoiling could participate in positive autoregulation. Thus we see that both local and global supercoiling can change. We next consider how supercoiling changes might affect promoter activities.

6. DNA twist and transcription initiation

Since negative supercoiling facilitates strand separation, opening of the duplex has long been thought to be a major transcriptional event facilitated by supercoiling (96). However, the role of supercoiling must be more complex, since initiation passes through an optimum as supercoiling becomes more negative (97, 98). This might be explained by the effect of supercoiling on DNA twist, the number of times one strand of a duplex crosses the other in solution. It has been argued that twist influences relative axial rotation and may affect DNA binding by transcription proteins that recognize two or more nearby sequence motifs (99). Many *E. coli* promoters do contain two distinct RNA polymerase binding sites, the –10 and –35 regions. Factors that alter DNA twist are expected to rotate the –10 and –35 recognition sites either toward or away from an orientation optimal for initiation of transcription. Among factors that increase twist are increases in potassium ion concentration and decreases in negative supercoiling or temperature (55, 100). Thus environmental perturbations could selectively influence initiation of transcription for large numbers of genes through direct effects on DNA twist and indirect effects mediated by topoisomerases.

Obtaining evidence for twist effects on transcription has not been straightforward. This is partially because twist is not easily measured. In addition, twist is expected to be one of several supercoil-sensitive factors whose effects on promoter activity are not easily distinguishable. Thus experimental support for the twist idea is still indirect and based on several observations being most simply interpreted in terms of twist effects. The *lac* promoter provides an example for supercoiling effects. The spacing between the –10 and –35 regions of the P^S *lac* promoter is 18 bp, and maximal initiation of transcription *in vitro* occurs when superhelical density is –0.065 (97). If the spacing is reduced to 17 bp, axial rotation between the two promoter regions is reduced. Thus attaining optimal rotation with the 17 bp promoter should require more twist than needed for the 18 bp promoter. Since lowering negative supercoiling increases twist, optimal twist for the 17-bp spacer template is expected to occur at a lower level of supercoiling. This is the observed result, as the optimum for the 17-bp spacer occurs at a superhelical density of –0.046 (97).

The effect of salts on initiation of transcription can also be explained in terms of changes in DNA twist and axial rotation of promoter elements. Potassium ion concentration affects initiation of transcription in a parabolic way that produces a transcription initiation optimum for several promoters (101). In one comparison, optima occurred at lower potassium concentrations for promoters with longer spacers, consistent with the twist hypothesis. Another indirect observation concerns a comparison of sodium, rubidium, and potassium ion effects on transcription initiation from the *rep* promoter of pBR322. With a supercoiled template, increasing sodium causes a decline in transcription initiation over a concentration range at which potassium causes an increase and rubidium causes a bell-shaped response (102). Over the same range of salt concentration, all three cations decrease transcription initiation from a linear template. Thus potassium and rubidium appear to affect transcription from the supercoiled template via DNA conformation. These transcription differences correlate with potassium and rubidium having much greater effects on DNA twist than sodium. Thus the case for DNA twist having a role in transcription initiation is gradually building, although the idea still rests on the untested assumption that RNA polymerase is too inflexible to readily accomodate small changes in axial rotation between the –10 and –35 sequence motifs.

7. Bent DNA

Another way for changes in supercoiling to be sensed by promoters is through bends and loops. Bends (curves) in DNA were initially detected by abnormally slow electrophoretic migration of short DNA fragments when the bend was in the middle of the fragment (103). Bends in purified DNA (intrinsic bending) are often associated with short (about 5 bp), repeated stretches of poly(A) arranged on the same side of the double helix. Bends can also be generated by protein binding (induced bending). Both intrinsic and induced bending have been connected to initiation of transcription (104), with intrinsically bent DNA often being located immediately upstream from the –35 region of promoters. Indeed, many *E. coli* promoters have a poly(A) stretch predicted to confer an intrinsic bend centered around position –44 (105), and a survey of non-linear structures in bacterial DNA data bases reveals that about half of the sharpest bends are located near promoter sequences at approximately position –50 (106). Examples include ribosomal and tRNA promoters of *E. coli*, the *bla* promoter of pUC19, the *alu156* promoter of the *B. subtilis* phage SP82, and the streptococcal plasmid pLS1 promoter PctII (for references see Reference 104).

Initiation of transcription is affected in several ways by DNA bending. In *E. coli*, crosslinking of RNA polymerase with the *lacUV5* promoter has shown that there are multiple contacts between DNA and the β, β′, and σ subunits (107), as if the DNA partially wraps around the polymerase. Bends would facilitate this process (108). Bends also serve as docking sites for proteins that stimulate or prevent RNA polymerase–promoter interactions. CAP (catabolite activator protein), which generates bend angles ranging between 70° and 96° (109), serves as an example. This protein activates RNA polymerase both through protein–protein interactions and through changes in DNA structure at the promoter. In the case of the *gal* promoter, DNA bending caused by CAP favors the contact of upstream DNA sequences with RNA polymerase, leading to open-complex formation. Repressors also bend the DNA at operators in a way that could interfere with RNA polymerase–promoter contacts (109). A third role of bends, especially when far upstream from a promoter, is to facilitate formation of loops in which repressors, bound to two distant operators, interact to block transcription. Two examples are discussed in the following section.

Negative supercoiling is likely to enhance processes utilizing bends, although a direct effect on bending is difficult to measure because bending is usually assayed with DNA fragments that cannot be supercoiled. Intrinsic bends in DNA tend to localize at terminal loops of interwound supercoils (110), so bent DNA will tend to place promoters near those loops. The loops would then facilitate binding of CAP as it further bends the DNA (111). When RNA polymerase binds, its interaction with CAP would lead to a tightening of the bend as duplex melting occurs. This bend tightening might then favor polymerase escape as initiation occurs, since tightly bent DNA would have a tendency to spring back (111).

8. DNA looping

Looping refers to an extreme bend in DNA that allows two distant regions to come into close proximity. Proteins bound at remote sites can then directly interact in a way that affects transcriptional status (112, 113). A feature often associated with looped DNA is the spacing of the distant protein binding sites by integral helical turns. Deletion or addition of half a turn of the helix places the sites out of phase, drastically reducing the biological effect. As pointed out above, interwound supercoils contain loops at their termini, and so supercoils are expected to facilitate loop formation (110). Loop formation is also influenced by DNA bending proteins, as described in later sections. Below we discuss the *lac* and *ara* operons as examples in which loops play important roles in initiation of transcription.

The *lac* operon contains a primary operator (O_1) that overlaps the promoter; consequently, binding of the repressor interferes with transcription initiation. The repressor also binds at two nearby pseudo-operators that exhibit lower affinity for repressor than O_1. One pseudo-operator (O_2) is located 401 bp downstream from O_1 in *lacZ*, the first structural gene of the operon; the other (O_3) is positioned 92 bp upstream from O_1 in *lacI*, the gene encoding the repressor (114 and references therein). *In-vitro* studies indicate that two operator sites can be occupied simultaneously (115), and intracellular studies show that pairs of operator sequences give greater repression than a single operator (116, 117): removal of either O_2 or O_3 lowers repression by roughly threefold while removal of both lowers it by 50-fold (114). These results are explained by formation of O_1– O_2 and O_1– O_3 loops such that repressor molecules bound at two operators can interact. This general idea of a repression loop is supported by visualization of repressor-dependent loops by electron microscopy (115). Moreover, a mutant repressor has been found that can bind to an operator but cannot form a looping interaction. This repressor fails to fully repress (114, 118). A contribution of negative supercoiling was demonstrated *in vitro*: DNA–repressor complex half-life increased greatly when supercoiled DNA was compared to linear (119, 120).

The *ara* system (121) provides a more complex example. The *araBAD* operon is involved in the transport and metabolism of arabinose as an energy source. The key regulator of this system is AraC, which activates the promoter of the *araBAD* operon in the presence of arabinose and negatively regulates the operon in the absence of the sugar. The regulatory region includes two adjacent binding sites, I_1 and I_2, located in the promoter of *araBAD*, and two sites, O_1 and O_2, positioned about 200 bp upstream. In the absence of arabinose, a repression loop forms between O_2 and I_1. In the presence of arabinose, this loop is disrupted, and AraC binds both I_1 and I_2. This activates *araBAD* transcription. The O_2–I_1 loop also represses expression from *araC*, which is located upstream from *araBAD*. An additional element in the AraC system is CAP. This protein activates *araBAD* transcription in a distance- and orientation-dependent manner, perhaps bending DNA into a conformation that precludes repression loop formation (122). As with *lac*, supercoiling facilitates loop formation *in vitro* (123, 124).

The *lac* and *ara* systems probably differ in how looping acts. With *lac*, loop formation serves to block RNA polymerase access to the promoter by increasing the binding of repressor to the primary operator. In the *ara* system, loop formation helps keep the repressor from serving as a transcription activator (113). Loops seem to play a more active role in *gal* regulation by distorting the promoter and interferring with the promoter–RNA polymerase interaction. (125). Thus precisely how loops contribute to repression differs considerably from one system to another.

9. DNA bending proteins

The DNA bending proteins, which are discussed next, recognize, exacerbate, or even create bends in DNA, often influencing expression of particular genes. Since the action of the bending proteins can be affected by supercoiling, these proteins provide another set of ways in which changes in supercoiling can be exploited to help regulate patterns of gene expression. In some cases they even influence supercoiling itself. The bending proteins also have the potential to respond to environment directly, either through their action or their level of expression. Thus the bending proteins add a level of topological influence beyond that exerted by supercoiling.

9.1 HU

HU is a small (20 kDa) dimeric DNA-binding protein that is highly conserved among eubacteria. *In vitro*, HU distorts DNA in a way reminiscent of the wrapping generated by eukaryotic histones (126, 127), and it facilitates circularization of short DNA fragments (128). This DNA bending ability, plus the presence of the protein at tens of thousands of copies per cell, led to the suggestion that HU might play an important role in packaging the bacterial chromosome. Consistent with this idea, HU shows little nucleotide sequence preference in binding to DNA. However, the protein does bind preferentially to kinked DNA (129) and cruciforms (130); consequently DNA distortions could help focus HU to specific spots and allow its general bending activity to stimulate a variety of activities in which distant DNA elements need to be brought together to form specific nucleoprotein complexes. Examples in which HU appears to play a facilitating role include replicative transposition of bacteriophage Mu (131, 132), gene inversion (133, 134), and initiation of DNA replication (135). In the case of Mu transposition, HU binding to DNA maps between two binding sites of Mu transposase (136). An important feature of this type of HU action is the apparent insignificance of specific protein contacts: the activity of HU can be replaced by eukaryotic HMG (high mobility group) proteins, which also exhibit DNA bending activity (137, 138).

HU seems to play a similar bending role in transcription. For example, the protein stimulates binding of *lac* repressor and CAP to cognate sites on linear DNA (139). As pointed out in the previous section, *lac* repressor and CAP binding involve looping and bending, respectively. Neither is expected for *trp* repressor, and HU has no effect on *trp* repressor binding. Moreover, the effect of HU can be replaced by barium chloride, a salt that stabilizes bends in DNA. These results are consistent with HU increasing DNA bending.

HU also appears to have an effect on DNA supercoiling. *In vitro* it facilitates gyrase action (140, 141), and strains deficient in HU have subnormal levels of supercoiling (41, 142, 143). Since gyrase wraps DNA during the supercoiling reaction (reviewed in Reference 144), it would not be surprising if the wrapping were facilitated by HU. Indeed, we found that a deficiency of HU is readily suppressed by mutations that map in *gyrB* or by expression of *gyrB* from a plasmid (145). We also noticed that a deficiency of HU leads to an increase in topoisomerase I-mediated relaxing activity and that overexpression of HU causes a decrease in relaxing activity (146). Thus HU has two actions that facilitate supercoiling.

HU may also prevent excess negative supercoils from accumulating: a *topA-10* mutant, which exhibits hypernegative supercoiling (12, 147), cannot accept a deficiency of HU introduced by transduction (146). In this case HU prevents excess supercoiling. Since HU also prevents relaxation (described above), it appears to act as a supercoiling buffer. The molecular interactions responsible for these effects have yet to be defined.

9.2 IHF

The integration host factor (IHF) of *E. coli* is another small (20 kDa) dimeric protein. Its subunits are closely related to those of HU, but unlike HU, IHF binds strongly to specific sites on DNA. The sites are roughly 30 bp long and contain a 13 bp consensus sequence AATCAANNNNTTA in which N is any nucleotide (148–150). Binding is associated with a DNA bend of approximately 140° (151), which effectively creates a hairpin in DNA. IHF was originally described as a host factor required for site-specific integration of bacteriophage lambda; the protein is now known to function as an accessory component of a wide variety of site-specific protein–DNA interactions including DNA replication, site-specific recombination, and initiation of transcription (152, 153). IHF is generally viewed as an architectural factor that facilitates the construction of complex DNA–protein structures (154). Strongly supporting this idea are experiments in which substitution of distorted DNA for IHF binding sites alleviates the IHF requirement for site-specific recombination (155, 156).

With respect to transcription initiation, IHF frequently acts as a bending element. Examples are seen in IHF-dependent stimulation of transcription from σ^{54}-dependent promoters. These promoters are often activated by factors that bind far upstream or downstream from the promoter (157), and cases have been found in which IHF-binding sites lie between the activator binding site and the promoter. It is generally thought that the bend introduced by IHF brings the two together (158).

IHF can also activate gene expression through short-range interactions. For three known cases, Pe of bacteriophage Mu (159–161), P_L1 of bacteriophage lambda (162), and P_G2 of the *ilvGMEDA* operon of *E. coli* (163, 164), IHF binds just upstream from the promoter. In the case of *ilvGMEDA*, the IHF binding site is about 70 bp upstream from P_G2 in a second promoter called P_G1. Binding at P_G1 severely bends DNA and leads to a fourfold, phase-independent activation of P_G2. This activation depends on negative supercoiling (164). Presumably the bend and supercoiling facilitate polymerase binding and helix opening.

When an IHF binding site overlaps that of RNA polymerase, IHF can act as a negative regulator of transcription. This occurs in the *ompF* operon where IHF inhibits transcription indirectly through modulation of a regulator protein, OmpR (165, 166). IHF binding overlaps the OmpR-binding region and inhibits OmpR-mediated *ompF* activation. Another example of steric hindrance involves the P_G1 promoter of the *ilvGMEDA* operon (164, 167). As pointed out in the previous paragraph, IHF binds at P_G1. That binding prevents RNA polymerase interaction with P_G1 (164). Thus the site-specificity of IHF binding, as well as its DNA bending activity, are utilized in a number of ways to modulate transcription initiation.

9.3 FIS

The factor for inversion stimulation (FIS) is a third small (20 kDa), dimeric, multifunctional DNA-binding protein (168). FIS was originally described as a host factor that stimulates the bacteriophage lambda excision recombination system and the site-specific inversion systems called *gin* in phage Mu, *cin* in phage P1, and *hin* in *S. typhimurium* (9, 10, 169, 170). FIS was later found to be involved in initiation of replication at the chromosomal origin (171, 172). Like IHF, the action of FIS may be mediated by its ability to distort DNA into a bend. In the case of FIS, the bend can be as much as 90° (151, 173). With initiation of replication, FIS-mediated bending of *oriC* might facilitate unwinding of an AT-rich region of *oriC* that serves as the locus of DNA melting (174, 175).

Levels of FIS protein and mRNA in *E. coli* vary enormously during the course of growth in batch culture, peaking at about 40 000 dimers per cell in early exponential phase and being undetectable after entry into stationary phase. Indeed, dilution of stationary phase cells into fresh medium raises FIS mRNA by 1000-fold (176, 177). Regulation of the *tyrT* operon provides an example of how FIS activates stable RNA promoters during outgrowth of cells from stationary phase through its interaction with an upstream activating sequence (UAS). The UAS of the *tyrT* promoter contains three highly conserved FIS binding sites located between positions –60 and –150 with respect to the transcription start site (178–180). Upon saturation of the UAS sites, FIS and RNA polymerase form a nucleoprotein complex at the promoter, and initiation of transcription follows. Formation of this complex appears to involve binding of FIS to one side of the helix, since insertion of half a helical turn between UAS sites II and III impairs the cooperative interaction between FIS dimers, the capacity to form FIS–RNA polymerase–DNA complexes, and the ability to initiate transcription. A potentially important feature of the interaction is the ability of the σ^{70} subunit of RNA polymerase to promote the cooperative binding of FIS to UAS (181). This binding may allow the σ^{70} subunit released after initiation of transcription to be recaptured by the nucleoprotein complex and recycled in successive rounds of holoenzyme assembly. Such behavior would help focus transcription to genes involved in the protein synthesis machinery when RNA polymerase is limiting, as would be the case during recovery from a prolonged stationary phase.

9.4 H-NS

H-NS is a small, abundant, heat-stable protein associated with the bacterial nucleoid. One of the unusual features of H-NS is that it has a neutral p*I* even though 30 per cent of its amino acids are charged. The acidic and basic residues occur in patches (182), similar to the pattern in eukaryotic HMG proteins (183). Another characteristic of H-NS is avid binding to bent DNA and stretches of AT base pairs (184–188), which are often found at or near control regions such as promoters, origins of replication, and recombination sites. H-NS may also have global effects on DNA structure. For example, mutations in *hns*, the gene encoding H-NS, can alter DNA linking deficit (39, 58, 186, 189, 190), and biochemical studies suggest that H-NS compacts DNA (191, 192) and/or constrains supercoils (187). Thus H-NS can have a variety of effects, and it is not surprising that mutations in *hns* influence a wide variety of cellular processes including recombination, virulence, motility, chromosome deletion, gene expression, and transposition of bacteriophage Mu (193).

Unlike HU, IHF, and FIS, the effects of H-NS have not been clearly attributed to alterations in DNA structure. Instead, H-NS seems to act by simply binding to curved DNA at or near particular promoters. Examination of *virB* (194), *rrnB* (195), *proU* (186, 188), and *hns* (196–198) indicates that H-NS can block transcription by preventing RNA polymerase binding, although results from an abortive transcription initiation assay suggest that for the *lac* and *gal* promoters H-NS influences open complex formation rather than RNA polymerase binding (191, 199). H-NS may also strengthen specific repressor–DNA complexes, as seen with the bacteriophage Mu repressor (182).

Mutations in *hns* characteristically increase expression of many unrelated genes (listed in Reference 193, plus *rpoS* (200), *virF* (201), *spvR* (202), *mxiC* (203), *icsB* (203), *ilvIH* (204), *lysU* (205)). Such pleiotropy suggests that H-NS may participate in complex regulatory circuits. For example, during exponential growth H-NS decreases the expression of σ^s, a stationary phase sigma factor (200). H-NS also represses Lrp, the leucine regulatory protein (204), and SpvR (202), a member of the *lysR* family of transcription activators (206). Since each of these genes controls a set of genes (202, 207–210), H-NS may have global regulatory importance.

Some of the effects of H-NS are exerted on genes that are sensitive to changes in environmental conditions that also affect DNA supercoiling. Thus questions emerge concerning whether the binding and activity of H-NS are affected by changes in supercoiling. With respect to supercoiling, H-NS-dependent inhibition of transcription *in vitro* shows a broad optimum slightly above levels of supercoiling found inside cells (211). *In vivo,* creation of locally high negative supercoiling by induction of divergent transcription reduces the inhibitory effect of H-NS and activates *virB* transcription (94). Conversely, lowering supercoiling through gyrase mutations activates *bgl* (13), an operon negatively regulated by H-NS. Thus it appears that H-NS silencing may be reduced by both increases and decreases in supercoiling.

10. Concluding remarks

Changes in DNA supercoiling can alter expression of many genes (for examples, see References 147 and 212), and it is becoming increasingly clear that supercoiling is responsive to environmental influences. There are several ways in which the environment can affect supercoiling. One is through changes in the ratio of [ATP] to [ADP] affecting the activity of gyrase. Another is a direct effect on DNA itself; salt and temperature may have this type of action. A third is through changes in expression of genes: transcription produces local negative supercoils behind transcription complexes and positive supercoils ahead of them. Determining precisely how these changes influence a particular gene is difficult, since so many different factors can be affected. Nevertheless, a reasonably clear list of factors to check is beginning to emerge. That list now includes strand separation in the promoter, wrapping of DNA around RNA polymerase, twist effects on promoter recognition by RNA polymerase, formation of loops, and binding of bending proteins that can have either positive or negative effects.

It is not difficult to visualize an underlying role for supercoiling in transcription initiation if one assumes that the sensitivity of supercoiling to the environment evolved before the sophisticated regulatory mechanisms that modulate transcription. Then the promoters and regulatory elements would have evolved to take advantage of the supercoiling changes. For example, differences in promoter spacers could allow individual promoters to respond to supercoiling change differently through the effects DNA twist has on the axial rotation between their –10 and –35 elements. Likewise, promoters could differ in the sensitivity of open complex formation (denaturation) to negative supercoiling. DNA bending and looping would serve to amplify the signal generated by changes in supercoiling.

The general evolutionary idea about supercoiling may also apply to the DNA-bending proteins if their raison d'etre concerns chromosome packaging and if this role predates regulatory mechanisms controlling initiation of transcription. Then the effects of the DNA bending proteins on transcription would reflect specific promoters evolving to take advantage of these proteins, much as phages, plasmids, and transposons may have done. Although there is currently little evidence that HU, IHF, FIS, and H-NS contribute to chromosome folding in a general way, an interesting relationship has emerged between IHF and REP sequences (213–215). Several hundred of these repetitive extragenic palindromes are scattered around the bacterial chromosome, and one subset, called RIPs or RIBs, contain an IHF binding site flanked by conserved palindromic units. The IHF sites in RIPs are more highly conserved than IHF sites in general, as are the palindromic units. The presence of intrinsic bends, which are uncommon among IHF sites, the conservation of sequence, and the distribution of about one RIP per topological domain of the chromosome have led to the suggestion that RIPs are chromosomal folding sites. The next step is to design assays to detect chromosome folding facilitated by the bending proteins.

Acknowledgements

We thank Leroy Liu for critical discussions and Hyon Choy, Marila Gennaro, Amy Luttinger, and Anthony Maxwell for comments on the manuscript. This work was supported by NIH Grant AI 35257.

References

1. Roca, J. (1995) The mechanisms of DNA topoisomerases. *TIBS*, **20**, 156.
2. Gellert, M., O'Dea, M. H., Mizuuchi, K., and Nash, H. (1976) DNA gyrase: an enzyme that introduces superhelical turns into DNA. *Proc. Natl. Acad. Sci. USA.*, **73**, 3872.
3. Wang, J. C. (1971) Interaction between DNA and an *Escherichia coli* protein. *J. Mol. Biol.*, **55**, 523.
4. Kato, J.-I., Nishimura, Y., Imamura, R., Niki, H., Hiraga, S., and Suzuki, H. (1990) New topoisomerase essential for chromosome segregation in *E. coli*. *Cell*, **63**, 393.
5. Luttinger, A., Springer, A., and Schmid, M. (1991) A cluster of genes that affects nucleoid segregation in *Salmonella typhimurium*. *The New Biologist*, **3**, 687.
6. Khodursky, A. B., Zechiedrich, E. L., and Cozzarelli, N. R. (1995) Topoisomerase IV is a target of quinolones in *Escherichia coli*. *Proc. Nat. Acad. Sci. USA*, **92**, 11801.
7. Rouviere-Yaniv, J. and Gros, F. (1975) Characterization of a novel, low molecular weight DNA-binding protein from *Escherichia coli*. *Proc. Natl. Acad. Sci. USA*, **72**, 3428.
8. Kikuchi, A. and Nash, H. A. (1978) The bacteriophage lambda *int* gene product. *J. Biol. Chem.*, **253**, 7149.
9. Johnson, R. C. and Simon, M. I. (1985) Hin-mediated site-specific recombination requires two 26 bp recombination sites and a 60 bp recombinational enhancer. *Cell*, **41**, 781.
10. Kahmann, R., Rudt, F., Koch, C., and Martens, G. (1985) G inversion in bacteriophage Mu DNA is stimulated by a site within the invertase gene and a host factor. *Cell*, **41** , 771.
11. Crepin, M., Cukier-Kahn, R., and Gros, F. (1975) Effect of a low-molecular weight DNA-binding protein, H1 factor, on the *in vitro* transcription of the lactose operon in *Escherichia coli*. *Proc. Natl. Acad. Sci. USA*, **72**, 333.
12. Pruss, G. J., Manes, S. H., and Drlica, K. (1982) *Escherichia coli* DNA topoisomerase I mutants: increased supercoiling is corrected by mutations near gyrase genes. *Cell*, **31**, 35.
13. DiNardo, S., Voelkel, K., Sternglanz, R., Reynolds, A., and Wright, A. (1982) *Esherichia coli* DNA topoisomerase I mutants have compensatory mutations in DNA gyrase genes. *Cell*, **31**, 43.
14. Drlica, K. and Snyder, M. (1978) Superhelical *Escherichia coli* DNA: relaxation by coumermycin. *J. Mol. Biol.*, **120**, 145.
15. Menzel, R. and Gellert, M. (1983) Regulation of the genes for *E. coli* DNA gyrase: homeostatic control of DNA supercoiling. *Cell*, **34**, 105.
16. Tse-Dinh, Y.-C. (1985) Regulation of the *Escherichia coli* DNA topoisomerase I gene by DNA supercoiling. *Nucleic Acids Res.*, **13**, 4751.
17. Manes, S. H., Pruss, G. J., and Drlica, K. (1983) Inhibition of RNA synthesis by oxolinic acid is unrelated to average DNA supercoiling. *J. Bacteriol.*, **155**, 420.
18. Tse-Dinh, Y.-C. and Beran, R. (1988) Multiple promoters for transcription of the *E. coli* DNA topoisomerase I gene and their regulation by DNA supercoiling. *J. Mol. Biol.*, **202**, 735.

19. Franco, R. J. and Drlica, K. (1989) Gyrase inhibitors can increase *gyrA* expression and DNA supercoiling. *J. Bacteriol.*, **171**, 6573.
20. Goldstein, E. and Drlica, K. (1984) Regulation of bacterial DNA supercoiling: plasmid linking number varies with growth temperature. *Proc. Natl. Acad. Sci. USA*, **81**, 4046.
21. Esposito, F. and Sinden, R. R. (1987) Supercoiling in prokaryotic and eukaryotic DNA: changes in response to topological perturbation of plasmids in *E. coli* and SV40 *in vitro*, in nuclei, and in CV-1 cells. *Nucleic Acids Res.*, **15**, 5105.
22. Pruss, G., Franco, R., Chevalier, S., Manes, S., and Drlica, K. (1986) Effects of DNA gyrase inhibitors in *Escherichia coli* topoisomerase I mutants. *J. Bacteriol.*, **168**, 276.
23. Drlica, K. (1990) Bacterial topoisomerases and the control of DNA supercoiling. *Trends in Genetics*, **6**, 433.
24. Westerhoff, H., Aon, M., van Dam, K., Cortassa, S., Kahn, D., and vanWorkum, M. (1990) Dynamical and hierarchical coupling. *Biochim. Biophys. Acta*, **1018**, 142.
25. Dean, F., Krasnow, M., Otter, R., Matzuk, M., Spengler, S., and Cozzarelli, N. (1983) *Escherichia coli* type I topoisomerases: identification, mechanism and role in recombination. *Cold Spring Harbor Symp. Quant Biol.*, **47**, 769.
26. DiGate, R. and Marians, K. (1988) Identification of a potent decatenating enzyme from *Escherichia coli*. *J. Biol. Chem.*, **263**, 13366.
27. DiGate, R. and Marians, K. (1989) Molecular cloning and DNA sequence analysis of *Escherichia coli topB*, the gene encoding topoisomerase III. *J. Biol. Chem.*, **264**, 17924.
28. Schofield, M. A., Agbunag, R., Michaels, M., and Miller, J. (1992) Cloning and sequencing of *Escherichia coli mutR* shows its identity to *topB*, encoding topoisomerase III. *J. Bacteriol.*, **174**, 5168.
29. McNairn, E., Bhriain, N. N., and Dorman, C. (1995) Overexpression of the *Shigella flexneri* genes coding for DNA topoisomerase IV compensates for loss of DNA topoisomerase I: effect on virulence gene expression. *Mol. Microbiol.*, **15**, 507.
30. Raji, A., Zabel, D. J., Laufer, S., and Depew, R. E. (1985) Genetic analysis of mutations that compensate for loss of *Escherichia coli* DNA topoisomerase I. *J. Bacteriol.*, **162**, 1173.
31. Adams, D., Shekhtman, E., Zechiedrich, E., Schmid, M., and Cozzarelli, N. (1992) The role of topoisomerase IV in partitioning bacterial replicons and the structure of catenated intermediates in DNA replication. *Cell*, **71**, 277.
32. Peng, H. and Marians, K. (1993) Decatenation activity of topoisomerase IV during *oriC* and pBR322 DNA replication *in vitro*. *Proc. Natl. Acad. Sci. USA*, **90**, 8571.
33. Hiasa, H. and Marians, K. (1994) Topoisomerase IV can support oriC DNA replication *in vitro*. *J. Biol. Chem.*, **269**, 16371.
34. Steck, T. R. and Drlica, K. (1984) Bacterial chromosome segregation: evidence for DNA gyrase involvement in decatenation. *Cell*, **36**, 1081.
35. Gellert, M., O'Dea, M. H., Itoh, T., and Tomizawa, J.-I. (1976) Novobiocin and coumermycin inhibit DNA supercoiling catalyzed by DNA gyrase. *Proc. Natl. Acad. Sci. USA* , **73**, 4474.
36. Westerhoff, H., O'Dea, M., Maxwell, A., and Gellert, M. (1988) DNA supercoiling by DNA gyrase. A static head analysis. *Cell Biophysics*, **12**, 157.
37. Balke, V. and Gralla, J. D. (1987) Changes in the linking number of supercoiled DNA accompany growth transitions in *Escherichia coli*. *J. Bacteriol.*, **169**, 4499.
38. Karem, K. and Foster, J. (1993) The influence of DNA topology on the environmental regulation of a pH-regulated locus in *Salmonella typhimurium*. *Mol. Microbiol.*, **10**, 75.
39. Higgins, C. F., Dorman, C. J., Stirling, D. A., Waddell, L., Booth, I. R., May, G., and Bremer, E. (1988) A physiological role for DNA supercoiling in the osmotic regulation of gene expression in *S. typhimurium* and *E. coli*. *Cell*, **52**, 569.

40. McClellan, J., Boublikova, P., Palecek, E., and Lilley, D. (1990) Superhelical torsion in cellular DNA responds directly to environmental and genetic factors. *Proc. Natl. Acad. Sci. USA*, **87**, 8373.
41. Hsieh, L.-S., Rouviere-Yaniv, J., and Drlica, K. (1991) Bacterial DNA supercoiling and [ATP]/[ADP]: changes associated with salt shock. *J. Bacteriol.*, **173**, 3914.
42. Dorman, C., Barr, G., NiBhriain, N., and Higgins, C. (1988) DNA supercoiling and the anaerobic growth phase regulation of *tonB* gene expression. *J. Bacteriol.*, **170**, 2816.
43. Hsieh, L.-S., Burger, R. M., and Drlica, K. (1991) Bacterial DNA supercoiling and [ATP]/[ADP]: changes associated with a transition to anaerobic growth. *J. Mol. Biol.*, **219**, 443.
44. Jensen, P., Loman, L., Petra, B., van der Weijden, C., and Westerhoff, H. (1995) Energy buffering of DNA structure fails when *Escherichia coli* runs out of substrate. *J. Bacteriol.*, **177**, 3420.
45. Grossman, A. D., Erikson, J. W., and Gross, C. A. (1984) The *htpR* gene product of *Escherichia coli* is a sigma factor for heat-shock promoters. *Cell*, **38**, 383.
46. Hartl, F. V. (1994) Molecular chaperones in cellular protein folding. *Nature*, **381**, 571.
47. Mizushima, T., Natori, S., and Sekimizu, K. (1993) Relaxation of supercoiled DNA associated with induction of heat shock proteins in *Escherichia coli*. *Mol. Gen. Genet.*, **238**, 1.
48. Kaneko, T., Mizushima, T., Ohtsuka, Y., Kurokawa, K., Kataoka, K., Miki, T., and Sekimizu, K. (1996) Co-induction of DNA relaxation and synthesis of DnaK and GroEL proteins in *Escherichia coli* by expression of LetD (CcdB) protein, an inhibitor of DNA gyrase encoded by the F factor. *Mol. Gen. Genet.*, **250**, 593.
49. Ogata, Y., Mizushima, T., Kataoka, K., Miki, T., and Sekimizu, K. (1994) Identification of DNA topoisomerases involved in immediate and transient relaxation induced by heat shock in *Escherichia coli*. *Mol. Gen. Genet.*, **244**, 451.
50. Camacho-Carranza, R., Membrillo-Hernandez, J., Ramirez-Santos, J., Castro-Dorantes, J., Sanchez, V. C., and Gomez-Eichelmann, M. C. (1995) Topoisomerase activity during the heat shock response in *Escherichia coli* K-12. *J. Bacteriol.*, **177**, 3619.
51. Tse-Dinh, Y.-C. (1994) Biochemistry of bacterial type I DNA topoisomerases. *Adv. Pharmacol.*, **29A**, 21.
52. Qi, H., Menzel, R., and Tse-Dinh, Y.-C. (1996) Effect of the deletion of the sigma-32-dependent promoter (P1) for the *Escherichia coli* topoisomerase I gene on thermotolerance. *Mol. Microbiol.*, **21**, 703.
53. Lesley, S. A., Jovanovich, S. B., Tse-Dinh, Y.-C., and Burgess, R. R. (1990) Identification of a heat shock promoter in the *topA* gene of *E. coli*. *J. Bacteriology*, **172**, 6871.
54. Rohde, J. R., Fox, J. M., and Minnich, S. A. (1994) Thermoregulation in *Yersinia enterocolitica* is coincident with changes in DNA supercoiling. *Molec. Microbiol.*, **12**, 187.
55. Depew, R. E. and Wang, J. C. (1975) Conformational fluctuations of DNA helix. *Proc. Natl. Acad. Sci. USA*, **72**, 4275.
56. Pulleyblank, D., Shure, M., Tang, D., Vinograd, J., and Vosberg, H.-P. (1975) Action of nicking-closing enzyme on supercoiled and nonsupercoiled closed circular DNA: formation of a Boltzmann distribution of topological isomers. *Proc. Natl. Acad. Sci. USA*, **72**, 4280.
57. Grau, R., Gardiol, D., Gilkin, G., and Mendoza, D. d. (1994) DNA supercoiling and thermal regulation of unsaturated fatty acid synthesis in *Bacillus subtilis*. *Mol. Microbiol.*, **11**, 933.
58. Dorman, C. J., NiBhriain, N., and Higgins, C. F. (1990) DNA supercoiling and environmental regulation of virulence gene expression in *Shigella flexneri*. *Nature*, **344**, 789.
59. Mojica, F., Charbonnier, F., Juez, G., Rodriguez-Valera, F., and Forterre, P. (1994) Effects

of salt and temperature on plasmid topology in the halophilic archaeon *Haloferax volcanii*. *J. Bacteriol.*, **176**, 4968.

60. Forterre, P., Bergerat, A., and Lopez-Garcia, P. (1996) The unique DNA topology and DNA topoisomerases of hyperthermophilic archaea. *FEMS Microbiology Reviews*, **18**, 237.
61. Forterre, P. (1996) A hot topic: The origin of hyperthermophiles. *Cell*, **85**, 789.
62. Oshima, T., Hamasaki, N., Uzawa, T., and Friedman, S. M. (1990) Biochemical functions of unusual polyamines found in the cells of extreme thermophiles. In *The biology and chemistry of polyamines* (ed. S. H. Goldemberg and I. D. Algranati), p. 1. Oxford University Press, New York.
63. Baumann, H., Knapp, S., Londback, T., Ladenstein, R., and Hard, T. (1994) Solution structure and DNA-binding properties of a small thermostable protein from the archaea *Sulfolobis sulfatarius*. *Nature Struct. Biol.*, **1**, 808.
64. Searcy, D. G. (1975) Histone-like protein in the prokaryote *Thermoplasma acidophilum*. *Biochim. Biophys. Acta*, **395**, 535.
65. Starich, M. R., Sandman, K., Reeve, J. N., and Summers, M. F. (1996) NMR structure of HMfB from the hyperthermophile, *Methanothermus fervidus*, confirms that this archaeal protein is a histone. *J. Mol. Biol.*, **255**, 187.
66. Kikuchi, A. and Asai, K. (1984) Reverse gyrase: a topoisomerase which introduces positive superhelical turns into DNA. *Nature*, **309**, 677.
67. Bouthier de la Tour, C., Portemer, C., Huber, R., Forterre, P., and Duguet, M. (1991) Reverse gyrase in thermophilic eubacteria. *J. Bacteriol.*, **173**, 3921.
68. Jaxel, C., Nadal, M., Mirambeau, G., Forterre, P., Takahashi, M., and Duguet, M. (1989) Reverse gyrase binding to DNA alters the double helix structure and produces single-strand cleavage in the absence of ATP. *EMBO J.*, **8**, 3135.
69. Confaloniere, F., Elie, C., Nadal, M., Bouthier de la Tour, C., Forterre, P., and Duget, M. (1993) Reverse gyrase: a helicase-like domain and a type I topoisomerase in the same polypeptide. *Proc. Natl. Acad. Sci. USA*, **90**, 4753.
70. Kozyavkin, S. A., Krah, R., Gellert, M., Stetter, K., Lake, J., and Slesarev, A. (1994) A reverse gyrase with an unusual structure. A type I DNA topoisomerase form the hyperthermophile *Methanopyrus kandleri* is a two-subunit protein. *J. Biol. Chem.*, **269**, 11081.
71. Friedman, S. M., Malik, M., and Drlica, K. (1995) DNA supercoiling in a thermotolerant mutant of *Escherichia coli*. *Mol. Gen. Genet.*, **248**, 417.
72. Pruss, G. and Drlica, K. (1986) Topoisomerase I mutants: the gene on pBR322 that encodes resistance to tetracycline affects plasmid DNA supercoiling. *Proc. Natl. Acad. Sci. USA*, **83**, 8952.
73. Pruss, G. J. (1985) DNA Topoisomerase I mutants: Increased heterogeneity in linking number and other replicon-dependent changes in DNA supercoiling. *J. Mol. Biol.*, **185**, 51.
74. Lockshon, D. and Morris, D. (1983) Positively supercoiled plasmid DNA is produced by treatment of *Escherichia coli* with DNA gyrase inhibitors. *Nucleic Acids Res.*, **11**, 2999.
75. Liu, L. and Wang, J. (1987) Supercoiling of the DNA template during transcription. *Proc. Natl. Acad. Sci. USA*, **84**, 7024.
76. Wu, H.-Y., Shyy, S.-H., Wang, J. C., and Liu, L. F. (1988) Transcription generates positively and negatively supercoiled domains in the template. *Cell*, **53**, 433.
77. Koo, H.-S., Wu, H.-Y., and Liu, L. F. (1990) Effects of transcription and translation on gyrase mediated DNA cleavage in *Escherichia coli*. *J. Biol. Chem.*, **265**, 12300.
78. Lodge, J., Kazic, T., and Berg, D. (1989) Formation of supercoiling domains in plasmid pBR322. *J. Bacteriol.*, **171**, 2181.
79. Lynch, A. and Wang, J. C. (1993) Anchoring of DNA to the bacterial cytoplasmic membrane

through cotranslational synthesis of polypeptides encoding membrane proteins or proteins for export; a mechanism of plasmid hypernegative supercoiling in mutants deficient in DNA topoisomerase I. *J. Bacteriol.*, **175**, 1645.

80. Tsao, Y.-P., Wu, H.-Y., and Liu, L. F. (1989) Transcription-driven supercoiling of DNA: direct biochemical evidence from *in vitro* studies. *Cell*, **56**, 111.
81. Yang, L., Jessee, C. B., Lau, K., Zhang, H., and Liu, L. F. (1989) Template supercoiling during ATP-dependent DNA helix tracking: studies with simian virus 40 large tumor antigen. *Proc. Natl. Acad. Sci. USA*, **86**, 6121.
82. Koo, H.-S., Claassen, L., Grossman, L., and Liu, L. F. (1991) ATP-dependent partitioning of the DNA template into supercoiled domains by *E. coli UvrAB*. *Proc. Natl. Acad. Sci. USA*, **88**, 1212.
83. Rahmouni, A. R. and Wells, R. D. (1989) Stabilization of Z DNA *in vivo* by localized supercoiling. *Science*, **246**, 358.
84. Rahmouni, A. R. and Wells, R. D. (1992) Direct evidence for the effect of transcription on local DNA supercoiling *in vivo*. *J. Mol. Biol.*, **223**, 131.
85. Drolet, M., Phoenix, P., Menzel, R., Masse, E., Liu, L. F., and Crouch, R. J. (1995) Overexpression of RNase H partially complements the growth defect of an *Escherichia coli* Δ*topA* mutant: R-loop formation is a major problem in the absence of DNA topoisomerase I. *Proc. Natl. Acad. Sci.*, **92**, 3526.
86. Drolet, M., Bi, X., and Liu, L. F. (1994) Hypernegative supercoiling of the DNA template during transcription elongation *in vitro*. *J. Biol. Chem.*, **269**, 2068.
87. Mukai, F. H. and Margolin, P. (1963) Analysis of unlinked suppressors of an O^0 mutation in *Salmonella*. *Proc. Natl. Acad. Sci. USA*, **50**, 140.
88. Trucksis, M. and Depew, R. (1981) Identification and localization of a gene that specifies production of *Escherichia coli* DNA topoisomerase I. *Proc. Natl. Acad. Sci. USA*, **78**, 2164.
89. Pruss, G. J. and Drlica, K. (1985) DNA supercoiling and suppression of the *leu-500* promoter mutation. *J. Bacteriol.*, **164**, 947.
90. Richardson, S. M. H., Higgins, C., and Lilley, D. (1988) DNA supercoiling and the *leu-500* promoter mutation of *Salmonella typhimurium*. *EMBO J.*, **7**, 1863.
91. Chen, D., Bowater, R., Dorman, C. J., and Lilley, D. M. (1992) Activity of a plasmid-borne *leu-500* promoter depends on the transcription and translation of an adjacent gene. *Proc. Natl. Acad. Sci. USA*, **89**, 8784.
92. Chen, D., Bowater, R., and Lilley, D. M. (1993) Topological promoter coupling in *Escherichia coli*: Δ*topA*-dependent activation of the *leu-500* promoter on a plasmid. *J. Bacteriol.*, **176**, 3757.
93. Tan, J., Shu, L., and Wu, H.-Y. (1994) Activation of the *leu-500* promoter by adjacent transcription. *J. Bacteriol.*, **176**, 1077.
94. Tobe, T., Yoshikawa, M., and Sasakawa, C. (1995) Thermoregulation of *virB* transcription in *Shigella flexneri* by sensing of changes in local DNA superhelicity. *J. Bacteriol.*, **177**, 1094.
95. Cook, D., Ma, D., Pon, N., and Hearst, J. (1992) Dynamics of DNA supercoiling by transcription in *Escherichia coli*. *Proc. Natl. Acad. Sci. USA*, **89**, 10603.
96. McClure, W. R. (1985) Mechanism and control of transcription initiation in prokaryotes. *Ann. Rev. Biochem.*, **54**, 171.
97. Borowiec, J. A. and Gralla, J. D. (1987) All three elements of the *lac* P^s promoter mediate its transcriptional response to DNA supercoiling. *J. Mol. Biol.*, **195**, 89.
98. Brahms, J., Dargouge, O., Brahms, S., Ohara, Y., and Vagner, V. (1985) Activation and inhibition of transcription by supercoiling. *J. Mol. Biol.*, **181**, 455.

99. Wang, J.-Y. and Syvanen, M. (1992) DNA twist as a transcriptional sensor for environmental changes. *Mol Microbiol.*, **6**, 1861.
100. Anderson, P. and Bauer, W. (1978) Supercoiling in closed circular DNA: dependence upon ion type and concentration. *Biochemistry*, **17**, 594.
101. Prince, W. and Villarejo, M. R. (1990) Osmotic control of *proU* transcription is mediated through direct action of potassium glutamate on the transcription complex. *J. Biol. Chem.*, **265**, 17673.
102. Wang, J.-Y., Drlica, K., and Syvanen, M. (1997) Monovalent cations differ in their effects on transcription initiation from a sigma-70 promoter of *Escherichia coli*. *Gene*, **196**, 95.
103. Wu, H.-M. and Crothers, D. (1984) The locus of sequence-directed and protein-induced DNA bending. *Nature*, **308**, 509.
104. Peres-Martin, J., Rojo, F., and deLorenzo, V. (1994) Promoters responsive to DNA bending: a common theme in prokaryotic gene expression. *Microbiol. Rev.*, **58**, 268.
105. Galas, D. J., Eggert, M., and Waterman, M. S. (1985) Rigorous pattern-recognition methods for DNA sequences. *J. Mol. Biol.* , **186**, 117.
106. Van Wye, D., Bronson, E. C., and Anderson, J. N. (1991) Species-specific patterns of DNA bending and sequence. *Nucleic Acids Res.*, **19**, 5253.
107. Chenchik, A., Beabealashvili, R., and Mirzabekov, A. (1981) Topography of interaction of *E. coli* RNA polymerase subunits with *lacUV5* promoter. *FEBS Lett.*, **128**, 46.
108. Khunke, G., Theres, C., Fritz, K.-J., and Ehring, R. (1989) RNA polymerase and *gal* repressor bind simultaneously and with DNA bending to the control region of the *Escherichia coli* galactose operon. *EMBO J.*, **8**, 1247.
109. Kim, J., Zwieb, C., Wu, C., and Adhya, S. (1989) Bending of DNA by gene-regulatory proteins: construction and use of a DNA bending vector. *Gene*, **85**, 15.
110. Laundon, C. and Griffith, J. (1988) Curved helix segments can uniquely orient the topology of supertwisted DNA. *Cell*, **52**, 545.
111. Zinkel, S. S. and Crothers, D. M. (1991) Catabolite activator protein-induced DNA bending in transcription initiation. *J. Mol. Biol.*, **219**, 201.
112. Matthews, K. S. (1992) DNA looping. *Microbiol. Rev.*, **56**, 123.
113. Choy, H. and Adhya, S. (1996) Negative control. In Escherichia coli *and* Salmonella typhimurium (ed. F. Neidhardt), p. 1287. American Society for Microbiology, Washington, DC.
114. Oehler, S., Eismann, E. R., Kramer, H., and Muller-Hill, B. (1990) The three operators of the *lac* operon cooperate in repression. *EMBO J.*, **9**, 973.
115. Kramer, H., Niemoller, M., Amouyal, M., Revet, B., Wilchen-Bergmann, B. v., and Muller-Hill, B. (1987) Lac repressor forms loops with linear DNA carrying two suitably spaced operators. *EMBO J.*, **6**, 1481.
116. Besse, M., Wilcken-Bergmann, B.v., and Muller-Hill, B. (1986) Synthetic *lac* operator mediates repression through *lac* repressor when introduced upstream and downstream from *lac* promoter. *EMBO J.*, **5**, 1377.
117. Mossing, M. C. and Record, M. T. (1986) Upstream operators enhance repression of the *lac* promoter. *Science*, **233**, 889.
118. Brenowitz, M., Mandal, N., Pickar, A., Jamison, E., and Adhya, S. (1991) DNA-binding properties of a *lac* repressor mutant incapable of forming tetramers. *J. Biol. Chem.*, **266**, 1281.
119. Borowiec, J., Zhang, L., Sasse-Dwight, S., and Gralla, J. (1987) DNA supercoiling promotes formation of a bent repression loop in *lac* DNA. *J. Mol. Biol.*, **196**, 101.
120. Whitson, P. A. and Matthews, K. S. (1986) Dissociation of lactose repressor–operator DNA

complex: effects of size and sequence context of operator-containing DNA. *Biochemistry*, **25**, 3845.

121. Dunn, T. M., Hahn, S., Ogden, S., and Schleif, R. f. (1984) An operator at –280 base pairs that is required for repression of *araBAD* operon promoter: Addition of DNA helical turns between the operator and promoter cyclically hinders repression. *Proc. Natl. Acad. Sci. USA*, **81**, 5017.
122. Lobell, R. B. and Schleif, R. F. (1991) AraC–DNA looping: orientation and distance-dependent loop breaking by the cyclic AMP receptor protein. *J. Mol. Biol.*, **218**, 45.
123. Lobell, R. B. and Schleif, R. F. (1990) DNA looping and unlooping by AraC protein. *Science*, **250**, 528.
124. Hahn, S., Hendrickson, W., and Schleif, R. (1986) Transcription of *Escherichia coli ara in vitro*: the cyclic AMP receptor protein requirement for P_{BAD} induction that depends on the presence and orientation of the *araO2* site. *J. Mol. Biol.*, **188**, 355.
125. Choy, H., Park, S.-W., Parrack, P., and Adhya, S. (1995) Transcription regulation by inflexibility of promoter DNA in a looped complex. *Proc. Natl. Acad. Sci. USA*, **92**, 7327.
126. Rouviere-Yaniv, J., Germond, J.-E., and Yaniv, M. (1979) *E. coli* DNA binding protein HU forms nucleosome-like structure with circular double-stranded DNA. *Cell*, **17**, 265.
127. Broyles, S. and Pettijohn, D. E. (1986) Interaction of the *E. coli* HU protein with DNA: evidence for formation of nucleosome-like structures with altered DNA helical pitch. *J. Mol. Biol.*, **187**, 47.
128. Hodges-Garcia, Y., Hagerman, P., and Pettijohn, D. (1989) DNA ring closure mediated by protein HU. *J. Biol. Chem.*, **264**, 14621.
129. Pontiggia, A., Negri, A., and Beltrame, M. (1993) Protein HU binds specifically to kinked DNA. *Mol. Microbiol.*, **7**, 343.
130. Bonnefoy, E., Takahashi, M., and Rouviere-Yaniv, J. (1994) DNA-binding parameters of the HU protein of *Escherichia coli* to cruciform DNA. *J. Mol. Biol.*, **242**, 116.
131. Craigie, R., Arndt-Jovin, D., and Mizuuchi, K. (1985) A defined system for the DNA strand-transfer reaction at the initiation of bacteriophage Mu transposition: protein and DNA substrate requirements. *Proc. Natl. Acad. Sci. USA*, **82**, 7570.
132. Huisman, O., Faelen, M., Girard, D., Jaffe, A., Toussaint, A., and Rouviere-Yaniv, J. (1989) Multiple defects in *Escherichia coli* mutants lacking HU protein. *J. Bacteriol.*, **171**, 3704.
133. Johnson, R., Bruist, M., and Simon, M. (1986) Host protein requirements for *in vitro* site-specific DNA inversion. *Cell*, **46**, 531.
134. Wada, M., Kutsukake, K., Komano, T., Imamoto, F., and Kano, Y. (1989) Participation of the *hup* gene product in specific DNA inversion in *Escherichia coli*. *Gene*, **76**, 345.
135. Hwang, D. and Kornberg, A. (1992) Opening of the replication origin of *Escherichia coli* DNA by DnaA protein with protein HU or IHF. *J. Biol. Chem.*, **267**, 23083.
136. Lavoie, B. D. and Chaconas, G. (1993) Site-specific HU binding in the Mu transpososome: conversion of a sequence-independent DNA-binding protein into a chemical nuclease. *Genes Dev.*, **7**, 2510.
137. Paull, T. T., Haykinson, M. J., and Johnson, R. C. (1993) The nonspecific DNA-binding and -bending proteins HMF1 and HMG2 promote the assembly of complex nucleoprotein structures. *Genes Dev.*, **7**, 1521.
138. Paull, T. T. and Johnson, R. C. (1995) DNA looping by *Saccharomyces cerevisiae* high mobility group proteins NHP6A/B. *J. Biol. Chem.*, **270**, 8744.
139. Flashner, Y. and Gralla, J. (1988) DNA dynamic flexibility and protein recognition: differential stimulation by bacterial histone-like protein HU. *Cell*, **54**, 713.

140. Marians, K. (1987) DNA gyrase-catalyzed decatenation of multiply linked DNA dimers. *J. Biol. Chem.*, **262**, 10362.
141. Yang, Y. and Ames, G.(1990) The family of repetitive extragenic palindromic sequences: interaction with DNA gyrase and histonelike protein HU. In *The bacterial chromosome* (ed. K. Drlica and M. Riley), p. 211. American Society for Microbiology, Washington, DC.
142. Hillyard, D. R., Edlund, M., Hughes, K., Marsh, M., and Higgins, N. P. (1990) Subunit-specific phenotypes of *Salmonella typhimurium* HU mutants. *J. Bacteriol.*, **172**, 5402.
143. Rouviere-Yaniv, J., Kiseleva, E., Bensaid, A., Almeida, A., and Drlica, K. (1992) Protein HU and bacterial DNA supercoiling. In *Prokaryotic structure and function* (ed. S. Mohan, C. Dow, and J. Cole), p. 17. Cambridge University Press, Cambridge, UK.
144. Reece, R. and Maxwell, A. (1991) DNA gyrase: structure and function. *Crit. Rev. Biochem. and Molec. Biol.*, **26**, 335.
145. Malik, M., Bensaid, A., Rouviere-Yaniv, J., and Drlica, K. (1996) Histone-like protein HU and bacterial DNA topology: suppression of an HU deficiency by gyrase mutations. *J. Mol. Biol.*, **256**, 66.
146. Bensaid, A., Almeida, A., Drlica, K., and Rouviere-Yaniv, J. (1996) Cross-talk between topoisomerase I and HU in *Escherichia coli*. *J. Mol. Biol.*, **256**, 292.
147. Steck, T., Franco, R., Wang, J.-Y., and Drlica, K. (1993) Topoisomerase mutations affect the relative abundance of many *Escherichia coli* proteins. *Mol. Microbiol.*, **10**, 473.
148. Craig, N. and Nash, H. (1984) *E. coli* integration host factor binds to specific sites in DNA. *Cell*, **39**, 707.
149. Goodrich, J. A., Schwartz, M. L., and McClure, W. A. (1990) Searching for and predicting the activity of sites for DNA binding proteins: compilation and analysis of the binding sites for *Escherichia coli* integration host factor (IHF). *Nucleic Acids Res.*, **18**, 4993.
150. Leong, J. M., Nunes-Dudy, S., Lesser, C. F., Youderian, P., Susskind, M., and Landy, A. (1985) The Φ80 and P22 attachment sites. *J. Biol. Chem.*, **260**, 4468.
151. Thompson, J. and Landy, A. (1988) Emperical estimation of protein-induced DNA bending angles: application to lambda site-specific recombination complexes. *Nucleic Acids Res.*, **16**, 9687.
152. Freundlich, M., Ramani, N., Mathew, W., Sirko, A., and Tsui, P. (1992) The role of integration host factor in gene expression in *Escherichia coli*. *Mol. Microbiol.*, **6**, 2557.
153. Goosen, N. and Putte, P. v. d. (1995) The regulation of transcription initiation by integration host factor. *Mol. Microbiol.*, **16**, 1.
154. Grosschedl, R., Giese, K., and Pagel, J. (1994) HMG domain proteins: architectural elements in the assembly of nucleoprotein structures. *Trends in Genetics*, **10**, 94.
155. Goodman, S. D. and Nash, H. A. (1989) Functional replacement of a protein-induced bend in a DNA recombination site. *Nature*, **341**, 244.
156. Goodman, S. D., Nicholson, S. C., and Nash, H. A. (1992) Deformation of DNA during site-specific recombination of bacteriophage lambda replacement of IHF protein by HU protein or sequence-directed bends. *Proc. Natl. Acad. Sci. USA*, **89**, 11910.
157. Kustu, S., Santero, E., Keener, J., Popham, D., and Weiss, D. (1989) Expression of sigma-54 (*ntrA*)-dependent genes is probably united by a common mechanism. *Microbiol. Rev.*, **53**, 367.
158. Hoover, T. R., Santero, E., Porter, S., and Kustu, S. (1990) The integration host factor stimulates interaction of RNA polymerase with NIFA, the transcriptional activator for nitrogen fixation operons. *Cell*, **63**, 11.
159. Goosen, N., van Heuvel, M., Moolenaar, G., and van de Putte, P. (1984) Regulation of Mu transposition. II. The *Escherichia coli* HimD protein positively controls two repressor promoters and the early promoter of bacteriophage Mu. *Gene*, **32**, 419.

160. Goosen, N. and van de Putte, P. (1984) Regulation of Mu transposition I. Localization of the presumed recognition sites for HimD and Ner functions controlling bacteriophage Mu transcription. *Gene*, **30**, 41.
161. Krause, H. and Higgins, N. P. (1986) Positive and negative regulation of the Mu operator by Mu repressor and *Escherichia coli* integration host factor. *J. Biol. Chem.*, **261**, 3744.
162. Giladi, H., Gottesman, M., and Oppenheim, A. B. (1990) Integration host factor stimulates the phage lambda pL promoter. *J. Mol. Biol.*, **213**, 109.
163. Pagel, J. M. and Hatfield, G. W. (1991) Integration host factor-mediated expression of the *ilvGMEDA* operon of *Escherichia coli*. *J. Biol. Chem.*, **266**, 1985.
164. Pagel, J. M., Winkelman, J. W., Adams, C. W., and Hatfield, G. W. (1992) DNA topology-mediated regulation of transcription initiation from the tandem promoters of the *ilvGMEDA* operon of *Escherichia coli*. *J. Mol. Biol.*, **224**, 919.
165. Slauch, J. M. and Silhavy, T. J. (1991) *cis*-Acting *ompF* mutations that result in OmpR-dependent constitutive expression. *J. Bacteriol.*, **173**, 4039.
166. Ramani, N., Huang, L., and Freundlich, M. (1992) *In vitro* interactions of integration host factor with the *ompF* promoter-regulatory region of *Escherichia coli*. *Mol. Gen. Genet.*, **231**, 248.
167. Pereira, R. F., Ortuno, M. J., and Lawther, R. P. (1988) Binding of integration host factor (IHF) to the *ilvGp1* promoter of the *ilvGMEDA* operon of *Escherichia coli* K12. *Nucleic Acids Res.*, **16**, 5973.
168. Finkel, S. E. and Johnson, R. C. (1992) The Fis protein: it's not just for DNA inversion anymore. *Mol. Microbiol.*, **6**, 3257.
169. Thompson, J. F., deVargas, L. M., Koch, C., Kahmann, R., and Landy, A. (1987) Celllular factors couple recombination with growth phase: Characterization of a new component in the lambda site-specific recombination pathway. *Cell*, **50**, 901.
170. Haffter, P. and Bickle, T. A. (1987) Purification and DNA-binding properties of FIS and Cin, two proteins required for the bacteriophage P1 site-specific recombination system, *cin*. *J. Mol. Biol.*, **198**, 579.
171. Gille, H., Egan, J. B., Roth, A., and Messer, W. (1991) The FIS protein binds and bends the origin of chromosomal DNA replication, *oriC*, of *Escherichia coli*. *Nucleic Acids Res.*, **19**, 4167.
172. Filutowicz, M. F., Ross, W., Wild, J., and Gourse, R. L. (1992) Involvement of Fis protein in replication of the *Escherichia coli* chromosome. *J. Bacteriol.*, **174**, 398.
173. Pan, C. Q., .Feng, J. A., Finkel, S. E., Landgraf, R., Sigman, D., and Johnson, R. C. (1994) Structure of the *Escherichia coli* Fis-DNA complex probed by protein conjugated with 1,10-phenanthroline copper (I) complex. *Proc. Natl. Acad. Sci. USA*, **91**, 1721.
174. Gille, H. and Messer, W. (1991) Localized DNA melting and structural perturbation in the origin of replication, *oriC*, of *Escherichia coli in vitro* and *in vivo*. *EMBO J.*, **10**, 1579.
175. Roth, A., Urmoneit, B., and Messer, W. (1994) Functions of histone-like proteins in the initiation of DNA replication at *oriC* of *Escherichia coli*. *Biochemie*, **76**, 917.
176. Ball, C. A., Osuna, R., and Ferguson, K. C. (1992) Dramatic changes in Fis levels upon nutrient upshift in *Escherichia coli*. *J. Bacteriol.*, **174**, 8043.
177. Ninnemann, O., Koch, C., and Kahmann, R. (1992) The *E. coli* fis promoter is subject to stringent control and autoregulation. *EMBO J.*, **11**, 1075.
178. Nilsson, L., Vanet, A., Vijgenboom, E., and Bosh, L. (1990) The role of FIS in trans-activation of stable RNA operon of *E. coli*. *EMBO J.*, **9**, 727.
179. Zacharias, M., Goringer, H., and Wagner, R. (1992) Analysis of the FIS-dependent and FIS-independent transcription activation mechanism of the *Escherichia coli* ribosomal RNA P1 promoter. *Biochemistry*, **31**, 2621.

180. Lazaru, L. and Travers, A. (1993) The *Escherichia coli* FIS protein is not required for the activation of *tyrT* transcription on entry into exponential growth. *EMBO J.*, **12**, 2483.
181. Muskhelishvili, G., Travers, A., Heumann, H., and Kahmann, R. (1995) FIS and RNA polymerase holoenzyme form a specific nucleoprotein complex at a stable RNA promoter. *EMBO J.*, **14**, 1446.
182. Falconi, M., McGovern, V., Gualerzi, C., Hillyard, D., and Higgins, N. P. (1991) Mutations altering chromosomal protein H-NS induce mini-Mu transposition. *The New Biologist*, **3**, 615.
183. Haggren, W. and Kolodrubetz, D. (1988) The *Saccharomyces cerevisiae* ACP2 gene encodes an essential HMG1-like protein. *Mol. Cell. Biol.*, **8**, 1282.
184. Bracco, L., Kotlarz, D., Kolb, A., Diekmann, S., and Buc, H. (1989) Synthetic curved DNA sequences can act as transcriptional activators in *Escherichia coli*. *EMBO J.*, **8**, 4289.
185. Yamada, H., Muramatsu, S., and Mizuno, T. (1990) An *Escherichia coli* protein that preferentially binds to sharply curved DNA. *J. Biochem.*, **108**, 420.
186. Owen-Hughes, T., Pavitt, G., Santos, D., Sidebotham, J., Hulton, C., Hinton, J., and Higgins, C. (1992) The chromatin-associated protein H-NS interacts with curved DNA to influence DNA topology and gene expression. *Cell*, **71**, 255.
187. Tupper, A. E., Owen-Hughes, T. A., Ussery, D. W., Santos, D. S., Ferguson, D. J. P., Sidebotham, J. M., Hinton, J. C. D., and Higgins, C. F. (1994) The chromatin-associated protein H-NS alters DNA topology *in vitro*. *EMBO J.*, **13**, 258.
188. Lucht, J. M., Dersch, P., Kempf, B., and Bremer, E. (1994) Interactions of the nucleoid-associated DNA-binding protein H-NS with the regulatory region of the osmotically controlled *proU* operon of *Escherichia coli*. *J. Biol. Chem.*, **269**, 6578.
189. Hulton, C., Seirafi, A., Hinton, J., Sidebotham, J., Waddell, L., Pavitt, G., Owen-Hughes, T., Spassky, A., Buc, H., and Higgins, C. (1990) Histone-like protein H1 (H-NS), DNA supercoiling, and gene expression in bacteria. *Cell*, **63**, 631.
190. Hinton, J. C., Santos, D. S., Seirafi, A., Hulton, C. S., Pavitt, G. D., and Higgins, C. F. (1992) Expression and mutational analysis of the nucleoid-associated protein H-NS of *Salmonella typhimurium*. *Mol. Microbiol.*, **6**, 2327.
191. Spassky, A., Rimsky, S., Garreau, H., and Buc, H. (1984) H1a, an *E. coli* DNA-binding protein which accumulates in stationary phase, strongly compacts DNA *in vitro*. *Nucleic Acids Res.*, **12**, 5321.
192. Spurio, R., Durrenberger, M., Falconi, M., LaTeana, A., Pon, C. L., and Gualerzi, C. O. (1992) Lethal overproduction of the *Escherichia coli* nucleoid protein H-NS: Ultramicroscopic and molecular autopsy. *Mol. Gen. Genet.*, **231**, 201.
193. Ussery, D. W., Hinton, J. C. D., Jordi, B. J. A. M., Granum, P. E., Seirafi, A., Stephen, R. J., Tupper, A. E., Berridge, G., Sidebotham, J. M., and Higgins, C. F. (1994) The chromatin-associated protein H-NS. *Biochimie*, **76**, 968.
194. Tobe, T., Yoshikawa, M., Mizuno, T., and Sasakawa, C. (1993) Transcriptional control of the invasion regulatory gene *virB* of *Shigella flexneri*: Activation by VirF and repression by H-NS. *J. Bacteriol.*, **175**, 6142.
195. Tippner, D., Afflerbach, H., Bradaczek, C., and Wagner, R. (1994) Evidence for a regulatory function of the histone-like *Escherichia coli* protein H-NS in ribosomal RNA synthesis. *Mol. Microbiol.*, **11**, 589.
196. Ueguchi, C., Kakeda, M., and Mizuno, T. (1993) Autoregulatory expression of the *Escherichia coli hns* gene encoding a nucleoid protein: H-NS functions as a repressor of its own transcription. *Mol. Gen. Genet.*, **236**, 171.
197. Falconi, M., Higgins, N., Spurio, R., Pon, C., and Gualerzi, C. O. (1993) Expression of the

gene encoding the major bacterial nucleoid protein H-NS is subject to transcriptional auto-repression. *Mol. Microbiol.*, **10**, 273.
198. Dersch, P., Schmidt, K., and Bremer, E. (1993) Synthesis of the *Escherichia coli* K-12 nucleoid-associated DNA-binding protein H-NS is subjected to growth-phase control and autoregulation. *Mol. Microbiol.*, **8**, 875.
199. Rimsky, S. and Spassky, A. (1990) Sequence determinants for H1 binding on *Escherichia coli lac* and *gal* promoters. *Biochemistry*, **29**, 3765.
200. Yamashino, T., Ueguchi, C., and Mizuno, T. (1995) Quantitative control of the stationary phase-specific sigma factor, σ^s, in *Escherichia coli*: involvement of the nucleoid protein H-NS. *EMBO J.*, **14**, 594.
201. Colonna, B., Casalino, M., Fradiani, P., Zagaglia, C., Naitza, S., Leoni, L., Prosseda, G., Coppo, A., Ghelardini, P., and Nicoletti, M. (1995) H-NS regulation of virulence gene expression in enteroinvasive *Escherichia coli* harboring the virulence plasmid integrated into the host chromosome. *J. Bacteriol.*, **177**, 4703.
202. O'Byrne, C. P. and Dorman, C. J. (1994) Transcription of the *Salmonella typhimurium spv* virulence locus is regulated negatively by the nucleoid-associated protein H-NS. *FEMS Microbiol. Letters*, **121**, 99.
203. Porter, M. E. and Dorman, C. J. (1994) A role for H-NS in the thermo-osmotic regulation of virulence gene expression in *Shigella flexneri*. *J. Bacteriol.*, **176**, 4187.
204. Levinthal, M., Lejeune, P., and Danchin, A. (1994) The H-NS protein modulates the activation of the *ilvIH* operon of *Escherichia coli* K12 by Lrp, the leucine regulatory protein. *Mol. Gen. Genet.*, **242**, 736.
205. Ito, K., Oshima, T., Mizuno, T., and Nakamura, Y. (1994) Regulation of lysyl-tRNA synthetase expression by histone-like protein H-NS of *Escherichia coli*. *J. Bacteriol.*, **176**, 7383.
206. Pullinger, G. D., Baird, G. D., Williamson, C. M., and Lax, A. J. (1989) Nucleotide sequence of a plasmid gene involved in the virulence of salmonellas. *Nucleic Acids Res.*, **17**, 7983.
207. Newman, E. B., D'ari, R., and Lin, R. T. (1992) The leucine-Lrp regulon in *E. coli*—a global response in search of a raison d'etre. *Cell*, **68**, 617.
208. Ernsting, B. R., Atkinson, M. R., Ninfa, A. J., and Matthews, R. G. (1992) Characterization of the regulon controlled by the leucine-responsive regulatory protein in *Escherichia coli*. *J. Bacteriol.*, **174**, 1109.
209. Hengge-Aronis, R. (1993) Survival of hunger and stress: The role of *rpoS* in early stationary phase gene regulation in *E. coli*. *Cell*, **72**, 165.
210. Siegele, D. A. and Kolter, R. (1992) Life after log. *J Bacteriol.*, **174**, 345.
211. Ueguchi, C. and Mizuno, T. (1993) The *Escherichia coli* nucleoid protein H-NS functions directly as a transcriptional repressor. *EMBO J.*, **12**, 1039.
212. Jovanovich, S. and Lebowitz, J. (1987) Estimation of the effect of supercoiling on *Salmonella typhimurium* promoters by using random operon fusions. *J. Bacteriol.*, **169**, 4431.
213. Boccard, F. and Prentki, P. (1993) Specific interaction of IHF with RIBs, a class of bacterial repetitive DNA elements located at the 3' end of transcription units. *EMBO J.*, **12**, 5019.
214. Bachellier, S., Perrin, D., Hoffnung, M., and Gilson, E. (1993) Bacterial interspersed mosaic elements (BIMEs) are present in the genome of *Klebsiella*. *Mol. Microbiol.*, **7**, 537.
215. Oppenheim, A. B., Rudd, K. E., Mendelson, I., and Teff, D. (1993) Integration host factor binds to a unique class of complex repetitive extragenic DNA sequences in *Escherichia coli*. *Mol. Microbiol.*, **10**, 113.

7 Integration of control devices: A global regulatory network in *Escherichia coli*

REGINE HENGGE-ARONIS

1. Introduction

Until the present day, the fundamental concepts developed by F. Jacob and J. Monod in the 1960s during their studies of the *lac* operon (1), have mainly shaped our view of the regulation of gene expression. What is the structure and function of regulatory proteins, how do they bind to DNA, and how do they influence the rate of transcription of this DNA? Especially with *Escherichia coli*, these questions can be at least partially answered for many regulatory proteins. Based on this knowledge, we can now start to determine the roles of these single regulatory systems within the physiology of a cell. As single-cell organisms, bacteria process a multitude of environmental signals and use this information to optimally balance their metabolism, growth rate, DNA replication, and cell division. This requires the coordination and interaction of a large number of single regulatory systems in regulatory networks.

Natural environments of bacteria are usually characterized by more or less severe nutrient limitation. Nutrients tend to become available only sporadically and therefore bacteria exhibit a feast-and-famine life-style. Bacteria rapidly readjust their rates of growth and cell division in response to changes in the nutrient concentration and composition. Under conditions of total starvation for an essential nutrient, they enter a physiological state commonly referred to as the stationary phase which is characterized by a maintenance metabolism and a pronounced stress resistance (2, 3). Some bacterial species also differentiate morphologically in response to starvation (see Chapter 11).

In addition to reacting appropriately to starvation stress, bacteria must also be able to cope with numerous other stresses, for instance with rapid changes in the temperature and osmolarity of the surrounding medium, in the availability of oxygen, or in the concentration of toxic oxygen radicals. Growing cells react to such adverse conditions by rapidly inducing systems that eliminate the stress factors and/or repair cellular damage, i.e. by stress-specific adaptive responses. Starved and therefore also energy-limited cells do not have the capacity for rapid *de novo* synthesis of stress-protective proteins. In compensation, they develop multiple stress resistance during entry into stationary phase, even in the absence of specific stress signals. Moreover, it has become clear that various stress conditions, such as high osmolarity, acid pH, and also non-optimally high or low temperature, elicit a response in growing cells, which in its molecular details and physiological consequences is very similar to the stationary phase response. Even though the stationary phase response has received most attention, it is now emerging as a part of broad general stress response that allows survival under the many different often highly adverse environmental conditions that prevail in the natural habitats of bacteria.

Although not identical, the general stress response and the more limited stress-specific adaptive responses are connected at various levels. This network has been studied most thoroughly in *Escherichia coli*. Therefore, this review will focus on the general stress response in *E. coli* which involves numerous signals and regulatory components, and thus by itself may already serve as a paradigm for a regulatory network. In addition, its emerging relationship to the carbon source specific starvation response as well as to the adaptive responses against high osmolarity and oxidative stress will be explored.

2. Characteristics of regulatory networks

A schematic representation of a simple signal transduction and regulatory pathway is shown in Fig. 1A. An extracellular or intracellular signal is recognized by a sensor which determines the activity or expression of a global regulatory protein that controls the expression of various target genes. If one or several of these target genes have themselves regulatory functions, the result is a regulatory cascade. Within a single bacterial cell a large number of such pathways are operating, and interactions between these pathways are conceivable at every level resulting in a complex regulatory network (Fig. 1B).

Regulatory networks have the following characteristics:

- Target genes are often controlled by more than one regulator. The result is a target gene-specific fine regulation in response to several signals.
- A master regulator may be essential for the expression of a large number of target genes, whereas various additional regulators may be responsible for fine-modulation. Sigma subunits of RNA polymerase are ideally suited to play the role of a master regulator.
- Signal input transmitted by several pathways may be integrated to control the activity or expression of a single key regulatory factor.
- Also the activity or expression of secondary regulators and therefore of subfamilies of target genes may be subject to multiple control.
- Regulators are often controlled at many levels. Whereas their activities may be affected, for instance, by covalent modification, by oligomerization, or by inter-

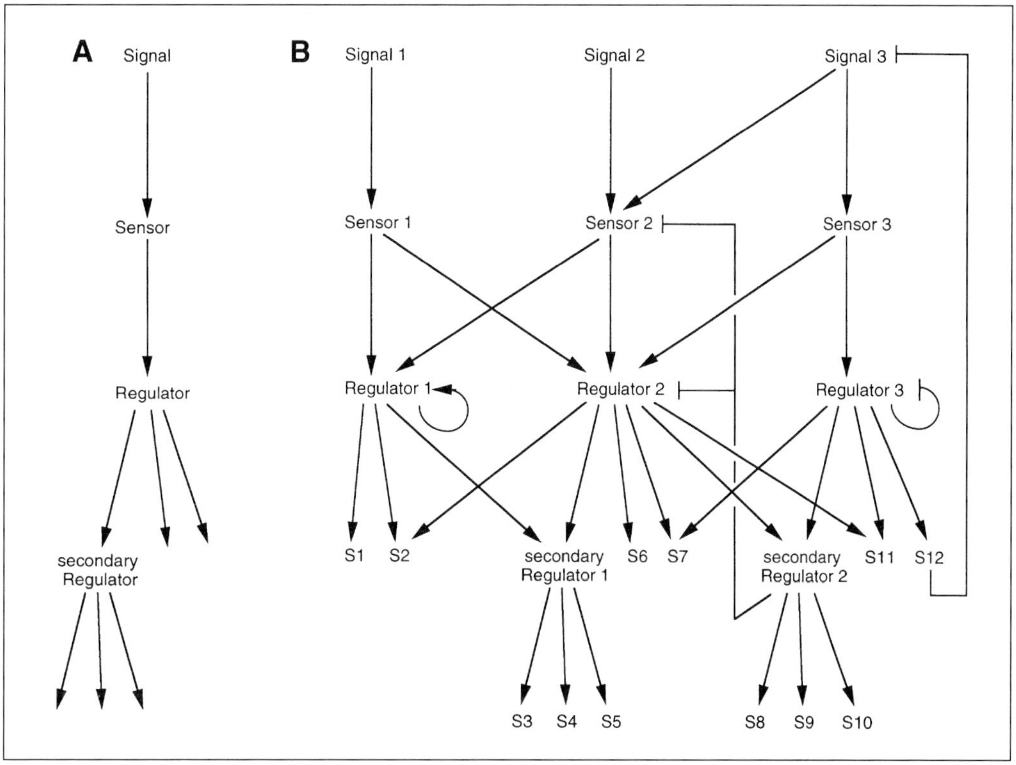

Fig. 1 Schematic representation of a regulatory network.

action with small signal molecules or other regulatory factors, their cellular levels may also be regulated by mechanisms that affect transcription, mRNA stability, translation, or protein turnover.

- Within regulatory networks, numerous positive and negative feedback mechanisms can operate that can potentiate responses or ensure that responses are not only switched on but also switched off appropriately.

3. The balance between the two primary sigma factors σ^{70} and σ^{S}

E. coli uses two primary sigma subunits of RNA polymerase, σ^{70} and σ^{S}, that control transcription from the large majority of promoters. σ^{70}, which is encoded by *rpoD*, is the major sigma subunit of RNA polymerase in growing *E. coli* cells (4) It is responsible for the recognition of the promoters of most 'house-keeping' genes, as well as of many genes that are induced in response to specific growth conditions (e.g. the presence of different carbon sources) or in response to several specific stresses. Induction of these genes is brought about by various types of activator proteins that interact with σ^{70}-containing RNA polymerase or by derepression. Under some stress conditions, however, alternative sigma factors are activated or induced that compete with σ^{70} for RNA polymerase core enzyme, and if present in sufficient amounts, at least transiently can massively redirect the transcriptional activity of a cell towards an alternative set of genes (see Chapter 3).

σ^S, which is encoded by the *rpoS* gene, plays an essential role in the expression of many stationary phase inducible and general stress-regulated genes (for recent reviews, see References 5–9). Sequence analysis indicated that σ^S is a very close relative of σ^{70}. Unlike alternative sigma factors, it contains an 'RpoD-box' (10) as well as two other motifs in the promoter-recognizing regions thought to be characteristic for vegetative or primary sigma factor (11). In *in-vitro* transcription assays with RNA polymerase reconstituted from purified components, many promoters are recognized by σ^S as well as by σ^{70} (12–14). The exact consensus sequence recognized by σ^S has been notoriously difficult to determine, but all studies indicate that at least in the –10 region it is probably nearly identical to that recognized by σ^{70} (TATACT and TATAAT, respectively (7)). Unlike σ^{70}-containing RNA polymerase, the σ^S-containing holoenzyme can activate transcription even in the absence of any functional –35 or upstream activating element (15); however, optimal function seems to involve contact in the –35 region which in many σ^S-dependent promoters contains CC instead of the TT(GACA) motif most common in σ^{70}-controlled promoters (16).

Whereas σ^{70} is indispensible, σ^S has been termed a non-essential primary sigma factor (11). While this is true for exponentially growing cells, i.e. under the usual laboratory conditions, σ^S is in fact essential for survival under conditions of prolonged nutrient starvation or of exposure to various life-threatening stress conditions (17, 18), i.e. conditions that prevail in the natural environments of bacteria.

In view of the similar promoter recognition specificities of σ^{70} and σ^S, the puzzling question arises of why *E. coli* needs two functionally so closely related sigma subunits of RNA polymerase. It is important to note that while *in vitro* σ^S and σ^{70} mostly recognize the same promoters, *in vivo* such a 'cross-recognition' does not appear to take place on a large scale. The expression of σ^S-dependent genes *in vivo* is often almost totally abolished in *rpoS* mutants, indicating that σ^{70} cannot just take over the function of σ^S in the expression of these genes, and the expression of a large majority of σ^{70}-controlled 'house-keeping' genes strongly decreases during entry into stationary phase and does not seem to be affected by the concomitant induction of σ^S (7). The reasons for this higher specificity for either σ^{70} or σ^S *in vivo* have not been understood in detail. It should be noted, however, that the standard conditions for *in vitro* transcription assays with purified reconstituted components are certainly not representative of the conditions inside the bacterial cell (see also below).

During entry into stationary phase, σ^{70} levels remain approximately constant (14, 19); on the other hand, the cellular σ^S content increases up to 20-fold (14, 19–21). A similar increase has also been observed in response to osmotic upshift in growing cells (19, 21, 22). Therefore the ratio between σ^S and σ^{70} is rather variable, and direct competition between the two sigmas may play a major role in directing RNA polymerase predominantly either to σ^{70}-dependent or σ^S-dependent promoters. In addition, intrinsic properties of RNA polymerase containing either σ^{70} and σ^S, such as a different potassium glutamate sensitivity (23), as well as for instance alterations in DNA superstructure in response to environmental changes (24) and histone-like proteins such as H-NS (25–27), Lrp (55, 57, 58) or IHF (55, 62) may influence the relative activities of the two holoenzyme forms of RNA polymerase at different promoters.

4. Control of the cellular level of σ^S

The cellular concentration of σ^S is a key variable controlled by numerous convergent signal transduction pathways. There are at least seven distinguishable environmental or physiological situations that result in an increase in the cellular σ^S content (Fig. 2). These are:

(1) a gradual reduction in growth rate (17, 21, 28, 29);

(2) late exponential phase (but not yet accompanied by a reduction in growth rate), provided the culture reaches a certain cell density (21);

(3) growth at low temperature (30);

(4) osmotic upshift or continuous growth under high osmolarity conditions (21, 22);

(5) starvation for either carbon, nitrogen or phosphorus sources (20, 21);

(6) heat shock (19, 31); and

(7) shift to acid pH (32).

The multitude of totally different signals that affect a single parameter, i.e. the cellular σ^S level, makes σ^S regulation a prime example of signal integration in *E. coli*. Moreover, the regulation of σ^S occurs at least at three different levels (Fig. 2). These include alterations in the rates of transcription (17, 28, 29) and translation (21, 22), as well as differentially controlled σ^S proteolysis (21, 22, 31, 33).

4.1 *rpoS* transcription

Three promoters contribute to *rpoS* transcription: the two closely spaced and non-growth-phase-controlled *nlpD* promoters (*nlpD* is located upstream of *rpoS* and encodes a lipoprotein (34, 35)) contribute to the basal rate of transcription in growing cells (35), whereas another promoter (*rpoS*p1) is located within the *nlpD* gene and its activity seems to be inversely growth rate regulated (17, 21, 33, 36). *rpoS*p1 exhibits –35 and –10 regions that conform relatively well to the consensus for σ^{70}-dependent promoters, and σ^S is not directly autoregulated (21, 29). So far, no regulatory factors have been found that specifically influence transcriptional initiation at *rpoS*p1. Moreover, all known stress conditions that result in increased σ^S levels, affect the post-transcriptional regulation of σ^S (see below).

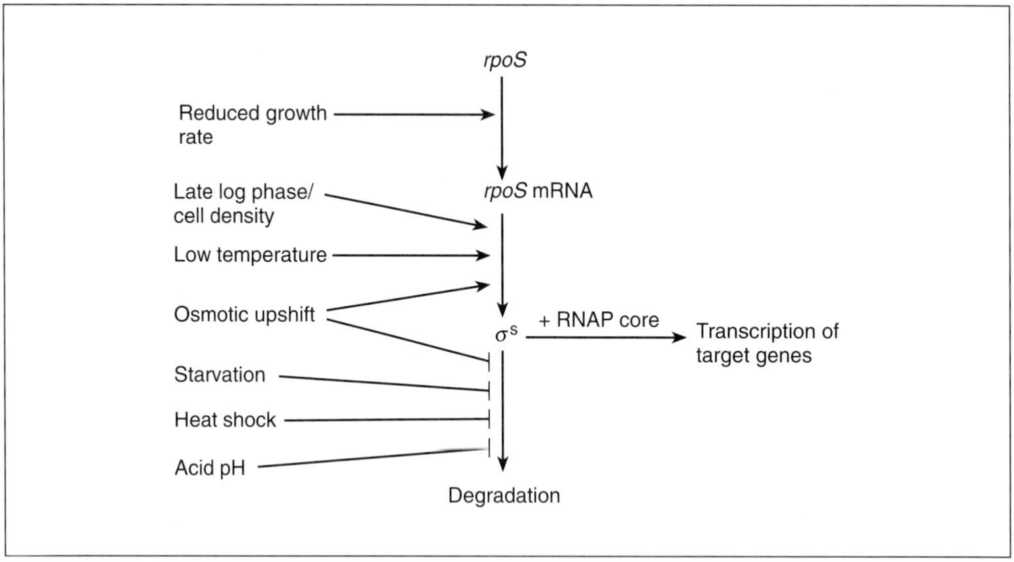

Fig. 2 Regulation of the cellular σ^S content (see text for details and references).

4.2 *rpoS* translation

Studies with transcriptional and translational *lacZ* fusions with identical points of insertion within the *rpoS* gene indicated that post-transcriptional control is more important for the regulation of the cellular σ^S content than the relatively moderate variations in the rate of *rpoS* transcription (21, 37). *rpoS* translation is stimulated during late exponential phase (provided the concentration of the carbon source is chosen such that the cells reach an optical density of at least 0.6) in response to an increase in the osmolarity of the medium (i.e. upon the addition of NaCl or sucrose) (21, 22) and during growth at low temperature (30).

rpoS mRNA has the potential to form complex and stable mRNA structures that include the translational initiation region (TIR). These secondary structures may interfere with translational initiation under non-inducing conditions and may be altered or opened up in response to inducing conditions (D. Traulsen and R. Hengge-Aronis, unpublished results). Recently, the RNA-binding Hfq (HF-I) protein was identified as essential for *rpoS* translation (38, 39). This protein has long been known as the host factor for phage Qβ RNA replication (40–42). Hfq protein seems to affect the stability of certain mRNA secondary structures in the *rpoS* TIR (S. Bouché and R. Hengge-Aronis, unpublished results). Its role in *rpoS* translation is the first cellular function identified for this protein in *E. coli*, but *rpoS* is obviously not its only target (43).

4.3 σ^S turnover

In growing cells, σ^S is an unstable protein with a half-life between 1.5 and 3 min, depending on the exact growth conditions (21, 22, 33). At the onset of glucose starvation, σ^S half-life changes to more than 10 min (21). Heat shock from 30 °C to 42 °C changes σ^S half-life to approximately 7 min (31). After osmotic upshift, a σ^S half-life of more than 45 min was observed (22). σ^S thus appears to be degraded by a protease whose activity or ability to recognize σ^S as a substrate is regulated in response to the nutrient supply and the osmolarity and temperature of the medium. The Clp protease (with the subunits ClpP and ClpX) has been implicated in the control of σ^S stability (44). Another component that is essential for σ^S turnover is the two-component response regulator RssB (45, 46, 143; see also below). Moreover, the DnaK chaperone seems to protect σ^S against degradation, and a role for DnaK in σ^S stabilization in response to starvation and heat shock has been proposed (31).

5. The stationary phase regulatory network

The regulatory patterns of σ^S-dependent genes are highly variable with respect to the time and extent of induction during transition into stationary phase. In addition, there are subsets of σ^S-controlled genes that respond differentially to various other stress conditions, such as high osmolarity, oxidative stress, or anaerobiosis. This differential regulation is accomplished by additional regulatory factors that together with σ^S as a master regulator constitute a complex regulatory network. For simplicity, the following will be restricted mostly to the stationary phase aspect of the general stress response mediated by σ^S.

As summarized in Fig. 3, different growth phase-regulated genes are controlled by different combinations of regulatory factors. The result is a characteristic fine-regulatory pattern for nearly every target gene (for examples, see Fig. 4). It should be

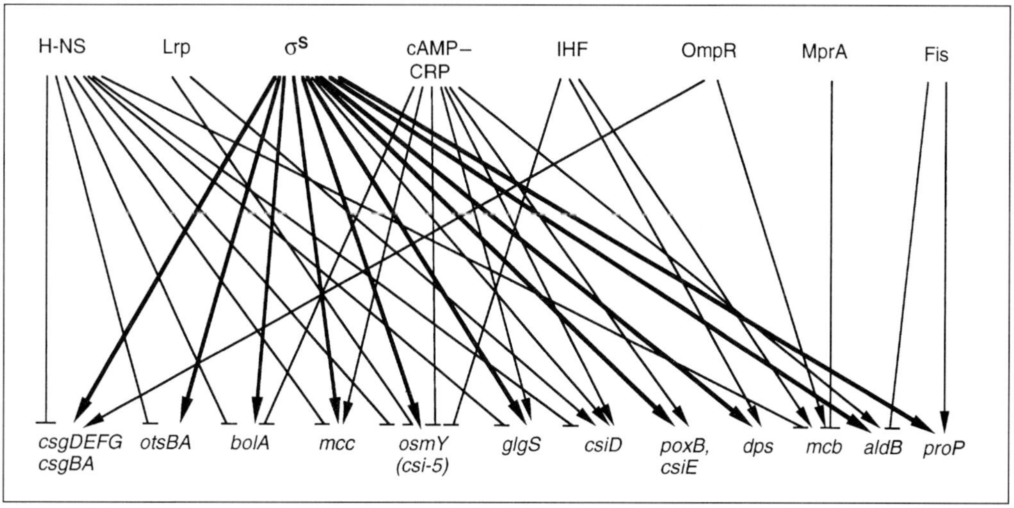

Fig. 3 The stationary phase regulatory network (for details and references, see text). This figure is a modified version of a previously published figure (8) and is shown here with permission.

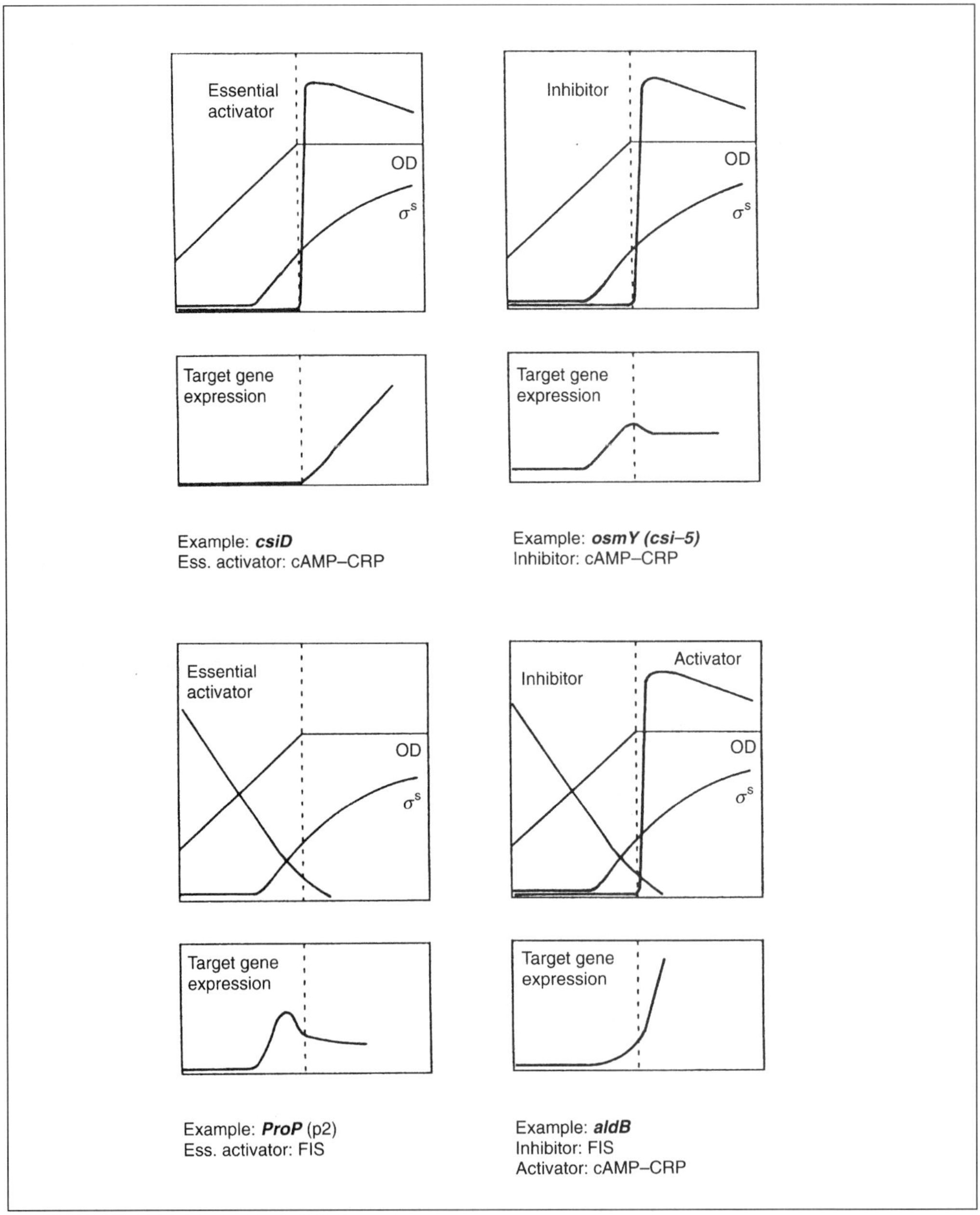

Fig. 4 The combined action of σ^S and various activators and inhibitors of gene expression produces strongly different growth phase related regulatory patterns. The upper panels show the typical idealized growth curve (OD) of a culture growing on glucose that becomes exhausted at the onset of starvation (indicated by a broken line), and the relative cellular levels of σ^S and of hypothetical activators or inhibitors that affect the expression of the target genes. The relative cellular levels of the gene products of the target genes are indicated in the lower panels. Examples of genes are given, whose expression in principle corresponds to the four different idealized patterns (*csiD* (50), *osmY* (55), *proPp2* (71), *aldB* (52)).

noted that a large variety of different regulatory patterns may be generated by combining a relatively small number of regulators, especially if these regulators can either activate or inhibit gene expression depending on the specific target genes. Many of these regulators are not classical activator or repressor proteins but more abundant histone-like proteins that bind to numerous sites on the chromosome and modulate gene expression as well as the structural organization of the chromosome.

5.1 Fine regulation by the cAMP–CRP complex, Lrp, IHF, and Fis

The involvement of the cAMP–CRP complex in the control of several σ^S-regulated genes provides a link between the general starvation or stationary phase response mediated by σ^S and the carbon starvation-specific response (see Reference 47 for a recent review on the role of cAMP and CRP protein). Positive control by cAMP–CRP was observed for several σ^S-dependent genes such as *glgS* (48), *csiD* and *csiE* (49, 50), *poxB* (51), *aldB* (52), and the microcin C7 operon (*mcc*) (53). CRP boxes are present upstream of the transcriptional start sites of *glgS* (48), *aldB* (52), and *csiD* at positions consistent with an activator function. *csiD* expression is almost totally dependent on high levels of cAMP–CRP, which restricts its expression to carbon starvation conditions (53a). In the cases of *csiE* (49) and *poxB* (51), cAMP–CRP may activate indirectly via an unknown intermediate regulator. On the other hand, several σ^S-dependent genes such as *bolA* (54), *osmY* (50, 55), *csiF* (50), and the *Salmonella spv* virulence genes (56) are under negative control by cAMP–CRP. This inhibition can be partially explained by the finding that the cAMP–CRP complex also plays a negative role in the expression of σ^S itself (17, 21).

Leucine-responsive regulatory protein (Lrp) plays a global regulatory role during a nutrient downshift from rich to minimal medium. In general, it represses genes required for the uptake and metabolism of nutrients present in rich medium and activates the expression of various biosynthetic genes (57). Moreover, a chromosome-organizing function has been suggested for Lrp (58). In addition, Lrp seems to have a function in the stationary phase regulatory network. In cells grown in rich medium, it down-modulates the expression of *osmY* during entry into stationary phase and thereby delays the time of *osmY* induction relative to σ^S induction. In minimal medium, it interferes with *osmY* expression throughout the growth cycle (55). In the regulation of *csiD*, Lrp acts as a positive modulator besides CRP (53a).

Integration host factor (IHF) is a sequence-specific histone-like protein involved in the formation of higher-order nucleoprotein complexes, in which the DNA-bending activity of IHF seems to be important for the proper arrangement of the other components (59–61). The cellular IHF content increases up to tenfold during entry into stationary phase (62). The transcription of *himA* and *hip* (*himD*), the structural genes for the two IHF subunits, is partly σ^S-dependent (63). IHF, together with σ^S, activates the stationary phase expression of *dps* (64, 65), which encodes an abundant

DNA-binding protein with regulatory and DNA-protective functions (66). IHF also acts as a positive regulatory factor in the stationary phase induction of the σ^S-independent *mcb* microcin B17 operon (53). In the control of *osmY*, IHF plays an inhibitory role with an effect similar to that of cAMP–CRP and Lrp (55).

Recently, a strong overlap between a set of genes that are negatively regulated by Fis protein (*frg* genes) and the σ^S regulon has been reported (67). Fis is a small DNA-binding protein involved in recombination, replication, and transcriptional control (68) that is approximately 500-fold induced when stationary phase cells are subcultured in rich medium. During the exponential growth phase of a batch culture, Fis levels rapidly decline and are extremely low in stationary phase (69, 70). In some cases, such as *aldB* (52), repression by Fis and activation by σ^S may act synergistically to produce strong induction during entry into stationary phase (see also Fig. 4). The *proP*p2 promoter, on the other hand, requires both Fis and σ^S for activity. Consequently, it is activated during late exponential phase, but increased expression does not last into stationary phase (71).

5.2 The role of the histone-like protein H-NS

H-NS (H1) is a small DNA-binding protein encoded by the *hns* gene that preferentially binds to AT-rich curved DNA regions (see Reference 72 for review; also Chapter 6). It can specifically influence the expression of various target genes (e.g. the *proU* operon (73–75)) by binding in the regulatory regions, but it also influences the degree of negative superhelicity of DNA, and when present in high concentrations, it can compact DNA and thereby may play a role in chromosome organization (75–77). Though already abundant in exponential phase, the cellular H-NS level may further increase during transition into stationary phase (78–80).

H-NS interferes with the expression of many σ^S-dependent genes during the exponential phase of growth (summarized in Fig. 3). These genes include *csgBA* and *hdeAB* (25), *mcc* (53), *osmY*, *otsBA*, *bolA* (26), *cbpA* (27), as well as 22 σ^S-controlled proteins discernible on two-dimensional O'Farrell gels (26). H-NS may directly bind to the regulatory regions of some of these genes, but others are indirectly controlled by H-NS, since H-NS also strongly inhibits the expression of the master regulator σ^S in growing cells. Surprisingly, H-NS acts at the post-transcriptional level of σ^S control, but details of this action are not known (26, 27).

Whereas the expression of the above-mentioned genes is strongly reduced in σ^S-deficient mutants, in some cases it has been observed that expression again improves when a secondary mutation in *hns* is introduced (25–27, 53). These results indicate that in *hns* mutants, σ^{70} can activate promoters that in a wild-type genetic background are almost exclusively recognized by σ^S, and that therefore either specific H-NS binding and/or DNA superhelicity and compactness as influenced by H-NS may be crucial factors that determine whether *in-vivo* transcription of a gene is initiated by RNA polymerase containing σ^S or also by σ^{70}-containing holoenzyme.

Taken together, H-NS has at least two functions in the stationary phase induction of gene expression: (i) by unknown post-transcriptional mechanisms, H-NS contributes to maintaining a low cellular level of the master regulator σ^S during exponential phase; and (ii) by direct binding and/or maintaining or producing a certain DNA superstructure, H-NS interferes with the recognition of σ^S-dependent promoters by σ^{70}-containing RNA polymerase, and thus couples activity of these promoters to the cellular level of σ^S. H-NS has thus emerged as a second key regulatory factor besides σ^S for stationary phase gene expression.

5.3 Molecular structure of stationary phase inducible promoter regions

So far, no obvious features exclusively characteristic for growth phase controlled promoters could be detected. Single transcriptional start sites are present upstream of *aldB* (52), *csiE* (49), *dps* (64, 65), *fic* (81), *katE* (82), *osmY* (55, 83), *poxB* (51), and *treA* (84). Other growth phase controlled genes have more than one promoter. Either all of them contribute to stationary phase induction as was observed for *glgS* (48), or only one of the promoters is stationary phase activated, and this promoter often is σ^S dependent as was found for *bolA* (54, 85, 86), *cfa* (87), *osmB* (88, 89), *proP* (71, 90), *pqi-5* (91), and *wrbA* (92). The other promoters may be subject to regulation by different signals and thus establish links to other regulatory circuits.

Sequence, deletion, and/or binding site analyses of the promoter regions of *dps* (64, 65), *osmY* (55, 83), *glgS* (48), and *aldB* (52) indicate that binding sites for the various regulatory factors involved in the control of the respective gene are present in the promoter region. There are, however, a few exceptions, such as an indirect cAMP–CRP dependence of *csiE* (49) and *poxB* (51), and indirect regulation by σ^S via regulatory cascades (see below). However, taken together, stationary phase inducible promoter regions often contain multiple binding sites for regulatory factors, indicating the formation of rather complex regulatory nucleoprotein structures.

5.4 Regulatory cascades within the σ^S regulon and connections to other regulatory circuits

Several σ^S-regulated genes have themselves regulatory function. These include *appY* (93–97), *bolA* (85, 98), *dps* (66), *csiD* (50), and *rob* (99–102) (a regulatory function for *dps* and *csiD* has been inferred from the altered patterns of protein synthesis in mutant strains, as shown by two-dimensional gel-electrophoresis). The expression of some secondary regulators and therefore their target genes is not only under σ^S control, but is also subject to an additional control in response to other stress conditions. The expression of certain subsets of σ^S-dependent genes is thus under oxygen control via AppY (103), under oxidative stress control via Dps (64, 65), and under carbon starvation-specific control via CsiD (50) (summarized in Fig. 5).

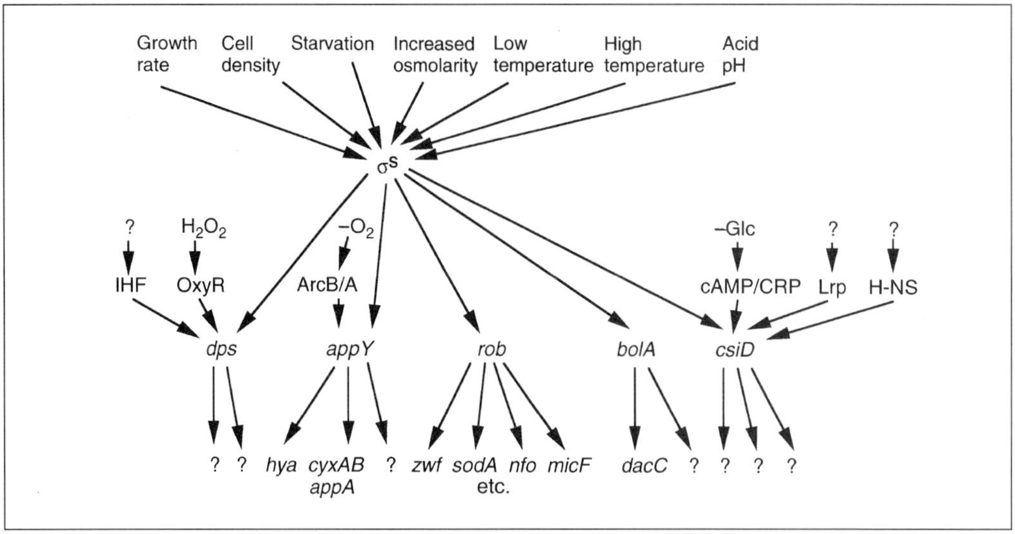

Fig. 5 Cascade regulation within the σ^S regulon. Secondary regulators often are the points of integration of additional environmental signals that differentially influence the expression of subsets of σ^S-controlled genes (see text for details and references). This figure is a modified version of a previously published figure (8) and is shown here with permission.

6. Connections between the responses to stationary phase, high osmolarity, and oxidative stress

In the following, the connections between the stationary phase response and the regulatory mechanisms that respond to osmotic and oxidative stress will be pointed out. However, no comprehensive survey of these regulatory mechanisms is intended here (these subjects are covered by various recent reviews (9, 104–110)).

The stationary phase response and the hyperosmotic response in growing cells are mainly connected by σ^S being a global regulator for both (5, 7, 9). The cellular σ^S content is as strongly increased in response to osmotic upshift as during transition into stationary phase. This osmotic induction is due to increased *rpoS* translation as well as increased σ^S stability (22). As a consequence, most σ^S-controlled genes are also hyperosmotically induced (summarized in Reference 9), although exceptions have been reported such as the *csg* (111), *csiE* (49), and *aldB* (52) genes. The expression of these latter genes may require additional factors besides σ^S that are not present in cells growing at increased osmolarity.

The 'classical' examples of osmoregulated systems include the *proU* operon and the OmpF/OmpC porin system which is under the control of the EnvZ/OmpR two-component system (for recent reviews, see References 112 and 113, respectively). While the osmotic regulation of these systems is certainly not dependent on σ^S, σ^S still appears to have a fine-modulatory role. *proU* has a minor promoter (p1) which is σ^S-dependent and hyperosmotically inducible (114, 115). OmpF levels decrease not only in response to high osmolarity but also during entry into stationary phase. σ^S is essential for this decrease that occurs at the level of *ompF* transcription by an unknown mechanism (18, 113, 116).

The histone-like protein H-NS seems to play a direct role in osmotic induction of the *proU* operon (73–75, 112). Since many proteins that are overproduced in *hns* mutants are osmotically regulated, a global osmoregulatory role has been ascribed to H-NS (72). However, the recent finding that most of these proteins are σ^S-dependent for expression, and that *hns* mutants show strongly increased levels of σ^S, indicates that often the osmoregulatory role of H-NS may be indirect and is actually due to its involvement in the control of σ^S (9, 26). In *hns* mutants, σ^S levels appear to be constitutively high, and no more osmotic regulation is observed (26), suggesting that H-NS may play an important role in the signal transducing and/or regulatory mechanisms that are responsible for hyperosmotic induction of σ^S. Strikingly, this regulation of σ^S is post-transcriptional (21, 26), whereas H-NS so far has been characterized as a DNA-binding histone-like protein.

A variety of connections also exist between the σ^S-dependent general stress response and the various adaptive responses to oxidative stress. *katG*, which encodes catalase HPI, *dps*, the structural gene for a DNA-binding protein (see above), and *gor*, which encodes glutathione reductase, have been found to be under the control of σ^S as well as of OxyR (64, 117, 118). OxyR is a transcriptional activator that acts together with σ^{70}-containing RNA polymerase. In OxyR, which is directly activated by hydrogen peroxide, sensory and regulatory functions are combined in a single protein (119, 120). *katG*, *dps*, and *gor* are thus under oxidative stress control, whereas

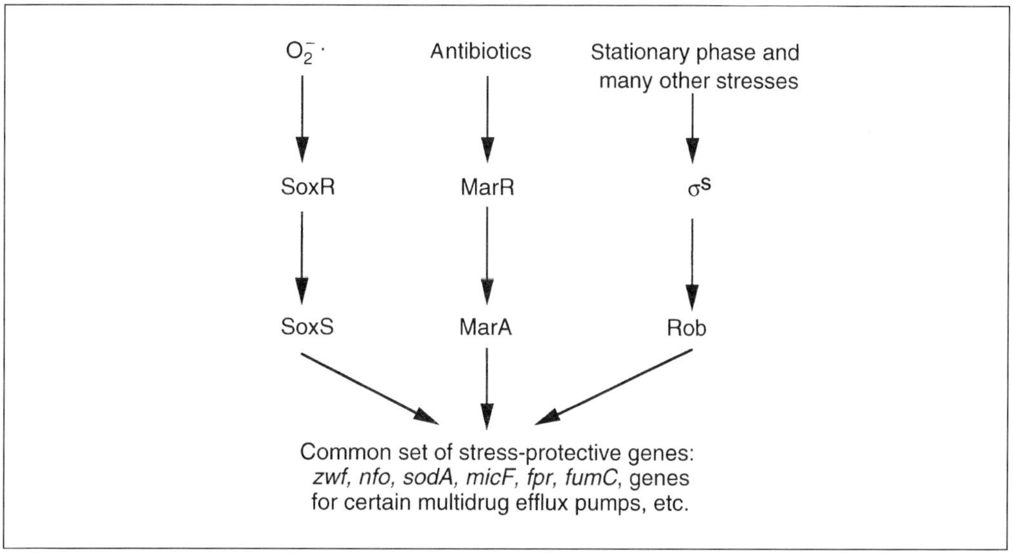

Fig. 6 Convergent regulation by the highly homologous regulators SoxS, MarA, and Rob. Different stress signals can activate a common set of stress-protective genes by inducing either SoxS, MarA, or Rob which recognize a single activating motif present in the promoter region of many target genes (for details and references, see text).

the basal level of expression of *katG* (117) and stationary phase induction of *dps* (64, 65) and *gor* (118) are σ^S-dependent. In addition, *oxyS* RNA is induced under conditions of oxidative stress via OxyR and interferes with *rpos* translation. H_2O_2 stress thus interferes with the induction of the general stress response (157).

SoxS, MarA, and Rob (Fig. 6) are highly homologous proteins that activate many stress-inducible genes by binding to a common activating motif present upstream of these genes (99, 100, 102, 121). While SoxS and MarA are induced by oxidative stress (superoxide anion) (122) and certain antibiotics (123), respectively, Rob is stationary phase induced and under σ^S control (101). Among the many proteins under the common control of these three factors are glucose-6-phosphate dehydrogenase (*zwf*), endonuclease IV (*nfo*), a superoxide dismutase (*sodA*) (99, 100), as well as certain multidrug efflux pumps (124). Also *micF* RNA, which interferes with OmpF expression by an antisense mechanism, is under the control of SoxS/MarA/Rob (100, 125), which explains downregulation of OmpF under conditions of oxidative stress (125) and possibly in stationary phase (113, 116). Signal integration in this system is thus not at the level of a single regulator being activated or induced by several signals, but different signals induce three highly related but distinct regulators that recognize a common motif present in the promoter regions of the target genes (Fig. 6).

7. Signal transduction

7.1 Small signal molecules

The concept of variations in the intracellular concentrations of small signal molecules, sometimes also termed second messengers or alarmones, that reflect changes outside the cell, is common in cell biology. In addition, the intracellular concentrations of 'ordinary' metabolic intermediates, that may vary according to the physiological and especially nutritional status of a cell, can also affect the regulation of gene expression. In *E. coli*, a variety of molecules influences the cellular level of σ^S and thereby affects the general stress and stationary phase responses.

A classical example is 3′,5′-cyclic-AMP (cAMP; for a recent review, see Reference 47). Its cellular level is moderately increased in response to carbon source downshift, whereas sudden total glucose starvation results in a rapid and strong increase (126–128). Changes in the cellular cAMP content are brought about by the control of adenylate cyclase activity by $EIIa^{Glc}$ (129), a component of the phosphoenolpyruvate: sugar phosphotransferase system, as well as by inhibition of cAMP phosphodiesterase (130) and cAMP excretion (126, 131, 132). Changes in the cellular cAMP content and therefore also in the concentration of the cAMP–CRP complex not only affect the expression of numerous genes under catabolite repression but also that of stationary phase inducible σ^S-dependent or σ^S-independent genes (48, 49, 54, 55, 133–136). In addition, cAMP–CRP partially interferes with σ^S expression during exponential growth by an unknown mechanism (17, 21).

The cellular content of acetyl phosphate varies with the nature of the carbon source provided or the growth phase of a culture (137, 138). Due to its ability to serve as a phospho-donor for the phosphorylation of two-component response regulators (139), changes in the cellular level of acetyl phosphate may in principle fine-modulate the expression of many systems (137, 140–142). A *pta ack* mutant that does not synthesize acetyl phosphate, exhibits increased stability and therefore higher levels of σ^S. This effect is dependent on the presence of the response regulator RssB which is essential for σ^S proteolysis (see below). Acetyl phosphate is not required for the σ^S system to respond to external stress signals, but seems to modulate it according to the metabolic, i.e. nutritional, status of the cell (143).

Guanosine 3′,5′-bispyrophosphate (ppGpp) is another small molecule with global effects on gene expression. Due to a differential activity of two ppGpp synthases (encoded by the *relA* and *spoT* genes), the cellular ppGpp content increases in response to starvation for amino acids (stringent response) as well as for sources of carbon (energy), nitrogen, and phosphorus (144–146). ppGpp interferes with transcriptional initiation at stringently controlled promoters (e.g. the *rrnB* promoter (147)), but also seems to reduce the rate of transcriptional elongation and thereby maintains coupling of transcription and translation under starvation conditions (148, 149). ppGpp is required for the expression of wild-type levels of σ^S and was shown to play a positive role in transcriptional elongation of *rpoS* (20, 36). In ppGpp-free *relA spoT* mutants the global pattern of protein expression during entry into stationary phase is strongly altered (116).

Evidence has been presented that a homoserine lactone may be involved in the control of σ^S. *thrA* mutants that are devoid of homoserine and homoserine phosphate exhibit strongly reduced σ^S levels, and this phenotype can be suppressed by adding homoserine lactone to the growth medium (150). In various other species, homoserine lactones are produced as signal molecules that accumulate in the medium and are used by the cells to monitor population density (151).

UDP-glucose is yet another small molecule that influences the cellular σ^S content. UDP-glucose pyrophosphorylase-deficient *galU* mutants show increased σ^S levels during exponential phase (152). UDP-glucose has a fine-modulatory role in the control of *rpoS* translation but does not seem to be crucial for osmotic or growth phase dependent signal transduction in the control of σ^S (A. Muffler and R. Hengge-Aronis, unpublished results).

As a rapid response to osmotic upshift and reduced turgor pressure, the uptake of K^+ is stimulated, and therefore an increased cellular K^+ concentration may serve as a primary intracellular high osmolarity signal (104, 105). As diverse activities as a relief of H-NS binding to a regulatory region in the osmoregulated *proU* operon (74) and the hyperosmotic activation of the *otsA*-encoded trehalose-6-phosphate synthase (153) have been ascribed to high levels of potassium ions. K^+ glutamate also affects sigma factor selectivity at some stress-regulated promoters by activating σ^S-mediated transcriptional initiation, while interfering with initiation by σ^{70}-containing RNA polymerase (23). In addition, K^+ might be involved in the hyperosmotic induction of σ^S, because glycine betaine, a potent osmoprotectant that reduces K^+ uptake in osmotically shifted cells, also alleviates osmotic induction of σ^S (22).

7.2 Two-component systems as signal transducers

Two-component systems are signal-transducing devices consisting of at least one sensory protein, that in response to some mostly environmental stimuli autophosphorylates and serves as a phospho-donor for a response regulator that displays an activity (usually, but not always, as a transcription factor) that is modulated by phosphorylation of its N-terminal receiver domain. In *E. coli,* the NtrB/NtrC and PhoR/PhoB systems are involved in the specific responses to nitrogen and phosphate limitation, respectively. The activity of the CreC(formerly PhoM)/CreB system seems to respond to variations in the carbon and energy source. The EnvZ/OmpR system controls the osmotic regulation of outer membrane porin composition in *E. coli.* All these systems have been recently reviewed (113, 154–156; and see also Chapter 8), and will not be further discussed here.

A two-component response regulator, RssB (alternatively termed SprE and MviA in *S. typhimurium*), has been found to be involved in the post-transcriptional regulation of σ^S. *rssB* null mutants show increased levels of σ^S during exponential phase growth, and are attenuated for osmotic and growth phase dependent σ^S induction. In contrast to wild-type strains, *rssB* mutants do not exhibit σ^S turnover (32, 45, 46). At present it is unclear, whether RssB itself has proteolytic activity or controls the activity of a protease (ClpXP (44)) or the recognition of σ^S as a substrate for the protease. Consistent with its unprecedented role in the control of proteolysis, the C-terminal output domain of RssB does not show significant similarity to any other protein of known function. So far, no corresponding sensor histidine kinase has been found for RssB, but phosphorylation of RssB at asp58 is obviously important, since a replacement of asp58 by several other amino acids has pronounced effects on the cellular σ^S level (D. Fischer and R. Hengge-Aronis, unpublished results). Since various stress signals trigger a stop of σ^S turnover, and the *rssB* mutant phenotype gives a loss of turnover, the regulated event in RssB activity control may be dephosphorylation rather than phosphorylation.

8. Perspectives

For *E. coli,* we possess detailed knowledge of the regulatory mechanisms involved in the control of many environmentally regulated genes. The corresponding signal transduction and integration pathways, however, have often been more refractory to molecular analysis (but might be most rewarding for future analysis). Nevertheless, even the combination of the presently available limited data has allowed us to recognize the first outlines of complex global regulatory networks. Knowing the position of a signal transducing or regulatory factor within this complex network is crucial for understanding its physiological role. The elucidation of regulatory networks thus brings back physiology to the analysis of living cells, which has long been dominated by the molecular analysis of single components. The prospect that at some time in the future we may, at least in principle, understand how a whole cell simultaneously senses a multitude of environmental signals and processes this information to optimally balance its rates of growth and cell division or to survive under the most adverse conditions is now becoming realistic.

References

1. Jacob, F. and Monod, J. (1961) Genetic regulatory mechanisms in the synthesis of proteins. *J. Mol. Biol.*, **3**, 318–356.
2. Kolter, R., Siegele, D. A., and Tormo, A. (1993) The stationary phase of the bacterial life cycle. *Ann. Rev. Microbiol.*, **47**, 855–874.
3. Siegele, D. A. and Kolter, R. (1992) Life after log. *J. Bacteriol.*, **174**, 345–348.
4. Gross, C. A., Lonetto, M., and Losick, R. (1992) Bacterial sigma factors. In *Transcriptional regulation* (ed. S. L. McKnight and K. R. Yamamoto), pp. 129–176. Cold Spring Harbor Laboratory Press, Cold Spring Harbor, NY.
5. Hengge-Aronis, R. (1993) Survival of hunger and stress: the role of *rpoS* in stationary phase gene regulation in *Escherichia coli*. *Cell*, **72**, 165–168.
6. Hengge-Aronis, R. (1993) The role of *rpoS* in early stationary phase gene regulation in *Escherichia coli* K12. In *Starvation in bacteria* (ed. S. Kjelleberg), pp. 171–200. Plenum Press, New York.
7. Loewen, P. C. and Hengge-Aronis, R. (1994) The role of the sigma factor σ^S (KatF) in bacterial global regulation. *Annu. Rev. Microbiol.*, **48**, 53–80.
8. Hengge-Aronis, R. (1996) Regulation of gene expression during entry into stationary phase. In Escherichia coli *and* Salmonella typhimurium*: Cellular and molecular biology* (ed. F. C. Neidhardt), pp. 1497–1512. American Society for Microbiology, Washington DC.
9. Hengge-Aronis, R. (1996) Back to log phase: σ^S as a global regulator in the osmotic control of gene expression in *Escherichia coli*. *Mol. Microbiol.*, **21**, 887–893.
10. Tanaka, K., Shiina, T., and Takahashi, H. (1988) Multiple principal sigma factor homologs in eubacteria: identification of the '*rpoD* Box'. *Science*, **242**, 1040–1042.
11. Lonetto, M., Gribskov, M., and Gross, C. A. (1992) The σ^{70} family: sequence conservation and evolutionary relationships. *J. Bacteriol.*, **174**, 3843–3849.
12. Kolb, A., Kotlarz, D., Kusano, S., and Ishihama, A. (1995) Selectivity of the *E. coli* RNA polymerase $E\sigma^{38}$ for overlapping promoters and ability to support CRP activation. *Nucl. Acids Res.*, **23**, 819–826.
13. Nguyen, L. H., Jensen, D. B., Thompson, N. E., Gentry, D. R., and Burgess, R. R. (1993) *In vitro* functional characterization of overproduced *Escherichia coli katF/rpoS* gene product. *Biochemistry*, **32**, 11112–11117.
14. Tanaka, K., Takayanagi, Y., Fujita, N., Ishihama, A., and Takahashi, H. (1993) Heterogeneity of the principal sigma factor in *Escherichia coli*: the *rpoS* gene product, σ^{38}, is a second principal sigma factor of RNA polymerase in stationary phase *Escherichia coli*. *Proc. Natl. Acad. Sci. USA*, **90**, 3511–3515.
15. Tanaka, K., Kusano, S., Fujita, N., Ishihama, A., and Takahashi, H. (1995) Promoter determinants for *Escherichia coli* RNA polymerase holoenzyme containing σ^{38} (the *rpoS* gene product). *Nucl. Acids Res.*, **23**, 827–834.
16. Wise, A., Brems, R., Ramakrishnan, V., and Villarejo, M. (1996) Sequences in the –35 region of *Escherichia coli rpoS*-dependent genes promote transcription by $E\sigma^S$. *J. Bacteriol.*, **178**, 2785–2793.
17. Lange, R. and Hengge-Aronis, R. (1991) Identification of a central regulator of stationary-phase gene expression in *Escherichia coli*. *Mol. Microbiol.*, **5**, 49–59.

18. McCann, M. P., Kidwell, J. P., and Matin, A. (1991) The putative σ factor KatF has a central role in development of starvation-mediated general resistance in *Escherichia coli*. *J. Bacteriol.*, **173**, 4188–4194.
19. Jishage, M., Iwata, A., Ueda, S., and Ishihama, A. (1996) Regulation of RNA polymerase sigma subunit synthesis in *Escherichia coli*: Intracellular levels of four species of sigma subunit under various growth conditions. *J. Bacteriol.*, **178**, 5447–5451.
20. Gentry, D. R., Hernandez, V. J., Nguyen, L. H., Jensen, D. B., and Cashel, M. (1993) Synthesis of the stationary-phase sigma factor σ^S is positively regulated by ppGpp. *J. Bacteriol.*, **175**, 7982–7989.
21. Lange, R. and Hengge-Aronis, R. (1994) The cellular concentration of the σ^S subunit of RNA-polymerase in *Escherichia coli* is controlled at the levels of transcription, translation and protein stability. *Genes Dev.*, **8**, 1600–1612.
22. Muffler, A., Traulsen, D. D., Lange, R., and Hengge-Aronis, R. (1996) Posttranscriptional osmotic regulation of the σ^S subunit of RNA polymerase in *Escherichia coli*. *J. Bacteriol.*, **178**, 1607–1613.
23. Ding, Q., Kusano, S., Villarejo, M., and Ishihama, A. (1995) Promoter selectivity control of *Escherichia coli* RNA polymerase by ionic strength: differential recognition of osmoregulated promoters by $E\sigma^D$ and $E\sigma^S$ holoenzymes. *Mol. Microbiol.*, **16**, 649–656.
24. Kusano, S., Ding, Q. Q., Fujita, N., and Ishihama, A. (1996) Promoter selectivity of *Escherichia coli* RNA polymerase $E\sigma^{70}$ and $E\sigma^{38}$ holoenzymes—Effect of DNA supercoiling. *J. Biol. Chem.*, **271**, 1998–2004.
25. Arnqvist, A., Olsén, A., and Normark, S. (1994) σ^S-dependent growth-phase induction of the *csgBA* promoter in *Escherichia coli* can be achieved *in vivo* by σ^{70} in the absence of the nucleoid-associated protein H-NS. *Mol. Microbiol.*, **13**, 1021–1032.
26. Barth, M., Marschall, C., Muffler, A., Fischer, D., and Hengge-Aronis, R. (1995) A role for the histone-like protein H-NS in growth phase-dependent and osmotic regulation of σ^S and many σ^S-dependent genes in *Escherichia coli*. *J. Bacteriol.*, **177**, 3455–3464.
27. Yamashino, T., Ueguchi, C., and Mizuno, T. (1995) Quantitative control of the stationary phase-specific sigma factor, σ^S, in *Escherichia coli*: involvement of the nucleoid protein H-NS. *EMBO J.*, **14**, 594–602.
28. Mulvey, M. R., Switala, J., Borys, A., and Loewen, P. C. (1990) Regulation of transcription of *katE* and *katF* in *Escherichia coli*. *J. Bacteriol.*, **172**, 6713–6720.
29. Schellhorn, H. E. and Stones, V. L. (1992) Regulation of *katF* and *katE* in *Escherichia coli* K-12 by weak acids. *J. Bacteriol.*, **174**, 4769–4776.
30. Sledjeski, D. D., Gupta, A., and Gottesman, S. (1996) The small RNA, DsrA, is essential for the low temperature expression of RpoS during exponential growth in *E. coli*. *EMBO J.*, **15**, 3993–4000.
31. Muffler, A., Barth, M., Marschall, C., and Hengge-Aronis, R. (1997) Heat shock regulation of σ^S turnover: a role for DnaK and the relationship between the stress responses mediated by σ^S and σ^{32} in *Escherichia coli*. *J. Bacteriol.*, **179**, 445–452.
32. Bearson, S. M. D., Benjamin Jr, W. H., Swords, W. E., and Foster, J. W. (1996) Acid shock induction of RpoS is mediated by the mouse virulence gene *mviA* of *Salmonella typhimurium*. *J. Bacteriol.*, **178**, 2572–2579.
33. Takayanagi, Y., Tanaka, K., and Takahashi, H. (1994) Structure of the 5′ upstream region and the regulation of the *rpoS* gene of *Escherichia coli*. *Mol. Gen. Genet.*, **243**, 525–531.
34. Ichikawa, J. K., Li, C., Fu, J., and Clarke, S. (1994) A gene at 59 minutes on the *Escherichia coli* chromosome encodes a lipoprotein with unusual amino acid repeat sequences. *J. Bacteriol.*, **176**, 1630–1638.

35. Lange, R. and Hengge-Aronis, R. (1994) The *nlpD* gene is located in an operon with *rpoS* on the *Escherichia coli* chromosome and encodes a novel lipoprotein with a potential function in cell wall formation. *Mol. Microbiol.*, **13**, 733–743.
36. Lange, R., Fischer, D., and Hengge-Aronis, R. (1995) Identification of transcriptional start sites and the role of ppGpp in the expression of *rpoS*, the structural gene for the σ^S subunit of RNA-polymerase in *Escherichia coli*. *J. Bacteriol.*, **177**, 4676–4680.
37. McCann, M. P., Fraley, C. D., and Matin, A. (1993) The putative σ factor KatF is regulated posttranscriptionally during carbon starvation. *J. Bacteriol.*, **175**, 2143–2149.
38. Muffler, A., Fischer, D., and Hengge-Aronis, R. (1996) The RNA-binding protein HF-I, known as a host factor for phage Qβ RNA replication, is essential for the translational regulation of *rpoS* in *Escherichia coli*. *Genes Dev.*, **10**, 1143–1151.
39. Brown, L. and Elliott, T. (1996) Efficient translation of the RpoS sigma factor in *Salmonella typhimurium* requires host factor I, an RNA-binding protein encoded by the hfq gene. *J. Bacteriol.*, **178**, 3763–3770.
40. Franze de Fernandez, M. T., Eoyang, L., and August, J. T. (1968) Factor fraction required for the synthesis of bacteriophage Qβ RNA. *Nature (London)*, **219**, 588–590.
41. Franze de Fernandez, M. T., Hayward, W. S., and August, J. T. (1972) Bacterial proteins required for replication of phage Qβ ribonucleic acid. Purification and properties of host factor I, a ribonucleic acid-binding protein. *J. Biol. Chem.*, **247**, 824–831.
42. Barrera, I., Schuppli, D., Sogo, J. M., and Weber, H. (1993) Different mechanisms of recognition of bacteriophage Qß plus and minus strand RNAs by Qß replicase. *J. Mol. Biol.*, **232**, 512–521.
43. Muffler, A., Traulsen, D. D., Fischer, D., Lange, R., and Hengge-Aronis, R. (1997) The RNA-binding protein HF-I plays a global regulatory role which is largely, but not exclusively, due to its role in expression of the σ^S subunit of RNA polymerase in *Escherichia coli*. *J. Bacteriol.*, **179**, 297–300.
44. Schweder, T., Lee, K.-H., Lomovskaya, O., and Matin, A. (1996) Regulation of *Escherichia coli* starvation sigma factor (σ^S) by ClpXP protease. *J. Bacteriol.*, **178**, 470–476.
45. Muffler, A., Fischer, D., Altuvia, S., Storz, G., and Hengge-Aronis, R. (1996) The response regulator RssB controls stability of the σ^S subunit of RNA polymerase in *Escherichia coli*. *EMBO J.*, **15**, 1333–1339.
46. Pratt, L. A. and Silhavy, T. J. (1996) The response regulator, SprE, controls the stability of RpoS. *Proc. Natl. Acad. Sci. USA*, **93**, 2488–2492.
47. Botsford, J. L. and Harman, J. G. (1992) Cyclic AMP in prokaryotes. *Microbiol. Rev.*, **56**, 100–122.
48. Hengge-Aronis, R. and Fischer, D. (1992) Identification and molecular analysis of *glgS*, a novel growth phase-regulated and *rpoS*-dependent gene involved in glycogen synthesis in *Escherichia coli*. *Mol. Microbiol.*, **6**, 1877–1886.
49. Marschall, C. and Hengge-Aronis, R. (1995) Regulatory characteristics and promoter analysis of *csiE*, a stationary phase-inducible σ^S-dependent gene under positive control of cAMP–CRP in *Escherichia coli*. *Mol. Microbiol.*, **18**, 175–184.
50. Weichart, D., Lange, R., Henneberg, N., and Hengge-Aronis, R. (1993) Identification and characterization of stationary phase-inducible genes in *Escherichia coli*. *Mol. Microbiol.*, **10**, 407–420.
51. Chang, Y.-Y., Wang, A.-Y., and Cronan Jr, J. E. (1994) Expression of *Escherichia coli* pyruvate oxidase (PoxB) depends on the sigma factor encoded by the *rpoS* (*katF*) gene. *Mol. Microbiol.*, **11**, 1019–1028.
52. Xu, J. M. and Johnson, R. C. (1995) *aldB*, an RpoS-dependent gene in *Escherichia coli*

encoding an aldehyde dehydrogenase that is repressed by Fis and activated by CRP. *J. Bacteriol.*, **177**, 3166–3175.

53. Moreno, F., San Millán, J. L., del Castillo, I., Gómez, J. M., Rodríguez-Sáinz, M. C., González-Pastor, J. E., and Díaz-Guerra, L. (1992) *Escherichia coli* genes regulating the production of microcins MccB17 and MccC7. In *Bacteriocins, microcins and antibiotics* (ed. R. James, C. Lazdunski, and F. Pattus), pp. 3–13. Springer-Verlag, Berlin Heidelberg.

53a. Marschall, C., Labrousse, V., Kreimer, M., Weichart, D., Kolb, A., and Hengge-Aronis, R. (1998) Molecular analysis of the regulation of *csiD*, a carbon-starvation-inducible gene in *Escherichia coli* that is exclusively dependent on σ^s and requires activation by cAMP-CRP. *J. Mol. Biol.*, **276**, 339–353.

54. Lange, R. and Hengge-Aronis, R. (1991) Growth phase-regulated expression of *bolA* and morphology of stationary phase *Escherichia coli* cells is controlled by the novel sigma factor σ^S (*rpoS*). *J. Bacteriol.*, **173**, 4474–4481.

55. Lange, R., Barth, M., and Hengge-Aronis, R. (1993) Complex transcriptional control of the σ^S-dependent stationary phase-induced *osmY* (*csi-5*) gene suggests novel roles for Lrp, cyclic AMP (cAMP) receptor protein–cAMP complex and integration host factor in the stationary phase response of *Escherichia coli*. *J. Bacteriol.*, **175**, 7910–7917.

56. O'Byrne, C. P. and Dorman, C. J. (1994) The *spv* virulence operon of *Salmonella typhimurium* LT2 is regulated negatively by the cyclic AMP (cAMP)–cAMP receptor protein system. *J. Bacteriol.*, **176**, 905–912.

57. Calvo, J. M. and Matthews, R. G. (1994) The leucine-responsive regulatory protein, a global regulator of metabolism in *Escherichia coli*. *Microbiol. Rev.*, **58**, 466–490.

58. D'Ari, R., Lin, R. T., and Newman, E. B. (1993) The leucine-responsive regulatory protein: more than a regulator? *Trends Biochem. Sci.*, **18**, 260–263.

59. Drlica, K. and Rouviere-Yaniv, J. (1987) Histone-like proteins of bacteria. *Microbiol. Rev.*, **51**, 301–319.

60. Friedman, D. I. (1988) Integration host factor: a protein for all reasons. *Cell*, **55**, 545–554.

61. Nash, H. A. (1996) The HU and IHF proteins: Accessory factors for complex protein–DNA assemblies. In *Regulation of gene expression in* Escherichia coli (ed. E. C. C. Lin and A. S. Lynch), pp. 149–179. R. G. Landes, Georgetown, TX.

62. Ditto, M. D., Roberts, D., and Weisberg, R. A. (1994) Growth phase variation of integration host factor level in *Escherichia coli*. *J. Bacteriol.*, **176**, 3738–3748.

63. Aviv, M., Giladi, H., Schreiber, G., Oppenheim, A. B., and Glaser, G. (1994) Expression of the genes coding for the *Escherichia coli* integration host factor are controlled by growth phase, *rpoS*, ppGpp and by autoregulation. *Mol. Microbiol.*, **14**, 1021–1031.

64. Altuvia, S., Almirón, M., Huisman, G., Kolter, R., and Storz, G. (1994) The *dps* promoter is activated by OxyR during growth and by IHF and σ^S in stationary phase. *Mol. Microbiol.*, **13**, 265–272.

65. Lomovskaya, O. L., Kidwell, J. P., and Matin, A. (1994) Characterization of the σ^{38}-dependent expression of a core *Escherichia coli* starvation gene, *pexB*. *J. Bacteriol.*, **176**, 3928–3935.

66. Almirón, M., Link, A., Furlong, D., and Kolter, R. (1992) A novel DNA binding protein with regulatory and protective roles in starved *Escherichia coli*. *Genes Dev.*, **6**, 2646–2654.

67. Xu, J. and Johnson, R. C. (1995) Identification of genes negatively regulated by Fis: Fis and RpoS comodulate growth-phase-dependent gene expression in *Escherichia coli*. *J. Bacteriol.*, **177**, 938–947.

68. Finkel, S. E. and Johnson, R. C. (1992) The Fis protein: it's not just for DNA inversion anymore. *Mol. Microbiol.*, **6**, 3257–3265.

69. Ball, C. A., Osuna, R., Ferguson, K. C., and Johnson, R. C. (1992) Dramatic changes in Fis levels upon nutrient upshift in *Escherichia coli*. *J. Bacteriol.*, **174**, 8043–8056.
70. Ninneman, O., Koch, C., and Kahmann, R. (1992) The *E. coli fis* promoter is subject to stringent control and autoregulation. *EMBO J.*, **11**, 1075–1083.
71. Xu, J. and Johnson, R. C. (1995) Fis activates the RpoS-dependent stationary phase expression of *proP* in *Escherichia coli*. *J. Bacteriol.*, **177**, 5222–5231.
72. Higgins, C. F., Hinton, J. C. D., Hulton, C. S. J., Owen-Hughes, T., Pavitt, G. D., and Seirafi, A. (1990) Protein H1: a role for chromatin structure in the regulation of bacterial gene expression and virulence? *Mol. Microbiol.*, **4**, 2007–2012.
73. Lucht, J. M., Dersch, P., Kempf, B., and Bremer, E. (1994) Interactions of the nucleoid-associated DNA-binding protein H-NS with the regulatory region of the osmotically controlled *proU* operon on *Escherichia coli*. *J. Biol. Chem.*, **269**, 6578–6586.
74. Mellies, J., Brems, R., and Villarejo, M. (1994) The *Escherichia coli proU* promoter element and its contribution to osmotically signaled transcription activation. *J. Bacteriol.*, **176**, 3638–3645.
75. Owen-Hughes, T. A., Pavitt, G. D., Santos, D. S., Sidebotham, J. M., Hulton, C. S., Hinton, J. C. D., and Higgins, C. F. (1992) The chromatin-associated protein H-NS interacts with curved DNA to influence DNA topology and gene expression. *Cell*, **71**, 255–265.
76. Spassky, A., Rimsky, S., Garreau, H., and Buc, H. (1984) H1a, an *E. coli* DNA-binding protein which accumulates in stationary phase, strongly compacts DNA *in vitro*. *Nucl. Acids Res.*, **12**, 5321–5340.
77. Tupper, A. E., Owen-Hughes, T. A., Ussery, D. W., Santos, D. S., Ferguson, D. J. P., Sidebotham, J. M., Hinton, J. C. D., and Higgins, C. F. (1994) The chromatin-associated protein H-NS alters DNA topology *in vitro*. *EMBO J.*, **13**, 258–268.
78. Dersch, P., Schmidt, K., and Bremer, E. (1993) Synthesis of the *Escherichia coli* K-12 nucleoid-associated DNA-binding protein H-NS is subjected to growth phase control and autoregulation. *Mol. Microbiol.*, **8**, 875–889.
79. Falconi, M., Higgins, N. P., Spurio, R., Pon, C. L., and Gualerzi, C. O. (1993) Expression of the gene encoding the major bacterial nucleoid protein H-NS is subject to transcriptional autorepression. *Mol. Microbiol.*, **10**, 273–282.
80. Ueguchi, C., Kakeda, M., and Mizuno, T. (1993) Autoregulatory expression of the *Escherichia coli hns* gene encoding a nucleoid protein: H-NS functions as a repressor of its own transcription. *Mol. Gen. Genet.*, **236**, 171–178.
81. Utsumi, R., Kusafuka, S., Nakayama, T., Tanaka, K., Takayanagi, Y., Takahashi, H., Noda, M., and Kawamukai, M. (1993) Stationary phase-specific expression of the *fic* gene in *Escherichia coli* K-12 is controlled by the *rpoS* gene product (σ^{38}). *FEMS Microbiol. Lett.*, **113**, 273–278.
82. von Ossowski, I., Mulvey, M. R., Leco, P. A., Borys, A., and Loewen, P. C. (1991) Nucleotide sequence of *Escherichia coli katE*, which encodes catalase HPII. *J. Bacteriol.*, **173**, 514–520.
83. Yim, H. H., Brems, R. L., and Villarejo, M. (1994) Molecular characterization of the promoter of *osmY*, an *rpoS* dependent gene. *J. Bacteriol.*, **176**, 100–107.
84. Repoila, F. and Gutierrez, C. (1991) Osmotic induction of the periplasmic trehalase in *Escherichia coli* K12: characterization of the *treA* promoter. *Mol. Microbiol.*, **5**, 747–755.
85. Aldea, M., Garrido, T., Hernández-Chico, C., Vicente, M., and Kushner, S. R. (1989) Induction of a growth-phase-dependent promoter triggers transcription of *bolA*, an *Escherichia coli* morphogene. *EMBO J.*, **8**, 3923–3931.
86. Bohannon, D. E., Connell, N., K., L., Tormo, A., Espinosa-Urgel, M., Zambrano, M. M., and Kolter, R. (1991) Stationary-phase-inducible 'gearbox' promoters: Differential effects of *katF* mutations and role of σ^{70}. *J. Bacteriol.*, **173**, 4482–4492.

87. Wang, A.-Y. and Cronan Jr, J. E. (1994) The growth phase-dependent synthesis of cyclopropane fatty acids in *Escherichia coli* is the result of an RpoS (KatF)-dependent promoter plus enzyme instability. *Mol. Microbiol.*, **11**, 1009–1017.
88. Hengge-Aronis, R., Klein, W., Lange, R., Rimmele, M., and Boos, W. (1991) Trehalose synthesis genes are controlled by the putative sigma factor encoded by *rpoS* and are involved in stationary phase thermotolerance in *Escherichia coli*. *J. Bacteriol.*, **173**, 7918–7924.
89. Jung, J. U., Gutierrez, C., Martin, F., Ardourel, M., and Villarejo, M. (1990) Transcription of *osmB*, a gene encoding an *Escherichia coli* lipoprotein, is regulated by dual signals. *J. Biol. Chem.*, **265**, 10574–10581.
90. Mellies, J., Wise, A., and Villarejo, M. (1995) Two different *Escherichia coli proP* promoters respond to osmotic and growth phase signals. *J. Bacteriol.*, **177**, 144–151.
91. Koh, Y.-S. and Roe, J.-H. (1996) Dual regulation of the paraquat-inducible gene *pqi-5* by SoxS and RpoS in *Escherichia coli*. *Mol. Microbiol.*, **22**, 53–61.
92. Yang, W., Ni, L., and Somerville, R. L. (1993) A stationary phase protein of *Escherichia coli* that affects the mode of association between the *trp* repressor protein and operator-bearing DNA. *Proc. Natl. Acad. Sci. USA*, **90**, 5796–5800.
93. Atlung, T., Nielsen, A., and Hansen, F. G. (1989) Isolation, characterization, and nucleotide sequence of *appY*, a regulatory gene for growth phase-dependent gene expression in *Escherichia coli*. *J. Bacteriol.*, **171**, 1683–1691.
94. Atlung, T. and Brøndsted, L. (1994) Role of the transcriptional activator AppY in regulation of the *cyx appA* operon of *Escherichia coli* by anaerobiosis, phosphate starvation and growth phase. *J. Bacteriol.*, **176**, 5414–5422.
95. Brøndsted, L. and Atlung, T. (1994) Anaerobic regulation of the hydrogenase 1 (*hya*) operon of *Escherichia coli*. *J. Bacteriol.*, **176**, 5423–5428.
96. Brøndsted, L. and Atlung, T. (1996) Effect of growth conditions on expression of the acid phosphatase (*cgx–appA*) operon and the *appY* gene, which encodes a transcriptional activator of *Escherichia coli*. *J. Bacteriol.*, **178**, 1556–1564.
97. Dassa, J., Fsihi, H., Marck, C., Dion, M., Kieffer-Bontemps, M., and Boquet, P. L. (1992) A new oxygen-regulated operon in *Escherichia coli* comprises the genes for a putative third cytochrome oxidase and for pH 2.5 acid phosphatase (*appA*). *Mol. Gen. Genet.*, **229**, 342–352.
98. Aldea, M., Hernandez-Chico, C., de la Campa, A. G., Kushner, S. R., and Vicente, M. (1988) Identification, cloning and expression of *bolA*, an *ftsZ*-dependent morphogene of *Escherichia coli*. *J. Bacteriol.*, **170**, 5169–5176.
99. Ariza, R. R. Z., Li, Z., Ringstad, N., and Demple, B. (1995) Activation of multiple antibiotic resistance and binding of stress-inducible promoters by *Escherichia coli* Rob protein. *J. Bacteriol.*, **177**, 1655–1661.
100. Jair, K.-W., Yu, X., Skarstad, K., Thöny, B., Fujita, N., Ishihama, A., and Wolf Jr, R. E. (1996) Transcriptional activation of promoters of the superoxide and multiple antibiotic resistance regulons by Rob, a binding protein of the *Escherichia coli* origin of chromosomal replication. *J. Bacteriol.*, **178**, 2507–2513.
101. Kakeda, M., Ueguchi, C., Yamada, H., and Mizuno, T. (1995) An *Escherichia coli* curved DNA-binding protein whose expression is affected by the stationary phase specific sigma factor σ^S. *Mol. Gen. Genet.*, **248**, 629–634.
102. Li, Z. Y. and Demple, B. (1996) Sequence specificity for DNA binding by *Escherichia coli* SoxS and Rob proteins. *Mol. Microbiol.*, **20**, 937–945.
103. Atlung, T., Knudsen, K., Heerfordt, L., and Brøndsted, L. (1997) Effect of σ^S and the transcriptional activator AppY on induction of the *Escherichia coli hya* and *cbdAB-appA* operons in response to carbon and phosphate starvation. *J. Bacteriol.*, **179**, in press.

104. Csonka, L. N. (1989) Physiological and genetic responses of bacteria to osmotic stress. *Microbiol. Rev.*, **53**, 121–147.
105. Csonka, L. N. and Epstein, W. (1996) Osmoregulation. In Escherichia coli *and* Salmonella typhimurium*: Cellular and molecular biology* (ed. F. C. Neidhardt). American Society for Microbiology, Washington DC, pp. 1210–1223.
106. Demple, B. (1991) Regulation of bacterial oxidative stress genes. *Annu. Rev. Genet.*, **25**, 315–337.
107. Demple, B. and Harrison, L. (1994) Repair of oxidative damage to DNA: enzymology and biology. *Annu. Rev. Biochem.*, **63**, 915–948.
108. Farr, S. B. and Kogoma, T. (1991) Oxidative stress responses in *Escherichia coli* and *Salmonella typhimurium*. *Microbiol. Rev.*, **55**, 561–585.
109. Hidalgo, E. and Demple, B. (1996) Adaptive responses to oxidative stress: The *soxRS* and *oxyR* regulons. In *Regulation of gene expression in* Escherichia coli (ed. E. C. C. Lin, and A. S. Lynch). R. G. Landes, Georgetown, TX, pp. 435–452.
110. Storz, G., Tartaglia, L. A., Farr, S. B., and Ames, B. N. (1990) Bacterial defenses against oxidative stress. *TIG*, **6**, 363–368.
111. Olsén, A., Arnqvist, A., Hammar, M., Sukupolvi, S., and Normark, S. (1993) The RpoS sigma factor relieves H-NS-mediated transcriptional repression of *csgA*, the subunit gene of fibronectin binding curli in *Escherichia coli*. *Mol. Microbiol.*, **7**, 523–536.
112. Lucht, J. M. and Bremer, E. (1994) Adaptation of *Escherichia coli* to high osmolarity environments: osmoregulation of the high-affinity glycine betaine transport system ProU. *FEMS Microbiol. Rev.*, **14**, 3–20.
113. Pratt, L. A. and Silhavy, T. J. (1995) Porin regulon of *Escherichia coli*. In *Two-component signal transduction* (ed. J. A. Hoch and T. J. Silhavy), pp. 105–127. ASM Press, Washington DC.
114. Rajkumaki, K., Kusano, S., Ishihama, A., Mizuno, T., and Gowrishankar, J. (1996) Effects of H-NS and potassium glutamate on σ^S- and σ^{70}-directed transcription *in vitro* from osmotically regulated P1 and P2 promoters of *proU* in *Escherichia coli*. *J. Bacteriol.*, **178**, 4176–4181.
115. Manna, D. and Gowrishankar, J. (1994) Evidence for involvement of proteins HU and RpoS in transcription of the osmoresponsive *proU* operon in *Escherichia coli*. *J. Bacteriol.*, **176**, 5378–5384.
116. Nyström, T. (1994) Role of guanosine tetraphosphate in gene expression and the survival of glucose or seryl-tRNA starved cells of *Escherichia coli* K12. *Mol. Gen. Genet.*, **245**, 355–362.
117. Ivanova, A., Miller, C., Glinsky, G., and Eisenstark, A. (1994) Role of *rpoS(katF)* in *oxyR*-independent regulation of hydroperoxidase I in *Escherichia coli*. *Mol. Microbiol.*, **12**, 571–578.
118. Becker-Hapak, M. and Eisenstark, A. (1995) Role of *rpoS* in the regulation of glutathione oxidoreductase (*gor*) in *Escherichia coli*. *FEMS Microbiol. Lett.*, **134**, 39–44.
119. Storz, G., Tartaglia, L. A., and Ames, B. N. (1990) Transcriptional regulator of oxidative stress-inducible genes: direct activation by oxidation. *Science*, **248**, 189–194.
120. Toledano, M. B., Kullik, I., Trinh, F., Baird, P. T., Schneider, T. D., and Storz, G. (1994) Redox-dependent shift of OxyR-DNA contacts along an extended DNA-binding site: A mechanism for differential promoter selection. *Cell*, **78**, 897–909.
121. Jair, K. W., Martin, R. G., Rosner, J. L., Fujita, N., Ishihama, A., and Wolf, R. E. (1995) Purification and regulatory properties of MarA protein, a transcriptional activator of *Escherichia coli* multiple antibiotic and superoxide resistance promoters. *J. Bacteriol.*, **177**, 7100–7104.

122. Nunoshiba, T., Hidalgo, E., Amábile-Cuevas, C. F., and Demple, B. (1992) Two-stage control of an oxidative stress regulon: the *Escherichia coli* SoxR protein triggers redox-inducible expression of the *soxS* regulatory gene. *J. Bacteriol.*, **174**, 6054–6060.
123. Hächler, H., Cohen, S. P., and Levy, S. B. (1991) *marA*, a regulated locus which controls expression of chromosomal multiple antibiotic resistance in *Escherichia coli. J. Bacteriol.*, **173**, 5532–5538.
124. Nikaido, H. (1996) Multidrug efflux pumps of Gram-negative bacteria. *J. Bacteriol.*, **178**, 5853–5859.
125. Chou, J. H., Greenberg, J. T., and Demple, B. (1993) Posttranscriptional repression of *Escherichia coli* OmpF protein in response to redox stress: positive control of the *micF* antisense RNA by the *soxRS* locus.
126. Buettner, M. J., Spitz, E., and Rickenberg, H. V. (1973) Cyclic 3′,5′-monophosphate in *Escherichia coli. J. Bacteriol.*, **114**, 1068–1073.
127. Epstein, W., Rothman-Denes, L. B., and Hesse, J. (1975) Adenosine 3′:5′-cyclic monophosphate as mediator of catabolite repression in *Escherichia coli. Proc. Natl. Acad. Sci. USA*, **72**, 2300–2304.
128. Joseph, E., Bernsley, C., Guiso, N., and Ullmann, A. (1982) Multiple regulation of the activity of adenylate cyclase in *Escherichia coli. Mol. Gen. Genet.*, **185**, 262–268.
129. Peterkofsky, A., Svenson, I., and Amin, N. (1989) Regulation of *Escherichia coli* adenylate cyclase activity by the phosphoenolpyruvate:sugar phosphotransferase system. *FEMS Microbiol. Rev.*, **63**, 103–108.
130. Amin, N. and Peterkofsky, A. (1995) A dual mechanism for regulating cAMP levels in *Escherichia coli. J. Biol. Chem.*, **270**, 11803–11805.
131. Wright, L. F., Milne, D. P., and Knowles, C. J. (1979) The regulatory effects of growth rate and cyclic AMP levels on carbon catabolism and respiration in *Escherichia coli* K-12. *Biochim. Biophys. Acta*, **583**, 73–80.
132. Matin, A. and Matin, M. K. (1982) Cellular levels, excretion, and synthesis rates of cyclic AMP in *Escherichia coli* grown in continuous culture. *J. Bacteriol.*, **149**, 801–807.
133. Blum, P. H., Jovanovich, S. B., McCann, M. P., Schultz, J. E., Lesley, S. A., Burgess, R. R., and Matin, A. (1990) Cloning and *in vivo* and *in vitro* regulation of cyclic AMP-dependent carbon starvation genes from *Escherichia coli. J. Bacteriol.*, **172**, 3813–3820.
134. Groat, R. G., Schultz, J. E., Zychlinski, E., Bockman, A. T., and Matin, A. (1986) Starvation proteins in *Escherichia coli*: kinetics of synthesis and role in starvation survival. *J. Bacteriol.*, **168**, 486–493.
135. Schultz, J. E., Latter, G. I., and Matin, A. (1988) Differential regulation by cyclic AMP of starvation protein synthesis in *Escherichia coli. J. Bacteriol.*, **170**, 3903–3909.
136. Schultz, J. E. and Matin, A. (1991) Molecular and functional characterization of a carbon starvation gene of *Escherichia coli. J. Mol. Biol.*, **218**, 129–140.
137. McCleary, W. R. and Stock, J. B. (1994) Acetyl phosphate and the activation of two-component response regulators. *J. Biol. Chem.*, **269**, 31567–31572.
138. Prüß, B. M. and Wolfe, A. J. (1994) Regulation of acetyl phosphate synthesis and degradation, and the control of flagellar expression in *Escherichia coli. Mol. Microbiol.*, **12**, 973–984.
139. Lukat, G. S., McCleary, W. R., Stock, A. M., and Stock, J. B. (1992) Phosphorylation of bacterial response regulator proteins by low molecular weight phospho-donors. *Proc. Natl. Acad. Sci.*, **89**, 718–722.
140. McCleary, W. R., Stock, J. B., and Ninfa, A. J. (1993) Is acetyl-phosphate a global signal in *Escherichia coli*? *J. Bacteriol.*, **175**, 2793–2798.

141. Wanner, B. L. and Wilmes-Riesenberg, M. R. (1992) Involvement of phosphotransacetylase, acetate kinase, and acetyl-phosphate synthesis in the control of the phosphate regulon in *Escherichia coli*. *J. Bacteriol.*, **174**, 2124–2130.
142. Wanner, B. L. (1992) Is cross regulation by phosphorylation of two-component response regulator proteins important in bacteria? *J. Bacteriol.*, **174**, 2053–2058.
143. Bouché, S., Fischer, D., and Hengge-Aronis, R. (1997) Regulation of RssB-dependent proteolysis in *Escherichia coli*: a role for acetyl phosphate in a response regulator-controlled process. *Mol. Microbiol.*, **27**, 787–795.
144. Cashel, M., Gentry, D. R., Hernandez, V. J., and Vinella, D. (1996) The stringent response. In Escherichia coli *and* Salmonella typhimurium*: Cellular and molecular biology* (ed. G. C. Neidhardt), pp. 1458–1496. American Society for Microbiology, Washington, DC.
145. Irr, J. D. (1972) Control of nucleotide metabolism and ribosomal ribonucleic acid synthesis during nitrogen starvation of *Escherichia coli*. *J. Bacteriol.*, **110**, 554–561.
146. Spira, B., Silberstein, N., and Yagil, E. (1995) Guanosine 3′,5′-bispyrophosphate (ppGpp) synthesis in cells of *Escherichia coli* starved for P_i. *J. Bacteriol.*, **177**, 4053–4058.
147. Josaitis, C. A., Gaal, T., and Gourse, R. L. (1995) Stringent control and growth-rate-dependent control have nonidentical promoter sequence requirements. *Proc. Natl. Acad. Sci. USA*, **92**, 1117–1121.
148. Faxén, M. and Isaksson, L. A. (1994) Functional interactions between translation, transcription and ppGpp in growing *Escherichia coli*. *Biochim. Biophys. Acta*, **1219**, 425–434.
149. Vogel, U., Sørensen, M., Pedersen, S., Jensen, K. F., and Kilstrup, M. (1992) Decreasing transcription elongation rate in *Escherichia coli* exposed to amino acid starvation. *Mol. Microbiol.*, **6**, 2191–2200.
150. Huisman, G. W. and Kolter, R. (1994) Sensing starvation: a homoserine lactone-dependent signaling pathway in *Escherichia coli*. *Science*, **265**, 537–539.
151. Fuqua, W. C., Winans, S. C., and Greenberg, E. P. (1994) Quorum sensing in bacteria: the *luxR-luxI* family of cell density-responsive transcriptional regulators. *J. Bacteriol.*, **176**, 269–275.
152. Böhringer, J., Fischer, D., Mosler, G., and Hengge-Aronis, R. (1995) UDP-glucose is a potential intracellular signal molecule in the control of expression of σ^S and σ^S-dependent genes in *Escherichia coli*. *J. Bacteriol.*, **177**, 413–422.
153. Giaever, H. M., Styrvold, O. B., Kaasen, I., and Strøm, A. R. (1988) Biochemical and genetic characterization of osmoregulatory trehalose synthesis in *Escherichia coli*. *J. Bacteriol.*, **170**, 2841–2849.
154. Ninfa, A. J., Atkinson, M. R., Kamberov, E. S., Feng, J., and Ninfa, E. G. (1995) Control of nitrogen assimilation by the NR_I–NR_{II} two-component system of enteric bacteria. In *Two-component signal transduction* (ed. J. A. Hoch and T. J. Silhavy), pp. 67–88. ASM Press, Washington DC.
155. Porter, S. C., North, A. K., and Kustu, S. (1995) Mechanism of transcriptional activation by NtrC. In *Two-component signal transduction*. (ed. J. A. Hoch and T. J. Silhavy), pp. 147–158. ASM Press, Washington DC.
156. Wanner, B. L. (1995) Signal transduction and cross regulation in the *Escherichia coli* phosphate regulon by PhoR, CreC, and acetyl phosphate. In *Two-component signal transduction*. (ed. J. A. Hoch and T. J. Silhavy), pp. 203–221. ASM Press, Washington DC.
157. Zhang, A., Altuvia, S., Tiwari, A., Argamaa, L., Hengge-Aronis, R., and Storz, G. (1998) The OxyS regulatory RNA represses *rpoS* translation and binds the Hfg (HF–I) protein. *EMBO J.*, **17**, 6061–6068.

8 Two-component systems

MARIETTE R. ATKINSON and ALEXANDER J. NINFA

1. Introduction

In this chapter, we will discuss the signal transduction systems known as the two-component systems. This review covers the literature in this area through July 1995. Signal transduction by these systems requires at least one each of two different types of conserved protein domains, known as the histidine protein kinase domain (HPK) and response regulator domain (RR). Signal transduction by the HPK and RR domains involves the reversible phosphorylation of each of these domains and the transfer of phosphoryl groups between HPK and RR domains. Phosphorylation of the response regulator domain results in an altered conformation that permits a regulatory target to be engaged, either by the RR domain itself or, more frequently, by additional protein domains linked to the RR domain that are controlled by the conformational state of the RR domain. Various regulatory mechanisms, some of which act through the HPK domain and others acting directly on the RR domain, regulate the intracellular concentration of the phosphorylated form of the RR domain in response to stimuli. When the intracellular concentration of the phosphorlated form of the RR domain reaches a critical level, the regulatory target is engaged, triggering a physiological adaptation. Thus, the overall logic employed by the two-component systems is that a targeted physiological adaptation is regulated via the control of the concentration of the phosphorylated form of the RR domain. Different mechanisms of regulation of these phosphorylation and phosphotransfer reactions, in as much as these are currently known, will be discussed in this review.

The two-component signal transduction systems constitute the most common class of bacterial signal transduction protein, with well over 100 examples of such systems known. In Table 1 (1–125) we present the results of a literature search for two-component systems. Since Table 1 results from a literature search, as opposed to the analysis of sequences from protein sequence data bases, there are several potential problems that the reader should be aware of. First, in many cases, previous reviews cited unpublished personal communications or unpublished data from the authors' laboratories as the source of the sequence data; in these cases, we have simply listed the systems and cite the earlier review without access to the primary data. In some cases, open reading frames of unknown function are claimed to share homology with the HPK or RR domains, and these are also listed without access to the primary

Table 1 Prokaryotic two-component systems

System	Histidine protein kinase	Response regulator	Function	Organisms
		AdgA[b]	Unknown	*Rhodobacter capsulatus* (1)
Afs	AfsQ2	AfsQ1	Secondary metabolism	*Streptomyces coelicolor* (2)
Glp		AgmR	Glycerol metabolism	*Pseudomonas aeurginosa* (3)
Agr	AgrB	AgrA	Synthesis of virulence factors	*Staphylococcus aureus* (4)
Alg		AlgR AlgB	Alginate biosynthesis	*Pseudomonas aeruginosa* (5, 6)
Amf		AmfR	Aerial mycelium formation	*Streptomyces griseus* (7)
Arc	ArcB	ArcC	Anaerobic control TCA cycle	*Escherichia coli, Haemophilus influenzae* (8–10)
Asg	AsgA		Fruiting body development	*Myxococcus xanthus* (11)
Az	Off1[a,b]	Az[a]	Ornithine decarboxylase antizyme; unknown if transcriptional regulator	*Escherichia coli* (12)
Bae	BaeS	BaeR	Unknown	*Escherichia coli* (13)
Bar	BarA		Porin biosynthesis	*Escherichia coli* (9, 14)
Bas	BasS	BasR	Unknown	*Escherichia coli, Haemophilus influenzae* (10, 13)
Bvg	BvgS	BvgA	Virulence	*Bordetella pertussis, Bordetella parapertussis, Bordetella bronchiseptica* (15, 16)
Car	CarS	CarR	Global expression of carbohydrate catabolic pathways	*Azospirillum brasilense* (17)
Che	CheA	CheY CheB	Chemotaxis	*Escherichia coli, Salmonella typhimurium, Bacillus subtilis, Pseudomonas aeruginosa, Halobacterium salinarium, Rhizobium meliloti, Enterobacter aerogenes, Listeria monocytogenes* (1, 18–20)
		CheV		*Bacillus subtilis* (21)
Chv	ChvG	ChvI	Virulence	*Agrobacterium tumefaciens* (22)
Cia	CiaH	CiaR	Competence, penicillin resistance	*Streptococcus pneumoniae* (23)
Com	ComP	ComA	Competence	*Bacillus subtilis* (24)
Cop	CopS	CopR	Copper resistance	*Pseudomonas syringae* (25, 26)
Cpx	CpxA	CpxR	pH dependent virulence	*Escherichia coli, Shigella sonnei, Haemophilus influenzae* (10, 27, 28)
Cre	CreC	CreB	Unknown	*Escherichia coli* (29)
Cut	CutS[b]	CutR	Copper metabolism	*Streptomyces lividans* (26, 30)
Dct	DctB	DctD	C_4-dicarboxylate transport	*Rhizobium meliloti, Rhizobium leguminosarum* (31)
	DctS	DctR	C_4-dicarboxylate transport	*Rhodobacter capsulatus* (32)
Deg	DegS DegM	DegU	Exoenzyme synthesis	*Bacillus subtilis* (24, 33)
Dfr	DfrA[b]		Difunan resistance	*Synechococcus* sp. PCC 6803 (34)
Div	PleC/DivJ	DivK/PleD	Development and cell division	*Caulobacter crescentus* (35)
Dnr		DnrN	Daunorubicin biosynthesis	*Streptomyces peucetius* (36)
Evg	EvgS	EvgA	Unknown	*Escherichia coli* (37)
Fim		FimZ	Type I fimbrae expression	*Salmonella typhimurium* (38)
Fix	FixL	FixJ	Nitrogen fixation	*Rhizobium meliloti, Azorhizobium caulinodans,*
		ORF138		*Bradyrhizobium japonicum* (39, 40)
Flb		FlbD	Flagellin biosynthesis	*Caulobacter crescentus* (35, 41)
Frz	FrzE	FrzG FrzZ	Cellular development	*Myxococcus xanthus* (42)

Table 1 *Continued*

System	Histidine protein kinase	Response regulator	Function	Organisms
Gac	LemA	GacA	Virulence	*Pseudomonas syringae* (43, 44)
Gdh		GdhR		*Bacillus megaterium* (1, 45)
Hkn	HknA		Toxin production	*Bacillus thuringiensis* (46)
Hox	HoxX	HoxA	Hydrogen oxidation	*Alcaligenes eutrophus, Bradyrhizobium japonicum* (47–49)
Hnr		Hnr	Unknown	*Escherichia coli* (50)
Hup	HupT	HupR1	Hydrogenase gene expression	*Rhodobacter capsulatus* (51, 52)
Hyd	HydH	HydG	Hydrogenase gene expression	*Salmonella typhimurium, Escherichia coli* (53, 54)
Kpd	KpdD	KpdE	Potassium transport	*Escherichia coli* (55)
Lcr		LcrB	Unknown	*Rhizobium* sp. (1)
Lux	LuxN, LuxQ	LuxO	Expression of luminescence	*Vibrio harveyi* (56)
Mox	MxaY	MxaX	Methanol dehydrogenase expression	*Paracoccus denitrificans* (57, 58)
Mrk		MrkE	Type 3 fimbriae expression	*Klebsiella pneumoniae* (59)
Mry		Mry (VirR)[a]	Virulence	*Streptococcus pyogenes* (4, 60, 61)
Mtr		MtrA[b]	Unknown	*Mycobacterium tuberculosis* (62)
Nar	NarX	NarL	Nitrate nitrite utilization	*Escherichia coli* (63)
	NarQ	NarP		*Escherichia coli, Haemophilus influenzae* (10, 63)
Nis	NisK	NisR	Nisin biosynthesis	*Lactococcus lactis* (64, 65)
Nod	NodV	NodW	Nodulation	*Bradyrhizobium japonicum* (66, 67)
	NwsA	NwsB		
Ntr	NifR2	NifR1	Nitrogen assimilation	*Rhodobacter capsulatus* (68)
	NifL[b]			*Azobacter vinelandii* (1)
	NRII (NtrB)	NRI (NtrC)		*Escherichia coli, Klebsiella aerogenes, Klebsiella pnuemoniae, Vibrio alginolyticus, Agrobacterium tumefaciens, Bradyrhizobium sp., Rhizobium meliloti, Salmonella typhimurium* (69–74)
		GlnR		*Streptomyces coelicolor* (75)
	NtrY	NtrX		*Azorhizobium caulinodans* ORS571 (76)
Omp	EnvZ	OmpR	Porin biosynthesis	*Escherichia coli, Salmonella typhimurium* (77)
Pat		PatA	Heterocyst pattern formation	*Anabaena* 7120 (34, 78)
Pco	PcoS[b]	PcoR	Copper resistance	*Escherichia coli* (26)
Peh	PehR[b]	PehS[b]	Virulence	*Pseudomonas solanacearum* (79)
Pet		PetR	Respiratory and photosynthetic growth	*Rhodobacter capsulatus* (80)
Pfe	PfeS	PfeR	Expression of ferric enterobactin receptor	*Pseudomonas aeruginosa* (81)
Pgt	PgtB	PgtA	Phosphoglycerate transport	*Salmonella typhimurium* (82, 83)
Pho	PhoR	PhoB	Phosphate assimilation	*Escherichia coli, Psuedomonas aeruginosa, Shigella dysenteriae, Shigella flexneri, Klebsiella pneumoniae, Haemophilus influenzae* (10, 29, 84–86)
	PhoQ	PhoP	Virulence	*Salmonella typhimurium, Escherichia coli* (87, 88)
	PhoR	PhoP	Phosphate assimilation	*Bacillus subtilis* (89)
Pil	PilS	PilR	Type 4 pili	*Pseudomonas aeruginosa* (90, 91)
	PilB	PilA	Type 4 pili	*Neisseria gonorrhoeae* (4, 92)

Table 1 *Continued*

System	Histidine protein kinase	Response regulator	Function	Organisms
		PilG/PilH	Type 4 pili	*Myxococcus xanthus* (93)
Pln	PlnB	PlnC/PlnD	Plantaricin A biosynthesis	*Lactobacillus plantarum* C11 (94)
Pmr	PmrB	PmrA	Virulence	*Salmonella typhimurium* (87, 95)
Prr	PrrB	PrrA	Regulation of photosynthesis gene expression	*Rhodobacter sphaeroides* (96, 97)
Ram		RamR	Aerial mycelium formation	*Streptomyces coelicolor* (98)
Rca		RcaC	Chromatic adaption	*Fremyella diplosiphon* (99, 100)
Rcs	RcsC	RcsB	Capsule biosynthesis	*Escherichia coli, Erwinia stewartii, Erwinia amylovora, Klebsiella sp., Citrobacter freundii* (101)
Reg	RegB	RegA	Anaerobic control of photosynthetic genes	*Rhodobacter capsulatus* (102)
Res	ResE	ResD	Phosphate assimilation	*Bacillus subtilis* (89)
Rpf	RpfC		Pathogenesis	*Xanthomonas campestris* pathovar *campestris* (103)
Rpr	RprX	RprY	Tetracycline resistance	*Bacteroides fragilis* (104)
Rte	RteA	RteB	DNA transfer	*Bacteroides* (105)
Hmc		Rrf1	Synthesis of transmembrane redox protein complex	*Desulfovibrio vulgaris* subsp. *vulgaris* (106)
Sap	SapK	SapR	Immunity to sakacin A	*Lactobacillus sake* Lb706 (107)
Sas	SasA		Unknown	*Synechococcus* sp. PCC7942 (34, 108)
Spa	SpaK	SpaR	Lantibiotic biosynthesis	*Bacillus subtilis* (109)
Sph	SphS	SphR	Phosphate assimilation	*Synechococcus* sp. PCC7942 (34, 110, 111)
Spo	KinA KinB KinC	Spo0F Spo0A	Sporulation	*Bacillus subtilis* (112, 113) *Bacillus sp., Clostridium sp.* (114)
Srr		SrrA/SrrB[b]	Unknown	*Synechococcus* sp. PCC7942 (34)
Tct		TctD	Tricarboxylate transport	*Salmonella typhimurium* (115)
Tor		TorR	Trimethylamine *N*Ooxide reductase regulation	*Escherichia coli* (116)
Tus	UspT[b]	UrpT[b]	Unknown	*Escherichia coli* (1, 117, 118)
Uhp	UhpB	UhpA	Phosphosugar transport	*Escherichia coli, Salmonella typhimurium* (119, 120)
Van	VanS	VanR	Vancomycin resistance	*Enterococcus* (121)
Vir	VirA	VirG	Virulence	*Agrobacterium tumefaciens* (22)
Vir	VirS	VirR	Extracellular toxin production	*Clostridium perfringens* (122)
Vsr	VsrB VsrA	VsrC VsrD	Virulence	*Pseudomonas solanacearum* (79, 123)
		YbcA	Unknown	*Escherichia coli* (1)
	OrfH	OrfJ	Unknown	*Bacillus subtilis* (124)
	Orf2	Orf5	Unknown	*Pseudomonas fluorescens* (125)
		Orf[b]	Unknown	*Synechococcus* sp. PCC 7002 (1, 34)
		TrpX[b]	Unknown	*Pseudomonas aeruginosa* (1)
	Orf[b]		Unknown	*Bacillus brevis* (1)
	Orf[b]		Unknown	*Mycobacterium tuberculosis* (1)
		Orf[b]	Unknown	*Clostridium acetobutylicum* (1)
		Orf[b]	Unknown	*Leptospira interrogans* (1)
		Orf[b]	Unknown	*Synechocystis* sp. PCC 6803 (50)

[a] Limited homology.
[b] Complete sequence data not available.

sequence data. In Table 1, we attempted to list each different system as a single entry, with the organisms in which that system is found listed in the rightmost column. However, these designations are arbitrary, and systems that are considered to be the same by us may have subtly different function or structure in the various organisms. In other cases, we may have listed separate systems that are really analogous but are named differently in different organisms. In some cases, the designation of unlinked genes for HPK and RR domain-containing proteins as constituting a cognate pair rests solely on the observation that mutations in these genes affect a common physiological response; future work may show that these are members of distinct cognate pairs affecting a common physiological process. In some cases, multiple HPK and RR proteins are listed as a single system if a common physiological process is affected, even though these may act independently on separate targets. Finally, although we did not conduct a rigorous analysis of the sequence data for each system, in a few cases we were able to eliminate sequences cited in the literature as sharing significant homology with the HPK and RR domains, but which upon inspection by eye did not demonstrate convincing homology. The resulting Table 1 contains 149 examples of prokaryotic two-component systems, which we designated as constituting 85 different systems.

The number of reviews of the two-component systems is within an order of magnitude of the number of systems themselves (1, 50, 126–140), and a recent treatise covers the subject (141). The sequence relationships among the family members have been discussed in several recent reviews (1, 50, 129, 130, 133), and brief descriptions of the physiological roles of several two-component systems were provided in an earlier review (127). In addition, the structure function relationships and mechanisms of catalysis have been discussed in an excellent recent review by Stock and colleagues (128). We will focus on the reactions catalyzed by the HPK and RR domains and the differences exhibited by different two-component systems. Thus, the topics covered here overlap somewhat with the discussion presented by Stock and colleagues (128).

In the enteric bacteria *Escherichia coli* and *Salmonella typhimurium*, 28 different two-component systems are known to occur (Table 1). In contrast, the complete nucleotide sequencing of the *Hemophilus influenza* genome indicates that this organism contains only four different proteins containing the RR domain and five containing the HPK domain (Table 1; Reference 10). Thus the number of these systems is highly variable in the eubacteria. To a much lesser extent, two-component systems, or at least proteins with limited homology to the RR and HPK domains, have been identified in archaebacteria, plants, yeast, and mammalian mitochondria (133, 142–149). Thus, the phosphotransfer reactions catalyzed by the HPK and RR proteins apparently represents an ancient signalling mechanism.

An overview of this signalling mechanism is presented in Fig. 1. The HPK domain catalyzes the reversible phosphorylation of a conserved histidine residue within the HPK domain at the expense of ATP, producing ADP (150–157). The phosphorylated HPK domain is dephosphorylated by the RR domain by means of a reversible reaction in which the phosphoryl group is transfered to a conserved aspartate within

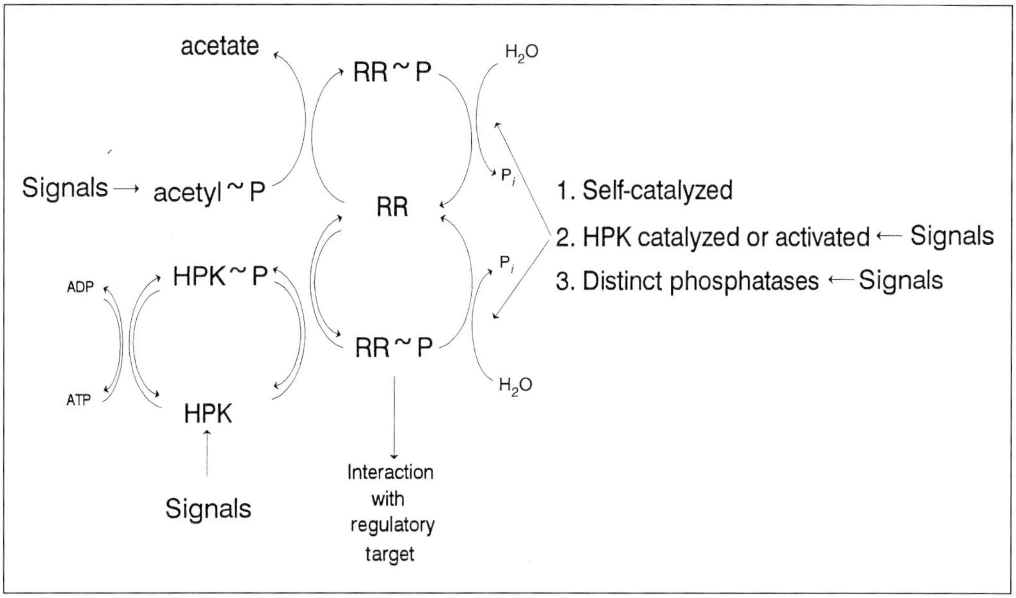

Fig. 1 Overview of two component system domain interactions. Abbreviations: HPK, histidine protein kinase domain; RR, response regulator domain; acetyl~P, acetyl phosphate.

the response regulator domain (152, 153, 157–174). The phosphorylated RR domain may be dephosphorylated by a number of mechanisms including (in addition to reversal of the phosphotransfer reaction with the HPK domain) a self-catalyzed 'autophosphatase' activity (151, 152), a regulated phosphatase activity of the HPK domain not involving phosphotransfer to the conserved histidine (151, 160, 163, 164, 171, 175–182), and dephosphorylation by distinct phosphatases (183–185).

Another route for the formation of the RR~P is by the direct RR-catalyzed transfer of phosphoryl groups from certain small molecule phosphoryl donors (169, 186–193; reviewed in References 192 and 193; Fig. 1). In Fig. 1, acetyl~P is depicted as serving this role, since for this compound there is clear evidence that the phosphotransfer reaction occurs *in vivo* and *in vitro* (186–189, 191). This reaction is apparently catalyzed by the RR domain, and is likely to occur by the same mechanism as phosphotransfer from the HPK~P.

In many cases, the intracellular concentrations of the proteins containing HPK and RR domains are regulated by the phosphorylation of the RR domain, providing for positive and/or negative autoregulatory circuits. In addition, the intracellular concentration of the RR~P may be controlled by signals acting at various points. For example, the autophosphorylation of the HPK domain may be regulated by the ATP/ADP ratio (which is itself a signal) and by signals impinging on the HPK domain directly or on linked regulatory domains which are controlled by association with small-molecule (182, 194, 195) or macromolecular (151, 160, 177, 179, 196) ligands. The interaction of the HPK~P with the RR domain may be regulated by control of protein concentration and/or allosteric control of the HPK domain. In addition, the formation of the RR~P is indirectly regulated by various signals that control the availability of small molecule phosphodonors (reviewed in References 192 and 193). Various signals may also control the rate of RR~P dephosphorylation. For example, the regulated phosphatase activity of the HPK domain and distinct phosphatase proteins may be regulated by association with small-molecule (182, 194, 195) or macromolecular (151, 160, 177, 179, 196) effectors. In addition, the biosynthesis of the distinct phosphatases may be regulated (181). Thus, for each isolated system the framework exists for the integration of distinct signals in the regulation of the intracellular concentration of the RR~P.

Interaction between different two-component systems permits the further integration of information in bringing about physiological regulation. In the simplest of cases, multiple HPK domains may directly interact with a single RR domain, bringing the regulation of the intracellular concentration of the RR~P under control of signals that affect either HPK domain (113, 163, 167; reviewed in Reference 197). These multiple HPK domains may act synergistically or antagonistically, depending on the system and conditions. In another case, two 'parallel' two-component systems, each containing proteins with HPK and RR domains, regulate a physiological adaptation (63, 176, 198); this arrangement serves to subject the physiological adaptation to the effects of signals controlling either HPK or RR protein. Subtle differences in the interaction of the RR domain-containing proteins with the regulatory targets may thus permit very fine control of these regulatory targets (63). Finally, a single HPK domain may interact with more than one RR domain (e.g. (167)), raising the possibility of branched signal transduction pathways from the common HPK domain.

In addition to these types of interactions, the phosphorylation of RR domains by small-molecule phosphorylated metabolic intermediates such as acetyl phosphate raises the possibility of indirect interactions between different two-component systems based on their effects on the intracellular pool(s) of the phosphodonor molecule(s). For example, draining of the intracellular pool of acetyl phosphate by the nitrogen-regulated (Ntr) two-component system during conditions of nitrogen excess has been observed (189). This is most likely due to the regulated phosphatase activity of the Ntr HPK domain-containing protein under conditions of nitrogen excess (160). Since acetyl phosphate is a phosphodonor for a number of RR domains, this action of the Ntr system indirectly affects the intracellular concentration of the phosphorylated form of these response regulator domains. Thus, both direct and indirect interactions between different two-component systems permits the coordination of different physiological adaptations in response to complex sets of signals.

Two-component systems of bacteria are involved in the regulation of a wide array of different biological processes (Table 1; reviewed in References 1 and 127). To illustrate the different requirements of these biological switches, three aspects of these biological switches will be considered here: whether the switch is reversible or not, whether the switch is rapid or not, and whether the signals that affect the switch are internal or external to the cell. As examples, we select the two-component systems that regulate chemotaxis (Che) and nitrogen assimilation (Ntr) in *E. coli*, and the Spo system, which controls the initiation of sporulation in *B. subtilis*. Due to the requirements of these different switches, there is considerable variation in the biochemical properties and regulatory mechanisms utilized.

The chemotaxis system is designed to continuously monitor the change in concentration of extracellular attractant or repellent compounds (reviewed in Reference 185). Thus, the switch must be capable of rapid evaluation of changes in the extracellular concentrations of these small-molecule compounds and be completely reversible; the time-scale of the 'adaptive response' is of the order of seconds. The environmental stimuli are a modest number of small molecules that bind to the periplasmic faces of transmembrane receptors. The receptors, in concert with another protein (CheW), control the rate of autophosphorylation of the HPK, CheA. Phosphotransfer is to two RR proteins, each of which has quite potent autophosphatase activity. In addition, a distinct phosphatase acts on one of these RR~P. Thus, as soon as the rate of HPK autophosphorylation is no longer stimulated by the receptors, the intracellular concentration of the two RR~P drops below the threshold required to engage the regulatory target; in this case the switch proteins of the flagellar motor.

In nitrogen regulation, the switch provided by the two-component system must respond to internal signals of nitrogen status, be continuously variable, and reversible. The two-component system is integrated into a bicyclic cascade containing two bifunctional enzymes; a signal transducing uridylyltransferase/uridylyl-removing enzyme and the HPK NRII. The reversible uridylylation of the PII protein regulates the phosphatase activity of NRII, permitting fine control of the RR~P concentration. The time-scale of the nitrogen-regulated adaptive response is somewhat longer than in the case of chemotaxis since at least part of the adaptive response involves the regulation of gene transcription, and the stability of the RR~P is greater than that seen with the chemotaxis RR proteins.

Sporulation of *B. subtilis* is not reversible, therefore the switch is not reversible. Instead, the switch is designed to integrate many different signals, some of which are extracellular signals and some of which are internal signals. The critical RR~P, Spo0A~P, is formed by the transfer of phosphoryl groups through a phosphorelay system containing another RR protein, Spo0F. A plethora of kinases and phosphatases act on the various components of the phosphorelay to control the concentration of Spo0A~P. The two RR~P in this system are considerably more stable than those in the Ntr and Che systems. Given the diversity of physiological processes regulated by two-component regulatory systems, we should expect that the common phosphotransfer reactions reflected in the homology shared by these systems should be modified and regulated to serve the needs of the particular regulatory systems. Thus, a single shared mechanism for the regulation of the intracellular concentration of the RR~P is not found.

2. Overview of holomogies and domain relationships of the HPK and RR proteins

This subject matter has been carefully reviewed (1, 50, 128, 129) and will only be briefly noted here in order to provide the reader with a framework for the discussion

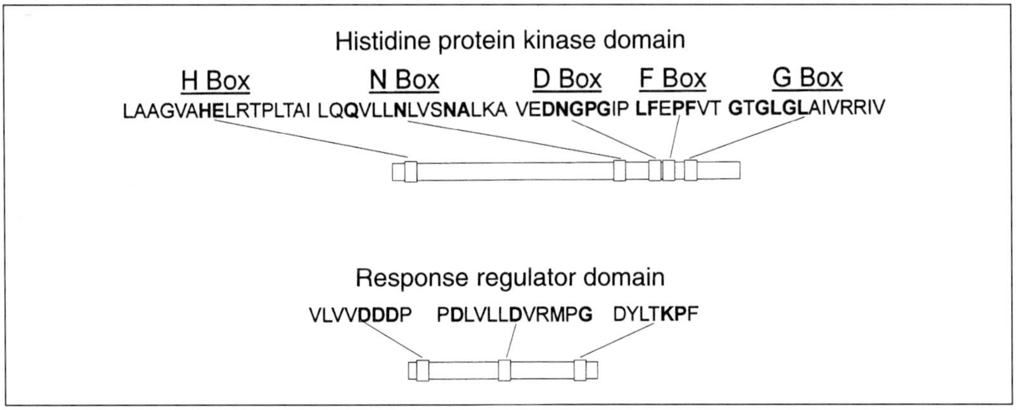

Fig. 2 Conserved amino acid motifs among histidine protein kinase and response regulator domains. The most highly conserved residues are shown in bold type. For a more complete sequence comparison see References 50, 128 and 129.

that follows. The HPK domain, consisting of about 225 amino acids, contains five conserved motifs that have been designated the H, N, D, F, and G motifs (Fig. 2). The RR domain, consisting of about 120 amino acids, contains three highly conserved motifs as depicted in Fig. 2. The most highly conserved residues within these motifs are designated in bold face in Fig. 2. Different arrangements of HPK and/or RR domains within proteins are depicted in Fig. 3.

Within the HPK domain, the H motif contains the conserved histidine residue that is the site of phosphorylation of the HPK domain (153–155). The precise role of the remaining conserved motifs has not been defined. On the basis of sequence comparisons to other types of protein kinases, Stock and colleagues have noted that the D motif is similar to the sequence of serine/threonine/tyrosine kinases responsible for coordinating Mg^{2+}, and the G motif is similar to glycine-rich sequences that comprise the nucleotide-binding sites of protein kinases (128). Mutations in the G motif of the *E. coli* HPK domain-containing proteins NRII (NtrB), EnvZ, and CheA have been observed to affect the binding of ATP (181, 199–202), supporting this view.

In several cases, one or more of the conserved HPK motifs are absent. For example, the SpoIIAB and RsbW proteins of *B. subtilis* lack the H and F motifs, and these proteins catalyze the phosphorylation of a serine residue in another protein that does not contain the response regulator domain (203–205). Thus, the H motif is not required for ATP binding or kinase activity; it should more properly be thought of as the substrate of the protein kinase activity. The UhpB HPK domain also lacks the F motif, and the FrzE HPK domain lacks the N motif, indicating that in these cases the absent motif is not required for histidine protein kinase activity.

In two cases, CheA and FrzE, the HPK domain is split, with the conserved H motif found within another protein domain (Fig. 3). Apparently, in the folded structure of these proteins, the displaced H motif is located near to the kinase active site formed by the rest of the HPK domain. A small peptide containing the isolated H motif of

CheA serves as a substrate for histidine phosphorylation by CheA (206).

The HPK domain is always found associated with other protein domains (Fig. 3). It is likely that in many cases these associated domains, which are not related to one another, are involved in the sensation of various stimuli and the regulation of the activities of the HPK domain. In many cases, the HPK domain is fused to an N-terminal segment that contains two or more membrane-spanning segments and a periplasmic domain, thus, these HPK domain-containing proteins are integral membrane proteins. This arrangement may facilitate the sensation of extracellular signals, with the presence of an extracellular ligand communicated to the cytoplasmic HPK domain via a conformational change in the transmembrane segments, which is brought about upon interaction of the ligand with the periplasmic domain (63). In other cases, the HPK domain is fused to an N-terminal segment with two or more membrane-spanning segments but no appreciable periplasmic domain (194). In these cases, a membrane localization may be required for optimal interaction with the cytoplasmic domains of other membrane proteins, or for sensation of stimuli found or generated at the membrane.

The CheA protein of *E. coli* and *S. typhimurium* is a soluble protein containing a 'split' HPK domain, as noted above, with the H motif found in an N-terminal domain. The domain relationships of CheA and the association of this protein with the other signal transduction components of the chemotaxis system have been studied genetically and biochemically (106–209; reviewed in Reference 185). CheA contains a short domain, designated the P2 domain, adjacent to the H motif near the N-terminus. This P2 domain has been implicated as playing an important role in the binding of one of the Che RR-containing proteins, designated CheY. Unfortunately, a similar domain has not yet been identified in other HPK-containing proteins, suggesting that the recognition of the cognate RR domain(s) may be different in those cases where the HPK domain is not 'split'. In addition to the P2 domain, CheA contains additional protein domains linked to the C-terminus of the HPK domain. Biochemical and genetic analysis of the roles of these C-terminal domains indicates that they are involved in the interaction of CheA with transmembrane receptor proteins and the CheW protein, which together serve to transmit environmental signals controlling the function of the HPK domain. Thus, CheA serves as an excellent example of how the HPK domain may be incorporated with non-conserved regulatory domains.

In quite a few cases, an HPK domain and RR domain are found fused in a single protein. In these proteins phosphoryl group transfer from the HPK domain to the associated RR domain is possible and in one case it seems that this HPK-associated RR domain serves to regulate the phosphorylation of another distinct RR domain (210). Among the set of proteins containing both HPK and RR domains, a subset have been identified (RteA, LemA, EvgS, BvgS, RpfC, BarA, and ArcB) that contain a conserved segment attached to the C-terminus of the RR domain (9). This C-terminal segment resembles the H motif of CheA and FrzE, and contains a conserved histidine moiety that can become phosphorylated by the HPK domain in an intermolecular reaction (9). Furthermore, the phosphorylated histidine at this site can serve as

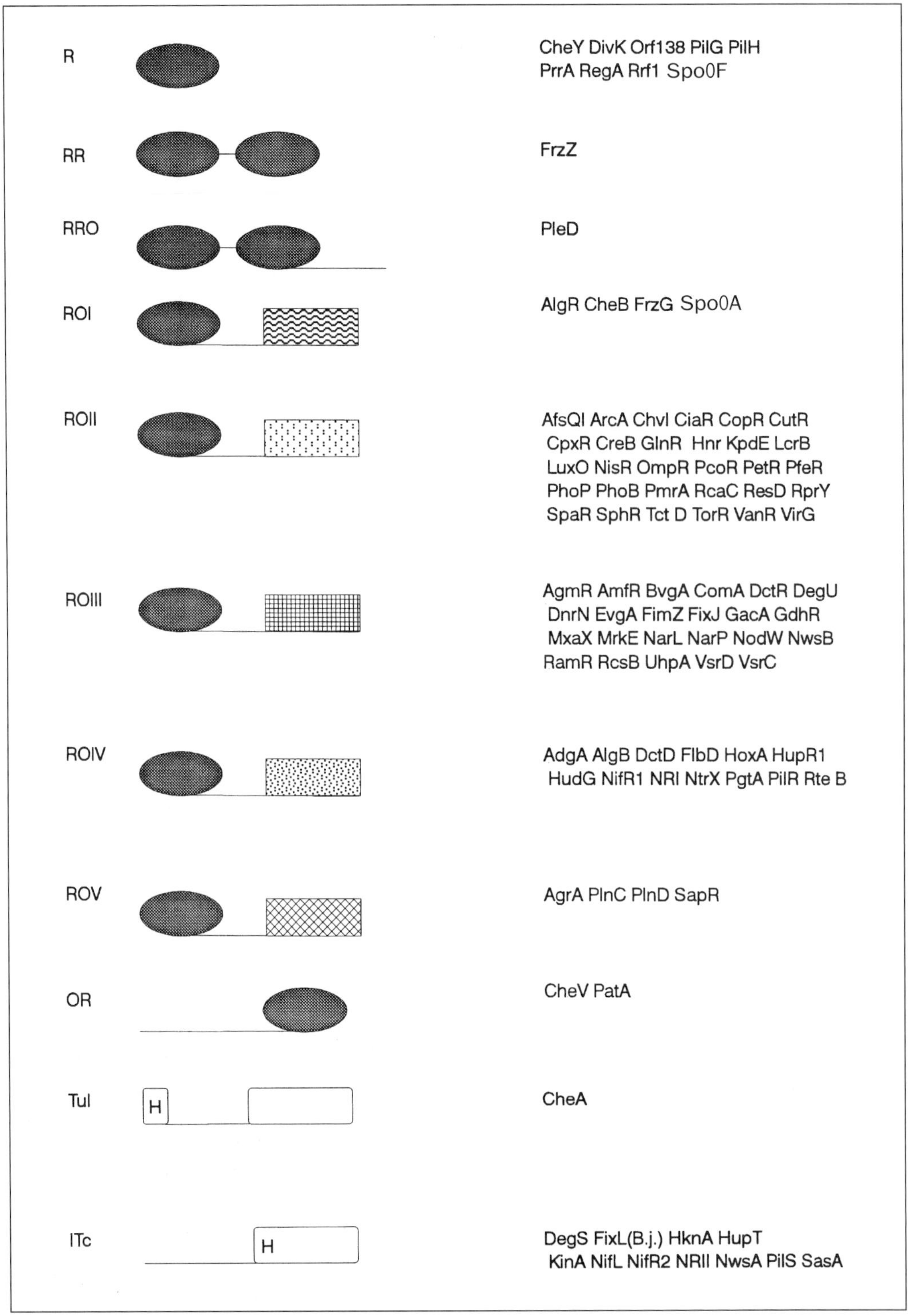
R
CheY DivK Orf138 PilG PilH
PrrA RegA Rrf1 Spo0F
RR
FrzZ
RRO
PleD
ROI
AlgR CheB FrzG Spo0A
ROII
AfsQI ArcA ChvI CiaR CopR CutR
CpxR CreB GlnR Hnr KpdE LcrB
LuxO NisR OmpR PcoR PetR PfeR
PhoP PhoB PmrA RcaC ResD RprY
SpaR SphR Tct D TorR VanR VirG
ROIII
AgmR AmfR BvgA ComA DctR DegU
DnrN EvgA FimZ FixJ GacA GdhR
MxaX MrkE NarL NarP NodW NwsB
RamR RcsB UhpA VsrD VsrC
ROIV
AdgA AlgB DctD FlbD HoxA HupR1
HudG NifR1 NRI NtrX PgtA PilR Rte B
ROV
AgrA PlnC PlnD SapR
OR
CheV PatA
TuI
H
CheA
ITc
H
DegS FixL(B.j.) HknA HupT
KinA NifL NifR2 NRII NwsA PilS SasA

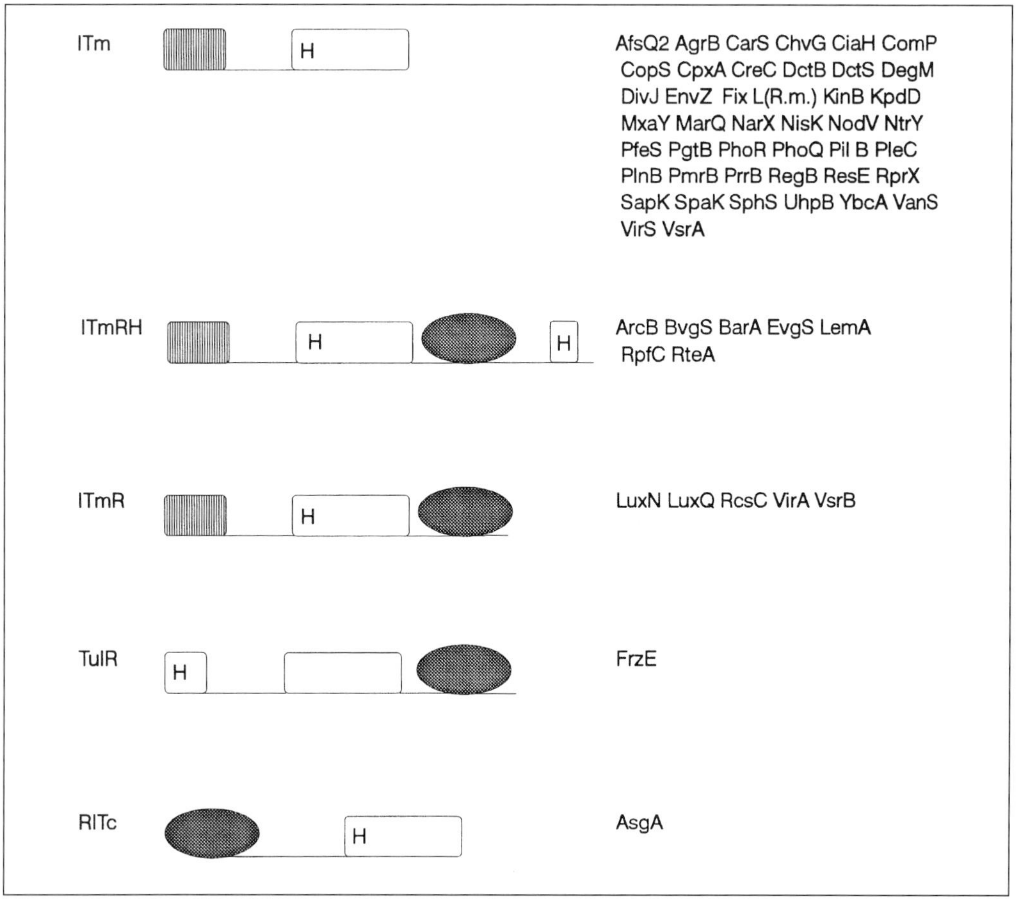

Fig. 3 Simplified domain arrangements of two-component system proteins. Body plan abbreviations: Tu, unorthodox transmitter domain; Tm, membrane associated transmitter (membrane spanning region indicated by striped box); Tc, cytoplasmic transmitter; I, sensory input domain; R, response regulator domain; H, histidine; O, output domain. ROI output domains are dissimilar. Other output domain groups have related sequences. These have been classified according to the scheme devised by Parkinson and Kofoid (1).

phosphodonor for the phosphorylation of a response regulator (9). Thus, it seems that this set of HPK-containing proteins contain both the standard H motif and a second copy of the H motif located distal to the associated RR domain. Interestingly, in the case of the ArcB protein, the distal H motif does not seem to be involved in the regulation of the intracellular concentration of the phosphorylated form of the cognate RR-containing ArcA protein (9). This distal H motif may be involved in the phosphorylation of another RR-containing protein *in vivo*.

The RR domain, typically about 125 amino acids in length, usually constitutes the cytoplasmic domain of a soluble protein. In addition, the RR domain may constitute entirely a separate protein, or be linked to an HPK domain as noted above. In those cases where the RR domain forms the cytoplasmic domain of a soluble protein, it seems likely that the phosphorylation of the RR domain, and the conformational change that is so stabilized, regulates the activities of the associated domains. In most cases, these associated domains are involved in transcriptional regulation, and at least three different types of 'transcriptional activation domains' have been recognized by sequence comparisons (1, 127). For each of these types of transcriptional activation domains, other proteins exist in which the transcriptional activation domain is found but the RR domain is absent (reviewed in Reference 127). Apparently, the activity of the transcriptional activation domain in those cases is regulated differently.

The RR domain contains several highly conserved motifs (reviewed in detail by Volz (50, 129)). Near the N-terminal end of the domain (~ position 12) a series of several acidic amino acids, usually aspartates, are found. An aspartic acid is typically found near position 55; this is the site of phosphorylation of the RR domain (173, 174). A threonine residue is typically found near position 87, and a lysine residue is typically found near position 105. The three-dimensional structure of the CheY protein, which consists of just the RR domain, has been determined by X-ray crystallography and NMR analysis, and the structure of the RR domain from the Nitrogen Regulator I (NtrC) protein of enteric bacteria has been determined by NMR analysis (211–214). These investigations revealed that the RR domain consists of a five-stranded β-sheet surrounded by α-helices. The highly conserved lysine and aspartate residues noted above are clustered at one end of the domain (reviewed by Volz (50, 129)).

Two examples in which the RR domain comprises the entire protein are the 'tumble generator' of the chemotaxis system, CheY, and an essential component of the system regulating sporulation in *B. subtilis*, Spo0F. There is compelling evidence from both genetic and biochemical studies that indicates that upon phosphorylation, CheY interacts with the flagellar switch proteins to bring about reversals or pauses in flagellar rotation (reviewed in Reference 185). In the case of Spo0F, the RR domain serves quite a different function (168), as noted earlier. The phosphorylated form of Spo0F serves as the phosphoryl donor for the phosphorylation of a protein, Spo0B, that contains neither HPK nor RR domains. The phosphorylated form of the Spo0B protein then serves as the phosphoryl donor for the phosphorylation of the RR domain found in the Spo0A protein. Thus, Spo0F serves as part of a phosphorelay system that generates as its end product the phosphorylated form of the RR domain contained in Spo0A (168).

How does the phosphorylation of the RR domain regulate the activity of associated protein domains? In at least some cases, it appears that the RR domain negatively regulates the activity of associated domains, and that phosphorylation and the attendant stabilization of an alternative conformation of the RR domain somehow results in relief from this negative regulation. Perhaps the clearest example comes from the study of the CheB protein, a methylesterase that acts on the chemotaxis receptor proteins (215). The N-terminal RR domain of CheB regulates the methylase activity, as clearly shown by measurements of esterase activity in the presence or absence of phosphorylation of this RR domain. Similarly, proteolytic cleavage of the

RR domain results in activation of the esterase activity, as does genetic deletion of the RR domain. Thus, the RR domain of CheB appears to inhibit the esterase activity intrinsic to the C-terminal domain of CheB. An additional example where the RR domain appears to negatively regulate the activity of associated domains comes from the study of the DctD transcriptional activator of *Rhizobium* (216). This protein consists of an N-terminal RR domain, a central transcriptional activation domain, and a C-terminal DNA-binding domain. Genetic deletion of the N-terminal RR domain results in a protein that activates transcription constitutively (216). Similar studies with the Spo0A and FixJ proteins also suggest that the RR domain negatively regulates transcriptional activation by these proteins (217, 218).

3. Overview of the phosphotransfer reactions involving HPK and RR proteins

The phosphotransfer reactions in which the HPK and RR domains are known to take part are listed below:

$$\text{ATP} + \text{HPK} \rightleftharpoons \text{ADP} + \text{HPK}{\sim}\text{P} \quad (8.1)$$

$$\text{HPK}{\sim}\text{P} + \text{RR} \rightleftharpoons \text{HPK} + \text{RR}{\sim}\text{P} \quad (8.2)$$

$$\text{RR}{\sim}\text{P} + \text{H}_2\text{O} \longrightarrow \text{RR} + \text{P}_i \quad (8.3)$$

$$\text{Sum of (8.1)–(8.3): ATP} \rightleftharpoons \text{ADP} + \text{P}_i$$

$$\text{HPK} + \text{RR}{\sim}\text{P} + \text{H}_2\text{O} \longrightarrow \text{HPK} + \text{RR} + \text{P}_i \quad (8.4a)$$

$$\text{HPK}{\sim}\text{P} + \text{RR}{\sim}\text{P} + \text{H}_2\text{O} \rightleftharpoons \text{HPK}{\sim}\text{P} + \text{RR} + \text{P}_i \quad (8.4b)$$

$$\text{RR} + \text{N}{\sim}\text{P} \longrightarrow \text{RR}{\sim}\text{P} + \text{N} \quad (8.5)$$

$$\text{RR}{\sim}\text{P} + \text{phosphatase} + \text{H}_2\text{O} \longrightarrow \text{RR} + \text{P}_i + \text{phosphatase} \quad (8.6)$$

$$\text{HPK-H}_2 + \text{ATP} \longrightarrow \text{HPK-H}_2{\sim}\text{P} + \text{ADP} \quad (8.7)$$

Equation (8.1) depicts the reversible autophosphorylation reaction in which the conserved histidine moiety in the H-motif becomes phosphorylated at the expense of ATP. The equilibrium constant for the reaction (in the direction drawn) is about 0.1 for CheA and 0.5 for NRII, corresponding to phosphodonor potentials for the phosphohistidine in CheA and NRII of about −9 to −10 kcal/M (M. Surrette, P. Park, J. B. Stock, E. G. Ninfa, A. J. Ninfa, unpublished data). The reaction is probably driven to the right (as drawn) *in vivo* by the very high concentration of ATP and by the rapid dephosphorylation of the HPK~P by the cognate RR domain. *In vitro*, the forward reaction for the CheA and NRII proteins is slow with turnovers of ~10/min (215; M. Surrette, P. Park, J. B. Stock, E. G. Ninfa, A. J. Ninfa, unpublished data). The reaction proceeds very rapidly in the reverse direction (as drawn) *in vitro* (219; E. Kamberov and A. Ninfa, unpublished data). Studies with NRII indicate that ATP-γ-S is a suitable substrate for the autophosphorylation reaction (8.1), and that the thiophosphate moiety is transferred to the RR-containing Ntr protein, NRI (reaction 8.2) (E. Kamberov and A. Ninfa, unpublished data). Furthermore, the thiophosphate moiety in NRI is hydrolyzed similar to phosphate (reaction 8.3). Similar results were also obtained with the ATP analog 8-azido-ATP. In contrast, AMP–PCP and AMP–PNP are competitive inhibitors for the NRII autophosphorylation reaction (E. Kamberov and A. J. Ninfa, unpublished data).

For the HPK domain-containing proteins CheA and NRII, isotope exchange kinetics of the reaction were used to estimate the dissociation constants for ADP and ATP from the phosphorylated and unphosphorylated forms of the enzyme (220). In addition, the ATP concentration dependence for phosphorylation of the response regulator has been determined for these enzymes (219; M. Surrette, P. Park, J. B. Stock, E. G. Ninfa, A. J. Ninfa, E. Kamberov, unpublished data), and the interaction of CheA with ATP has been measured by the equilibrium column method of Hummel and Dryer (219). These studies suggest that the dissociation of ATP for the unphosphorylated forms of the enzymes are in the range 100–300 μM, and that the dissociation of ADP from the unphosphorylated form of the enzyme occurs at a slightly higher concentration. Results from the isotope exchange studies suggests that the dissociation of ATP from the phosphorylated form of the enzyme occurs at a much higher concentration of ATP (0.5–1.0 mM), and the dissociation of ADP from the phosphorylated form of the enzymes occurs at a low concentration (20–100 μM). Since the ATP concentration *in vivo* is ~4.0 mM while the ADP concentration is about an order of magnitude lower, it has been suggested that regulation of the autophosphorylation activity is not likely to occur by regulation of the binding of the substrate, ATP, to the enzyme (219). However, several lines of data suggest that ATP and ADP stabilize the enzyme in two distinct conformations, as discussed below.

There are three clear examples of proteins that are related to the HPK domain, yet lack the H-motif histidine residue; these are two anti-sigma factors from *B. subtilis* and the branched chain keto acid dehydrogenase kinase from mammalian mitochondria (147, 203–205). All three of these proteins are serine protein kinases, and they apparently do not transfer phosphoryl groups via an enzyme~P intermediate. Studies with one of the *B. subtilis* anti-sigma proteins revealed that this protein is controlled by another protein (an anti-anti-sigma) via a sequestration mechanism (221). The affinity of the anti-sigma for the sigma and for the anti-anti-sigma is regulated by the nature of the nucleotide bound to the anti-sigma. In the presence of ATP, the anti-sigma binds preferentially to the sigma factor; in the presence of ADP it binds preferentially to the anti-anti-sigma protein. Thus, it seems that the conformational state of the anti-sigma factor may be regulated by the ATP/ADP ratio, which may constitute an important signal for the control of sporulation of *B. subtilis* (221). An additional indication that the conformation of the HPK domain may be regulated by nucleotide binding will be considered below with the discussion of the regulated phosphatase activity of the HPK proteins (reactions 8.4a and 8.4b).

The autophosphorylation reaction (reaction 8.1) proceeds only by a trans-intramolecular route. This was shown definitively with the NRII protein, which is a stable dimer of identical subunits (202). Hybrid dimers were formed *in vitro* from mutant subunits lacking either the site of autophosphorylation or the ability to bind to ATP; within the dimer, ATP bound to one subunit phosphorylates the H-motif histidine within the other subunit (202). Furthermore, a *cis*-transfer mechanism was excluded in experiments using hybrid dimers containing wild-type subunits and a subunit unable to bind ATP; in these dimers only the mutant subunit was phosphorylated to any appreciable extent upon the addition of ATP (202). Results with the EnvZ and CheA proteins support these conclusions (199–201). In the case of CheA, the protein is not a stable dimer, rather the subunits in a population are in a dynamic equilibrium. Thus, simple mixing of mutant subunits unable to bind ATP with mutant subunits lacking the site of autophosphorylation results in the phosphorylation of those subunits containing the intact histidine site (200, 201).

The stoichiometry of CheA autophosphorylation upon incubation with ATP at high concentration approaches 1 phosphate/subunit. However, due to the dynamic equilibrium of CheA subunits, this result is not unexpected and provides no information as to whether phosphorylation of the two potential sites of phosphorylation per dimer is coupled. Experiments with the stably dimeric NRII indicate that at high ATP concentration, a stoichiometry of $\sim$1 phosphate/subunit is obtained (E. Kamberov and A. Ninfa, unpublished data). Furthermore, hemiphosphorylated dimers were not observed during the time-course of autophosphorylation. Finally, preliminary experiments suggest that at low concentrations of ATP, the initial rate of phosphorylation of the NRI RR domain by NRII is stimulated by the ATP analogues AMP–PCP and AMP–PNP (E. Kamberov and A. Ninfa, unpublished). These data suggest that the phosphorylation of the two H-motif histidines within the NRII dimer may be coupled.

One of the most common misperceptions concerning the two-component systems is that the environmental stimulus regulates only the autophosphorylation of the HPK domain. In the case of the NRII protein, this seems to not be the case. For this enzyme, the autophosphorylation reaction seems to be unregulated, with regulation of the concentration of the phosphorylated RR domain occuring by the regulated dephosphorylation of the RR$\sim$P (see below). On the other hand, the autophosphorylation of CheA is apparently an important regulatory locus. CheA is regulated by interaction with the CheW signal transduction protein and transmembrane receptors known as the methyl-accepting chemotaxis proteins (MCP). In the presence of the receptors and CheW, the rate of CheA autophosphorylation is increased several hundred fold, and this stimulation is greatly diminished in the presence of the small-molecule ligands of the MCP proteins (reviewed in Reference 185). The autophosphorylation of the FixL protein of *Rhizobium meliloti* is also regulated; in this case the environmental stimulus is molecular oxygen, which is sensed by a non-covalently associated heme bound to the non-conserved N-terminal portion of FixL. In the presence of oxygen, the rate of FixL autophosphorylation is greatly diminished (182, 195).

In contrast to the autophosphorylation reaction (8.1), the transfer of phosphoryl groups from the HPK domain to the cognate RR domain (reaction 8.2) is very rapid. Despite claims to the contrary (222), convincing evidence for regulation of this reaction by any factor other than protein concentration has not been presented. Experiments in which various HPK and RR proteins from different systems were combined have demonstrated that the transfer of phosphoryl groups from HPK~P to (non-cognate) RR readily occurs, but requires considerably higher protein concentrations and is much slower that the corresponding reaction between cognate HPK and RR (171, 223). Apparently this reflects the weaker interaction between non-cognate protein pairs than between cognate protein pairs.

Experiments with the technique of surface plasmon resonance have suggested that CheA binds tightly to RR-protein CheY in the absence of ATP, and that ATP and the attendant phosphorylation of CheY results in dissociation of this complex (209). In contrast, no stable complexes of NRII with NRI or with the N-terminal RR domain of NRI were observed in the absence of ATP (E. Kamberov, P. Chandran, and A. J. Ninfa, unpublished). Instead, a complex between NRII and NRI could be detected, but only in the presence of ATP. This complex was barely detectable using the plasmon resonance technique, indicating that it is considerably less stable than the CheA:CheY complex. Apparently, the interactions of cognate HPK and RR pairs may differ considerably.

Reactions (8.1) and (8.2) lead to the simplified kinetic model presented in Fig. 4. The depiction in Fig. 4 assumes that the HPK protein is a stable dimer, and the autophosphatase activity of the RR~P is ignored. Other phosphatase activities affecting the RR~P are also ignored; thus the model represents only the forward reaction leading to the RR~P and its reversal. Furthermore, none of the mixed dimers are considered, and the model thus assumes that there will be no differences in dissociations or reaction rates due the status of the other protomer in the dimeric proteins, which is

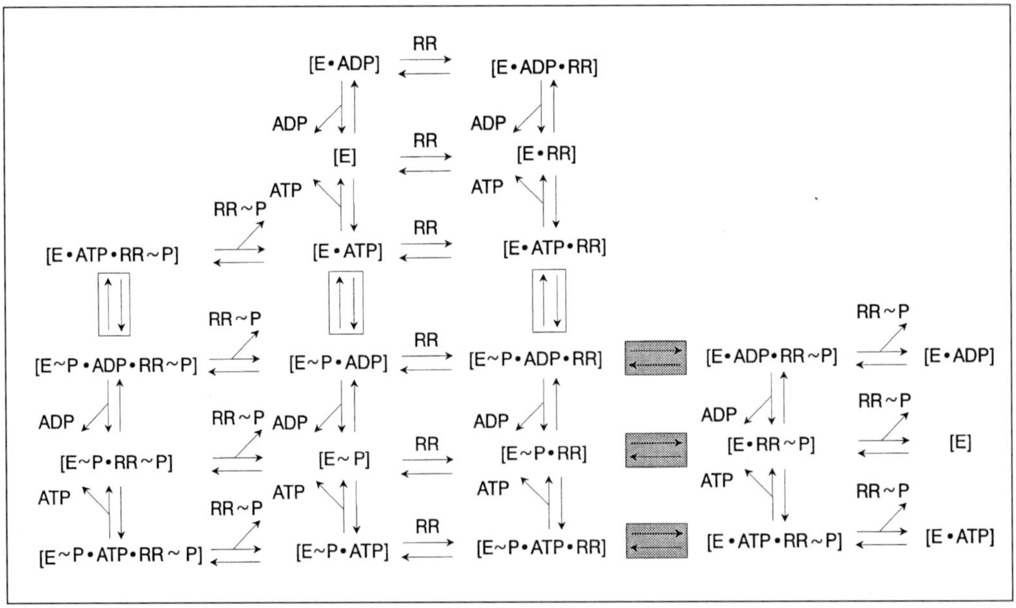

Fig. 4 Simplified kinetic model of phospho-transfer among two-component system protein domains. In this figure E represents a histidine protein kinase (HPK) domain. For further details, see text.

most unlikely. Finally, the model assumes that there is a single nucleotide binding site, i.e. no strictly allosteric nucleotide-binding sites. To present the model on a flat page, E, and three complexes, E.ATP, E.ADP, and E.ATP.RR~P, are shown twice. As shown, six enzyme complexes contain the RR~P, while six forms of the enzyme contain E~P (including E~P alone). Three enzyme forms may catalyze the phosphorylation of histidine within the H-motif (unfilled boxes), and three enzyme forms may take part in the phosphorylation of the active site aspartate in the RR domain (stippled boxes).

Little information is available as to relative rates of these reactions; the only useful data being the very high dissociation constant for the dissociation of ATP from the phosphorylated HPK domain (219, 220), which suggests that the pathway at the bottom of the figure may be relatively unimportant. It is important to realize that experiments designed to measure E~P (HPK~P) or RR~P in the presence of nucleotides measure the sum of these complexes. If such experiments are performed using similar concentrations of E and RR, the contribution by the various complexes may be considerable. As expected, the model predicts that the rate of RR~P production will be affected by the ATP, ADP, E, RR, and RR~P concentrations. ADP, RR~P, and ATP give rise to dead-end complexes: [E.ADP], [E.ADP.RR], [E~P.RR~P], [E~P.ATP.RR~P], and [E~P.ATP.RR~P]. Furthermore, [E~P.ADP.RR~P] can only produce ATP. Thus, ADP and RR~P, when high, should slow down the rate of production of RR~P. It has not been possible to do this experiment with very high concentrations of RR~P, owing to the instability of many RR~P (see below). However, the formation of RR~P in reaction mixtures containing ATP, excess RR, HPK, and various concentrations of ADP has been performed using the NRII and NRI proteins of the Ntr system. Surprisingly, ADP had little effect on the rate of formation of the RR~P (E. Kamberov and A. Ninfa, unpublished), but, since the experiment was performed at excess RR, this may only indicate that the rate of transfer from E~P to the RR is greater than the rate of transfer to ADP.

There is also potential inhibition of the rate of RR~P production by RR. For example, if the E.RR complex binds ATP less well than E, or if the rate of histidine phosphorylation by the E.ATP.RR complex is slower than the rate of histidine phosphorylation by the E.ATP complex, then high R will slow the rate by siphoning off the enzyme into less active complexes. This experiment has apparently not been performed.

The acylphosphate moiety in the RR~P has a characteristic instability at neutral pH that differs considerably depending on the protein (reviewed in Reference 130). The least stable of the characterized RR~P is CheB~P, which has a $t_{1/2}$ at neutral pH in the presence of Mg^{2+} of about 5–10 s; the most stable to date are OmpR~P and Spo0F~P, which have $t_{1/2}$ of several hours under these conditions. The chemical stability of an acylphosphate under these conditions is expected to be several hours. Thus, some of the RR~P seem to have an instability considerably greater than the expected chemical instability of the acylphosphate. Denaturation of the RR~P with SDS or chelation of the Mg^{2+} results in a great increase in the stability of these phosphoryl groups, thus the dephosphorylation of the native RR~P has been attributed to an 'autophosphatase' activity (151). This autophosphatase activity is greatly dimished by alteration of the conserved lysine near the C-terminus of the RR domain (224, 225; see Fig. 2). In the folded structure of the RR domain, this lysine lies adjacent to the phosphorylated aspartate residue. The rate of RR~P dephosphorylation will clearly affect the flux of phosphoryl groups through the HPK and RR domains (noted as the sum of reactions 8.1, 8.2, and 8.3). Also, since the RR~P may be involved in dead end complexes (Fig. 4), the autophosphatase activity may influence the rate of RR~P production.

The reactions listed in equations (8.4a) and (8.4b) depict the 'regulated phosphatase activity', by which the RR~P is dephosphorylated in a reaction stimulated by the HPK domain. The reaction is written in the forms shown in equations (8.4a) and (8.4b) because we do not know whether the phosphorylated HPK domain participates in this reaction. Many different HPK domains appear to have this activity (based on the results of physiological and genetic studies), and the activity is clearly present in the NRII, FixL, and EnvZ proteins (151, 160, 171, 175, 177, 179, 182). In the case of the NRII protein, it is known that the active site H-motif histidine is not required for the dephosphorylation of the RR~P, and that the reaction does not proceed by reversal of the phosphotransfer reaction (reaction 8.2) (179). Whether the regulated phosphatase activity represents the stimulation of the autophosphatase activity or occurs by another mechanism is unknown.

In two cases, the regulated phosphatase activity has been reconstituted in its properly regulated form *in vitro*; these are the FixL and NRII activities. In the former case, the activity is regulated by oxygen, which as noted earlier is sensed by a non-covalently associated heme moiety. In the latter case, the activity is controlled by the association of NRII with a regulatory subunit known as the PII protein (160). The availability of PII for this reaction is regulated in response to environmental signals by the covalent modification of PII. In its unmodified form, PII binds to NRII and elicits the regulated phosphatase activity (226). Several lines of evidence have indicated that PII is not itself a phosphatase but rather elicits this activity from NRII (160, 179).

An interesting feature of the NRII regulated phosphatase activity is that it is greatly stimulated by ATP but not by ADP (151, 177). Thus, these nucleotides may stabilize different conformations of the NRII protein. The phosphatase activity of the EnvZ protein was also stimulated by ATP; however, in this case, the activity was similarly stimulated by ADP (171). Thus, there may be a considerable difference in the way nucleotides affect conformation for the different HPK proteins. Another interesting feature of the PII-stimulated regulated phosphatase activity of NRII is that it apparently occurs simultaneously with the autophosphorylation and phosphotransfer (reactions 8.1 and 8.2) activities. That is, PII seems to have no effect on the autophosphorylation of NRII or on the transfer of phosphoryl groups to NRI (179). These results have led to the suggestion that the regulated phosphatase activity of NRII may be regulated independently from the other activities of NRII. In this regard NRII and FixL are clearly different, since in the case of FixL both the kinase and phosphatase activities of the protein are (reciprocally) regulated by oxygen (182).

In the case of the NRII and PII proteins, the regulated phosphatase activity is affected by the concentration of the unphosphorylated RR domain, suggesting that the RR forms a dead-end complex with the PII–NRII complex (179). Thus, the concentration of the RR, the concentration of the unmodified form of the PII protein, and the ATP/ADP ratio all have been observed to control the RR~P dephosphorylation rate *in vitro*.

The autophosphorylation of the RR domain at the expense of small molecule phosphodonors is depicted in reaction (8.5). The different RR domain-containing proteins have a characteristic spectrum of small-molecule phosphodonors that may serve as substrate (187, 190). So far, all RR proteins that have been tested may be phosphorylated by phosphoramidate. In addition, certain of the RR become phosphorylated when incubated with acetyl phosphate or carbamyl phosphate. In the case of the Ntr RR-containing protein NRI, it has been clearly demonstrated that the phosphorylation by small-molecule phosphodonors results in the same activated form that results upon phosphotransfer from the HPK~P (188). Furthermore, genetic and physiological studies of several two-component systems indicate that acetyl phosphate plays an important role in phosphorylating the RR *in vivo* (reviewed in References 192 and 193). Since the HPK proteins of most of these systems have a regulated phosphatase activity, mutations resulting in the accumulation of acetyl phosphate do not affect these systems in the presence of the HPK protein; mutations eliminating the HPK in these systems typically result in a low level constitutive activation of the regulatory target owing to the effect of acetyl phosphate. Thus, it has been an issue as to whether acetyl phosphate has a role in the signalling by two-component systems in cells that are not mutant for the cognate HPK. This issue appears to have been resolved by studies of the transcription of the flagellar control operon in *E. coli*; this operon is repressed by the OmpR response regulator protein only in cells able to synthesize acetyl phosphate, even in the presence of the cognate HPK protein (191).

The use of metabolic intermediates as signalling molecules, as is the case with acetyl phosphate, raises the possibility that the metabolic enzymes of the cell are also functioning as sensors of the environmental condition. Formally, the metabolite is acting both as metabolite and signalling molecule. In the case of acetyl phosphate, it is likely that what is being sensed is the acetyl-CoA pool, which, when elevated, may signify a condition of stress (reviewed in Reference 192).

It should also be noted that the use of a common phosphorylated metabolic intermediate, when coupled with the presence of regulated phosphatase activities, permits the indirect communication between distinct two-component systems. For example, it has been noted that in the presence of the Ntr HPK, NRII, the intracellular acetyl phosphate level is reduced about tenfold in a nitrogen-rich environment (189). This reduction may be due to the regulated phosphatase activity of NRII having the net effect of converting acetyl phosphate to inorganic phosphate by a combination of the reactions in equations (8.4a) and (8.5). Since acetyl phosphate affects flagellar gene expression via the phosphorylation of OmpR, it should be expected that flagellar expression is negatively regulated by NRII under conditions of nitrogen excess. This experiment has not yet been performed.

The dephosphorylation of the RR~P by distinct phosphatases is depicted in the reaction shown in equation (8.6). So far, there are three classes of such phosphatases known: the CheZ protein of the bacterial chemotaxis system, which is a phosphatase specific for CheY~P; the Spo0E gene product of the *B. subtilis* sporulation system, which is a phosphatase specific for Spo0A~P; and the Rap proteins of the sporulation system, which are phosphatases specific for Spo0F~P. Little is known concerning the regulation of these phosphatases, although the existence of such regulation seems likely. In the case of the Rap proteins, there is some evidence that the product of a partially overlapping gene brings about (directly or indirectly) regulation of the phosphatase activity after being secreted from the cell and then imported via a specific permease (J. Hoch and M. Perego, personal communication).

Finally, the phosphorylation of one class of HPK proteins at a second H-motif has been observed (equation 8.7). The HPK proteins of this class contain the orthodox HPK domain and a linked RR domain; the second H-motif is located downstream from the RR domain (Fig. 2). Preliminary studies as to the role of this second H-motif have suggested that phosphoryl groups may be transfered from this site to RR proteins, and that this site may be phosphorylated by HPK domains in an intermolecular reaction. As the discovery of this second H-motif is very recent, the physiological significance of this site of phosphorylation must await further studies.

4. Conclusions

It is not possible to cover the area of two-component systems in this short chapter, but the authors hope that we have conveyed a sense of the diversity and flexibility of these systems. Several key unresolved issues merit special attention, and we will briefly note these now. The most important of these is perhaps the mechanism and regulation of the regulated phosphatase activity, which plays a central regulatory role in many, if not most, of the systems. In particular, it is of interest whether this activity requires a conformation that excludes the autophosphorylation or phosphotransfer activities. We noted that the regulated phosphatase activity of the Ntr system appeared to occur in the presence of these other activities (179); however, since populations of protein molecules are involved in the *in vitro* assays, the possibility cannot be excluded that a small fraction of the molecules with very potent activity are responsible for the 'regulated phosphatase activity', and that in this small fraction of the molecules the other activities are switched off. In the FixL protein, the autophosphorylation activity and the regulated phosphatase activity are reciprocally regulated by oxygen (182). Thus, the issue of whether autophosphorylation and the regulated phosphatase activity require exclusive conformations is very much unresolved. Furthermore, whether this phosphatase activity represents a stimulation of the autophosphatase activity or is a separate activity remains to be resolved.

Another key question to be resolved is the role of ATP and ADP in regulating the conformational state of the HPK proteins. Although the dissociation constants for the few enzymes that have been studied so far suggest that nucleotide binding should not be a regulatory factor, it cannot be denied that nucleotides regulate the regulated phosphatase activity of NRII *in vitro* and also regulate the binding specificity of the closely related anti-sigma factors from *B. subtilis*.

Finally, the mechanisms of communication between different two-component systems should be of interest to work out, as these will reveal how the cell globally coordinates the expression of many systems in response to complex stimuli. These mechanisms of communication are likely to involve HPK and RR proteins with overlapping specificities, the interactions of various small-molecule phosphodonors with RR proteins, phosphatase activities specific for the various RR~P, and other, as yet undiscovered, factors.

References

1. Parkinson, J. S. and Kofoid, E. C. (1992) Communication modules in bacterial signaling proteins. *Annu. Rev. Genet.*, **26**, 71.
2. Ishizuka, H., Horinouchi, S., Kieser, H. M., Hopwood, D. A., and Beppu, T. (1992) A putative two-component regulatory sytem involved in secondary metabolism in Streptomyces spp. *J. Bacteriol.*, **174**, 7585.
3. Schweizer, H. P. (1991) The *agmR* gene, an environmentally responsive gene, complements defective *glpR*, which encodes the putative activator for glycerol metabolism in *Pseudomonas aeruginosa*. *J. Bacteriol.*, **173**, 6798.
4. Dziejman, M. and Mekalanos, J. J. (1995) Two-component signal transduction and its role in the expression of bacterial virulence factors. In *Two-component signal transduction* (ed. J. A. Hoch and T. J. Silhavy), p. 305. ASM Press, Washington, DC.
5. Deretic, V., Dikshit, R., Konyecsni, W. M., Chakrabarty, A. M., and Misra, T. K. (1989) The *algR* gene, which regulates mucoidy in *Pseudomonas aeruginosa*, belongs to a class of environmentally responsive genes. *J. Bacteriol.*, **171**, 1278.
6. Wozniak, D. J. and Ohman, D. E. (1991) *Pseudomonas aeruginosa* AlgB, a two-component response regulator of the NtrC family, is required for *algD* transcription. *J. Bacteriol.*, **173**, 1406.
7. Ueda, K., Miyake, K., Horinouchi, S., and Beppu, T. (1993) A gene cluster involved in aerial mycelium formation in *Streptomyces griseus* encodes proteins similar to the response regulators of two-component regulatory systems and membrane translocators. *J. Bacteriol.*, **175**, 2006.
8. Iuchi, S. and Lin, E. C. C. (1995) Signal transduction in the Arc system for control of operons encoding aerobic respiratory enzymes. In *Two-component signal transduction* (ed. J. A. Hoch, and T. J. Silhavy), p. 223. ASM Press, Washington, DC.
9. Ishige, K., Nagasawa, S., Tokishita, S., and Mizuno, T. (1994) A novel device of bacterial signal transducers. *EMBO J.*, **13**, 5195.
10. Fleischman, R. D., Adams, M. D., White, O., Clayton, R. A., Kirkness, E. F., and Kerlavage, A. R. (1995) Whole genome random sequencing and assembly of *Haemophilus influenzae* Rd. *Science.*, **269**, 496.
11. Plamann, L., Li, Y., Cantwell, B., and Mayor, J. (1995) The *Myxococcus xanthus asgA* gene encodes a novel signal transduction protein required for multicellular development. *J. Bacteriol.*, **177**, 2014.
12. Canellakis, E. S., Paterakis, A. A., Huang S.-C., Pandagiotidis, C. A., and Kyriakidis, D. A. (1993) Identification, cloning, and nucleotide sdequencing of the ornithine decarboxylase antizyme gene of *Escherichia coli*. *Proc. Natl. Acad. Sci. USA*, **90**, 7129.

13. Nagasawa, S.,Ishige, K., and Mizuno, T. (1993) Novel members of the 2-component signal-transduction genes in *Escherichia coli. J. Biochem.*, **114**, 350.
14. Nagasawa, S., Tokishita, S., Aiba, H., and Mizuno, T. (1992) A novel sensor-regulator protein that belongs to the homologous family of signal-transduction proteins involved in adaptive responses in *Escherichia coli. Mol. Microbiol.*, **6**, 799.
15. Uhl, M. A. and Miller, J. F. (1995) *Bordetella pertussis* BvgAS virulence by two-component regulatory systems. In *Two-component signal transduction* (ed. J. A. Hoch and T. J. Silhavy), p. 333. ASM Press, Washington, DC.
16. Arico', B., Scarlato, V., Monack, D. M., Falkow, S., and Rappuoli, R. (1991) Structural and genetic analysis of the *bvg* locus in *Bordetella* species. *Mol. Microbiol.*, **5**, 2481.
17. Chattopadhyay, S., Mukherjee, A., and Ghosh, S. (1994) Molecular cloning and sequencing of an operon, *carRS* of *Azospirillum brasilense*, that codes for a novel two-component regulatory system: Demonstration of a positive regulatory role of *carR* for global control of carbohydrate catabolism. *J. Bacteriol.*, **176**, 7484.
18. Amsler, C. D. and Matsumura, P. (1995) Chemotactic signal transduction in *Escherichia coli* and *Salmonella typhimurium*. In *Two-component signal transduction* (ed. J. A. Hoch and T. J. Silhavy), p. 89. ASM Press, Washington, DC.
19. Rudolph, J. and Oesterhelt, D. (1995) Chemotaxis and phototaxis require a CheA histidine kinase in the archaeon *Halobacterium salinarium. EMBO J.*, **14**, 667.
20. Dons, L., Olsen, J. E., and Rasmussen, O. F. (1994) Characterization of two putative *Listeria monocytogenes* genes encoding polypeptides homologous to the sensor protein CheA and the response regulator CheY of chemotaxis. *DNA Sequence*, **4**, 301.
21. Fredrick, K. L. and Helmann, J. D. (1994) Dual chemotaxis signaling pathways in *Bacillus subtilis*: a σ^D dependent gene encodes a novel protein with both cheW and cheY homologous domains. *J. Bacteriol.*, **176**, 2727.
22. Heath, J. D., Charles, T. C., and Nester, E. W. (1995) Ti plasmid and chromosomally encoded two-component systems important in plant cell transformation by *Agrobacterium* species. In *Two-component signal transduction* (ed. J. A. Hoch and T. J. Silhavy), p. 367. ASM Press, Washington, DC.
23. Guenzi, E., Gasc, A., Sicard, M. A., and Hakenbeck, R. (1994) A two-component signal-transducing system is involved in competence and penicillin susceptibility in laboratory mutants of *Streptococcus pneumoniae. Mol. Microbiol.*, **12**, 505.
24. Msadek, T., Kunst, F., and Rapoport, G. (1995) A signal transduction network in *Bacillus subtilis* includes the DegS/DegU and ComP/ComA two-component systems. In *Two-component signal transduction* (ed. J. A. Hoch and T. J. Silhavy), p 447. ASM Press, Washington, DC.
25. Mills, S. D., Jasalavich, C. A., and Cooksey, D. A. (1993) A two-component regulatory system required for copper-inducible expression of the copper resistance operon of *Pseudomonas syringae. J. Bacteriol.*, **175**, 1656.
26. Brown, N. L., Lee, B. T. O., and Silver, S. (1994) Bacterial transport of and resistance to copper. In *Metal ions in biological systems* (ed. H. Sigel and A. Sigel), Vol. 30, p. 405. Marcel Dekker, New York.
27. Dong, H., Iuchi, S., Kwan, H.-S., Lu, Z., and Lin, E. C. C. (1993) The deduced amino-acid sequence of the cloned *cpxR* gene suggests the protein is the cognate regulator for the membrane sensor, CpxA, in a two-component signal transduction system of *Escherichia coli. Gene*, **136**, 227.
28. Nakayama, S.-I. and Watanabe, H. (1995) Involvement of *cpxA*, a sensor of a two-component regulatory system, in the pH-dependent regulation of expression of *Shigella sonnei vir*F gene. *J. Bacteriol.*, **177**, 5062.

29. Wanner, B. L. (1995) Signal transduction and cross regulation in the *Esherichia coli* phosphate regulon by PhoR, CreC, and acetyl phosphate. In *Two-component signal transduction* (ed. J. A. Hoch and T. J. Silhavy), p. 203. ASM Press, Washington, DC.
30. Tseng, H.-C. and Chen, C. W. (1991) A cloned *ompR*-like gene of *Streptomyces lividans* 66 suppresses defective *melC1*, a putative copper-transfer gene. *Mol. Microbiol.*, **5**, 1187.
31. Gu, B., Lee, J. H., Hoover, T. R., Scholl, D., and Nixon, B. T. (1994) *Rhizobium meliloti* DctD, a σ^{54}-dependent transcriptional activator, may be negatively controlled by a subdomain in the C-terminal end of its two-component receiver module. *Mol. Microbiol.*, **13**, 51.
32. Hamblin, M. J., Shaw, J. G., and Kelly, D. J. (1993) Sequence analysis and interposon mutagenesis of a sensor-kinase (DctS) and response-regulator (DctR) controlling synthesis of the high-affinity C4-dicarboxylate transport system in *Rhodobacter capsulatus. Mol. Gen. Genet.*, **237**, 215.
33. Masui, N. H., Fujiwara, N., Takagi, M., and Imanaka, T. (1992) Cloning and nucleotide sequence of the regulatory gene, *degM*, for minor serine protease in *Bacillus subtilis. J. Ferment. Bioeng.*, **74**, 230.
34. Mann, N. H. (1994) Protein phosphorylation in cyanobacteria. *Microbiology*, **140**, 3207.
35. Lane, T., Benson, A., Hecht, G. B., Burton, G. J., and Newton, A. (1995) Switches and signal transduction networks in the *Caulobacter crescentus* cell cycle. In *Two-component signal trnsduction* (ed. J. A. Hoch and T. J. Silhavy), p. 403. ASM Press, Washington, DC.
36. Otten, S. L., Ferguson, J., and Hutchinson, C. R. (1995) Regulation of Daunorubicin production in *Streptomyces peucetius* by the dnrR_2 locus. *J. Bacteriol.*, **177**, 1216.
37. Utsumi, R., Katayama, S., Taniguchi, M., Horie, M., Ikeda, S., Igaki, H., Nakagawa, H., Miwa, A., Tanabe, H., and Noda, M. (1994) Newly identified genes involved in the signal transdution of *Escherichia coli* K-12. *Gene*, **140**, 73.
38. Swenson, D. L. and Clegg, S. (1992) Identification of ancillary *fim* genes affecting *fimA* expression in *Salmonella typhimurium*. *J. Bacteriol.*, **174**, 7697.
39. Agron, P. G., and Helinski, D. R. (1995) Symbiotic expression of *Rhizobium meliloti* nitrogen fixation genes is regulated by oxygen. In *Two-component signal transduction* (ed. J. A. Hoch and T. J. Silhavy), p. 275. ASM Press, Washington, DC.
40. Anthamatten, D. and Hennecke, H. (1991) The regulatory status of the *fixL*- and *fixJ*-like genes in *Bradyrhizobium japonicum* may be different from that in *Rhizobium meliloti. Mol. Gen. Genet.*, **225**, 38.
41. Wu, J., Benson, A. K., and Newton, A. (1995) Global regulation of a σ^{54}-dependent flagellar gene family in *Caulobacter crescentus* by the transcriptional activator FlbD. *J. Bacteriol.*, **177**, 3241.
42. Shi, W. and Zusman, D. R. (1995) The *frz* signal transduction system controls multicellular behavior in *Myxococcus xanthus*. In *Two-component signal transduction* (ed. J. A. Hoch and T. J. Silhavy), p. 419. ASM Press, Washington, DC.
43. Rich, J. J., Kinscherf, T. G., Kitten, T., and Willis, D. K. (1994) Genetic evidence that the *gacA* gene encodes the cognate response regulator for the *lemA* sensor in *Pseudomonas syringae. J. Bacteriol.*, **176**, 7568.
44. Mitamura, T., Ebora, R. V., Nakai, T., Makino, Y., Negoro, S., Urabe, I., and Okada, H. (1990) Structure of isoxyme genes of glucose dehydrogenase from *Bacillus megaterium* IAM1030. *J. Ferment. Bioeng.*, **70**, 363.
46. Malvar, T. and Baum, J. A. (1994) Tn*5401* disruption of the Spo0F gene, identified by direct chromosomal sequencing, results in CryIIIA overproduction in *Bacillus thuringiensis. J. Bacteriol.*, **176**, 4750.
47. Eberz, G. and Friedrich, B. (1991) Three *trans*-acting regulatory functions control hydrogenase synthesis in *Alcaligenes eutrophus. J. Bacteriol.*, **173**, 1845.

48. Lenz, O., Schwartz, E., Dernedde, J., Eitinger, M., and Friedrich, B. (1994) The *Alcaligenes eutrophus* H16 *hoxX* gene participates in hydrogenase regulation. *J. Bacteriol.*, **176**, 4385.
49. Van Soom, C., Verreth, C., Sampaio, M. K., and Vanderleyden, J. (1993) Identification of a potential transcriptional regulator of hydrogenase activity in free-living *Bradyrhizobium japonicum* strains. *Mol. Gen. Genet.*, **329**, 235.
50. Volz, K. (1993) Structural conservation in the CheY superfamily. *Biochemistry*, **32**, 11741.
51. Richaud, P., Colbeau, A., Toussaint, B., and Vignais, P. M. (1991) Identification and sequence analysis of the *hupR1* gene, which encodes a response regulator of the NtrC family required for hydrogenase expression in *Rhodobacter capsulatus. J. Bacteriol.*, **173**, 5928.
52. Elsen, S., Richaud, P., Colbeau, A., and Vignais, P. M. (1993) Sequence analysis and interposon mutagenesis of the *hupT* gene, which encodes a sensor protein involved in repression of hydrogenase synthesis in *Rhodobacter capsulatus. J. Bacteriol.*, **175**, 7404.
53. Chopra, A. K., Peterson, J. W., and Prasad, R. (1991) Cloning and sequence analysis of hydrogenase regulatory genes (*hydHG*) from *Salmonella typhimurium. Biochim. Biophys. Acta.*, **1129**, 115.
54. Stoker, K., Reijnders, W. N. M., Oltmann, L. F., and Stouthamer, A. H. (1989) Initial cloning and sequencing of *hydHG*, an operon homologous to *ntrBC* and regulating the labile hydrogenase activity in *Escherichia coli* K-12. *J. Bacteriol.*, **171**, 4448.
55. Walderhaug, M. P., Poarek, J. W., Voelkner, P., Daniel, J. M., Hesse, J. E., Altendorf, K., and Epstein, W. (1992) KpdD and KpdE, proteins that control expression of the *kpdABC* operon, are members of the two-component sensor-effector class of regulators. *J. Bacteriol.*, **174**, 2152.
56. Bassler, B. L. and Silverman, M. R. (1995) Intercellular communication in marine *Vibrio* species: Density-dependent regulation of the expression of bioluminescence. In *Two-component signal transduction* (ed. J. A. Hoch and T. J. Silhavy), p. 431. ASM Press, Washington, DC.
57. Harms, N., Reijnders, W. N. M., Anazawa, H., Van der Palen, C. J. N. M., van Spanning, R. J. M., Oltmann, L. F., and Stouthamer, A. H. (1993) Identification of a two-component regulatory system controlling methanol dehydrogenase synthesis in *Paracoccus denitrificans. Mol. Microbiol.*, **8**, 457.
58. Yang, H., Reijnders, W. N. M., van Spanning, R. J. M., Stouthamer, A. H., and Harms, N. (1995) Expression of the structural *mox* genes in *Paracoccus denitrificans* follows wild-type regulation in mutants with a deletion in *mxaY*, the gene encoding the signal sensor. *Microbiology*, **141**, 825.
59. Allen, B. L., Gerlach, F.-F., and Clegg, S. (1991) Nucleotide sequence and functions of *mrk* determinants necessary for expression of type 3 fimbrae in *Klebsiella pneumoniae. J. Bacteriol.*, **173**, 916.
60. Perez-Casal, J., Caparon, M. G., and Scott, J. R. (1991) Mry, a *trans*-acting positive regulator of the M protein gene of *Streptococcus pyogenes* with similarity to the receptor proteins of two-component regulatory systems. *J. Bacteriol.*, **173**, 2617.
61. McLandsborough, L. A. and Cleary, P. P. (1995) Insertional inactivation of *virR* in *Streptococcus pyogenes* M49 demonstrates that VirR functions as a positive regulator of ScpA, FcRA, OF, and M protein. *FEMS Microbiol. Lett.*, **128**, 45.
62. Curcic, R., Dhandayuthapani, S., and Deretic, V. (1994) Gene expression in mycobacteria: transcriptional fusions based on *xylE* and analysis of the promoter region of the response regulator *mtrA* from *Mycobacterium tuberculosis. Mol. Microbiol.*, **13**, 1057.
63. Stewart, V. and Rabin, R. S. (1995) Dual sensors and dual response regulators interact to

control nitrate- and nitrite-responsive gene expression in *Escherichia coli*. In *Two-component signal transduction* (ed. J. A. Hoch and T. J. Silhavy), p. 233. ASM Press, Washington, DC.
64. van der Meer, J. R., Polman, J., Beerthuyzin, M. M., Siezin, R. J., Kuipers, O. P., and De Vos, W. M. (1993) Characterization of the *Lactococcus lactis* nisin A operon genes *nisP*, encoding a subtilisin-like serine protease involved in precursor processing, and *nisR*, encoding a regulatory protein involved in nisin biosynthesis. *J. Bacteriol.*, **175**, 2578.
65. Engelke, G., Gutowski-eckel, Z., Kiesau, P., Siegers, K., Mannelmann, M., and Entian, K.-D. (1994) Regulation of nisin biosythesis and immunity in *Lactococcus lactis* 6F3. *Appl. Environ. Microbiol.*, **60**, 814.
66. Grob, P., Michel, P., Hennecke, H., and Gottfert, M. (1993) A novel response-regulator is able to suppress the nodulation defect of a *Bradyrhizobium japonicum nodW* mutant. *Mol. Gen. Genet.*, **241**, 531.
67. Grob, P., Hennecke, H., and Gottfert, M. (1994) Cross-talk between the two-component regulatory systems NodVW and NwsAB of *Bradyrhizobium japonicum. FEMS Microbiol Lett.*, **120**, 349.
68. Jones, R. and Haselkorn, R. (1989) The DNA sequence of the *Rhodobacter capsulatus ntrA*, *ntrB* and *ntrC* gene analogues required for nitrogen fixation. *Mol. Gen. Genet.*, **215**, 507.
69. Ninfa, A. J., Atkinson, M. R., Kamberov, E. S., Feng, J., and Ninfa, E. G. (1995) Control of nitrogen assimilation by the NR_I–NR_{II} two-component system of enteric bacteria. In *Two-component signal transduction* (ed. J. A. Hoch and T. J. Silhavy), p. 67. ASM Press, Washington, DC.
70. Maharaj, R., Rumbak, E., Jones, W. A., Robb, S. M., and Robb, F. T. (1989) Nucleotide sequence of the *Vibrio alginolyticus glnA* region. *Arch. Microbiol.*, **152**, 542.
71. Wardhan, H., McPherson, M. J., and Sastry, G. R. (1989) Identification, cloning and sequence analysis of the nitrogen regulation gene *ntrC* of *Agrobacterium tumefaciens. Mol. Plant Microb. Interact.*, **2**, 241.
72. Nixon, B. T., Ronson, C. W., and Ausubel, F. M. (1986) Two-component regulatory systems responsive to environmental stimuli share strongly conserved domains with the nitrogen assimilation regulatory genes *htrB* and *ntrC*. *Proc. Natl. Acad. Sci. USA*, **83**, 7850.
73. Szeto, W. W., Nixon, B. T., Ronson, C. W., and Ausubel, F. M. (1987) Identification and characterization of the *Rhizobium meliloti ntrC* gene: *R. meliloti* has separate regulatory pathways for activation of nitrogen fixation genes in free-living and symbiotic cells. *J. Bacteriol.*, **169**, 1423.
74. Klose, K. E., Weiss, D. S., and Kustu, S. (1993) Glutamate at the site of phophorylation of nitrogen-regulatory protein NTRC mimics aspartyl-phosphate and activates the protein. *J. Mol. Biol.*, **232**, 67.
75. Wray, L. V. and Fisher, S. H. (1993) The *Streptomyces coelicolor glnR* gene encodes a protein similar to other bacterial response regulators. *Gene*, **130**, 145.
76. Pawlowski, K., Klosse, U., and de Bruijn, F. J. (1991) Characterization of a novel *Azorhizobium caulinodans* ORS571 two-component regulatory system, NtrY/NtrX, involved in nitrogen fixation and metabolism. *Mol. Gen. Genet.*, **231**, 124.
77. Pratt, L. A. and Silhavy, T. J. (1995) Porin regulon of *Escherichia coli*. In *Two-component signal transduction* (ed. J. A. Hoch and T. J. Silhavy), p. 105. ASM Press, Washington, DC.
78. Liang, J., Scappino, L., and Haselkorn, R. (1992) The *patA* gene product, which contains a region similar to CheY of *Escherichia coli*, controls heterocyst pattern formation in the cyanobacterium *Anabaena* 7120. *Proc. Natl. Acad. Sci. USA*, **89**, 5655.
79. Huang, J., Carney, B. F., Denny, T. P., Weissinger, A. J., and Schell, M. A. (1995) A complex

network regulates expression of *eps* and other virulence genes of *Pseudomonas solanacearum*. *J. Bacteriol.*, **177**, 1259.
80. Tokito, M. K. and Daidal, F. (1992) *petR*, located upstream of the *fbcFBC* operon encoding the cytochrome bc_1 complex, is homologous to bacterial response regulators and necessary for photosynthetic and respiratory growth of *Rhodobacter capsulatus*. *Mol. Microbiol.*, **6**, 1645.
81. Dean, C. R. and Poole, K. (1993) Expression of the ferric enterobactin receptor (PfeA) of *Pseudomonas aeruginosa*: involvement of a two-component regulatory system. *Mol. Microbiol.*, **8**, 1095.
82. Yang, Y.-L., Goldrick, D., and Hong, J.-S. (1988) Identification of the products and nucleotide sequences of two regulatory genes involved in the exogenous induction of phosphoglycerate transport in *Salmonella typhimurium*. *J. Bacteriol.*, **170**, 4299.
83. Jiang, S.-Q., Yu, G.-Q., Li, Z.-G., and Hong, J.-S. (1988) Genetic evidence for modulation of the activator by two regulatory proteins involved in the exogenous induction of phosphoglycerate transport in *Salmonella typhimurium*. *J. Bacteriol.*, **170**, 4304.
84. Anba, J., Bidaud, M., Vasil, M. L., and Lazdunski, A. (1990) Nucleotide sequence of the *Pseudomonas aeruginosa phoB* gene, the regulatory gene for the phosphate regulon. *J. Bacteriol.*, **172**, 4685.
85. Lee, T.-Y., Makino, K., Shinagawa, H., Amemura, M., and Nakata, A. (1989) Phosphate regulon in members of the family *Enterobacteriaceae*: comparison of the *phoB-phoR* operons of *Escherichia coli, Shigella dysenteriae*, and *Klebsiella pneumoniae.J. Bacteriol.*, **171**, 6593.
86. Scholten, M., Janssen, R., Bogaarts, C., van Strein, J., and Tommassen, J. (1995) The *pho* regulon of *Shigella flexneri*. *Mol. Microbiol.*, **15**, 247.
87. Groisman, E. A. and Heffron, F. (1995) Regulation of *Salmonella* virulence by two-component regulatory systems. In *Two-component signal transduction* (ed. J. A. Hoch and T. J. Silhavy), p. 319. ASM Press, Washington, DC.
88. Kasahara, M., Nakata, A., and Shinagawa, H. (1992) Molecular analysis of the *Escherichia coli phoP-phoQ* operon. *J. Bacteriol.*, **174**, 492.
89. Hulett, F. M. (1995) Complex phosphate regulation by sequential switches in *Bacillus subtilis*. In *Two-component signal transduction* (ed. J. A. Hoch and T. J. Silhavy), p. 289. ASM Press, Washington, DC.
90. Boyd, J. M., Koga, T., and Lory, S. (1994) Identification and characterization of PilS, and essential retulator of pilin expression in *Pseudomonas aeruginosa*. *Mol. Gen. Genet.*, **243**, 565.
91. Ishimoto, K. S. and Lory, S. (1992) Identification of *pilR*, which encodes a transcriptional activator of the *Pseudomonas aeruginosa* pilin gene. *J. Bacteriol.*, **174**, 3514.
92. Taha, M.-K. and Giorgini, D. (1995) Phosphorylation and functional analysis of PilA, a protein involved in the transcriptional regulation of the pilin gene in *Neisseria gonorrhoeae*. *Mol. Microbiol.*, **15**, 667.
93. Darzins, A. (1994) Characterization of a *Pseudomonas aeruginosa* gene cluster involved in pilus biosynthesis and twitching motility: sequence similarity to the chemotaxis proteins of enterics and the gliding bacterium *Myxococcus xanthus*. *Mol. Microbiol.*, **11**, 137.
94. Diep, D. B., Havarstein, S., Nissen-Meyer, J., and Nes, I. F. (1994) The gene encoding plantaricin A, a bacteriocin from *Lactobacillus plantarum* C11, is located on the same transcription unit as an *agr*-like regulatory system. *Appl. Environ. Microbiol.*, **60**, 160.
95. Roland, K. L., Martin, L. E., Esther, C. R., and Spitznagel, J. K. (1993) Spontaneous *pmrA* mutants of *Salmonella typhimurium* LT2 define a new two-component regulatory system with a possible role in virulence. *J. Bacteriol.*, **175**, 4154.

96. Eraso, J. M. and Kaplan, S. (1994) *prrA*, a putative response regulator involved in oxygen regulation of photosynthesis gene expression in *Rhodobacter sphaeroides*. *J. Bacteriol.*, **176**, 32.
97. Eraso, J. M. and Kaplan, S. (1995) Oxygen-insensitive synthesis of the photosynthetic membranes of *Rhodobacter sphaeroides*: a mutant histidine kinase. *J. Bacteriol.*, **177**, 2695.
98. Ma, H. and Kendall, K. (1994) Cloning and analysis of a gene cluster from *Streptomyces coelicolor* that causes accelerated aerial mycelium formation in *Streptomyces lividans*. *J. Bacteriol.*, **176**, 3800.
99. Chiang, G. G., Schaefer, M. R., and Grossman, A. R. (1992) Complementation of a red-light-indifferent cyanobacterial mutant. *Proc. Natl. Acad. Sci. USA*, **89**, 9415.
100. Golden, S. S. (1995) Light-responsive gene expression in Cyanobacteria. *J. Bacteriol.*, **177**, 1651.
101. Gottesman, S. (1995) Regulation of capsule synthesis: modification of the two-component paradigm by an accessory unstable regulator. In *Two-component signal transduction* (ed. J. A. Hoch and T. J. Silhavy), p. 253. ASM Press, Washington, DC.
102. Mosley, C. S., Suzuki, J. Y., and Bauer, C. E. (1994) Identification and molecular genetic characterization of a sensor kinase responsible for coordinately regulating light harvesting and reaction center gene expression in response to anaerobiosis. *J. Bacteriol.*, **176**, 7566.
103. Tang, J.-L., Liu, Y.-N., Barber, C. E., Dow, J. M., Wootton, J. C., and Daniels, M. J. (1991) Genetic and molecular analysis of a cluster of *rpf* genes involved in positive regulation of synthesis of extracellular enzymes and polysaccharide in *Xanthomonas campestris* pathovar *campestris*. *Mol. Gen. Genet.*, **226**, 409.
104. Rasmussen, B. A. and Kovacs, E. (1993) Cloning and identification of a two-component signal-transducing regulatory system from *Bacteroides fragilis*. *Mol. Microbiol.*, **7**, 765.
105. Salyers, A. A., Shoemaker, N. B., and Stevens, A. M. (1995) Tetracycline regulation of conjugal transfer genes. In *Two-component signal transduction* (ed. J. A. Hoch and T. J. Silhavy), p. 393. ASM Press, Washington, DC.
106. Rossi, M., Brent, W., Pollock, R., Reji, M. W., Keon, R. G., Fu, R., and Voordouw, G. (1993) The *hmc* operon of *Desulfovivrio vulgaris* subsp. *vulgaris* Hildenborough encodes a potential transmembrane redox protein complex. *J. Bacteriol.*, **175**, 4699.
107. Axelsson, L., and Holck, A. (1995) The genes involved in production of an immunity to sakacin A, a bacteriocin from *Lactobacillus sake* Lb706. *J. Bacteriol.*, **177**, 2125.
108. Nagaya, M., Aiba, H., and Mizuno, T. (1993) Cloning of a sensory-kinase-encoding gene that belongs to the two-component regulatory family from the cyanobacterium *Synechococcus* sp. PCC7942. *Gene*, **131**, 119.
109. Klein, C., Kaletta, C., and Entian, K.-D. (1993) Biosynthesis of the lantibiotic subtilin is regulated by a histidine kinase/response regulator system. *Appl. Environ. Microbiol.*, **59**, 296.
110. Aiba, H., Nagaya, M., and Mizuno, T. (1993) Sensor and regulator proteins from the cyanobacterium *Synechocococcus* species PCC7942 that belong to the bacterial signal-transduction protein families: implication in the adaptive response to phosphate limitation. *Mol. Microbiol.*, **8**, 81.
111. Nagaya, M., Aiba, H., and Mizuno, T. (1994) The *sphR* product, a two-component system response regulator protein, regulates phosphate assimilation in *Synechococcus* sp. strain PCC7942 by binding to two sites upstream from the *phoA* promoter. *J. Bacteriol.*, **176**, 2210.
112. Hoch, J. A. (1995) Control of cellular development in sporulating bacteria by the

phosphorelay two-component signal transduction system. In *Two-component signal transduction* (ed. J. A. Hoch and T. J. Silhavy), p. 129. ASM Press, Washington, DC.

113. LeDeaux, J. R and Grossman, A. D. (1995) Isolation and characterization of *kinC*, a gene that encodes a sensor kinase homologous to the sporulation sensor kinases KinA and KinB in *Bacillus subtilis*. *J. Bacteriol.*, **177**, 166.
114. Brown, D. P., Ganova-Raeva, L., Green, B. D., Wilkinson, S. R., Young, M., and Youngman, P. (1994) Characterization of *spo0A* homologues in diverse *Bacillus* and *Clostridium* species identifies a probable DNA-binding domain. *Mol. Microbiol.*, **14**, 411.
115. Widenhorn, K. A., Somers, J. M., and Kay, W. W. (1989) Genetic regulation of the tricarboxylate transport operon (*tctI*) of *Salmonella typhimurium*. *J. Bacteriol.*, **171**, 4436.
116. Simon, G., Mejean, V., Jourlin, C., Chippaux, M., and Pascal, M.-C. (1994) The *torR* gene of *Escherichia coli* encodes a response regulator protein involved in the expression of the trimethylamine *N*-oxide reductase genes. *J. Bacteriol.*, **176**, 5601.
117. Roecklein, B., Pelletier, A., and Kuempel, P. (1991) The *tus* gene of *Escherichia coli*; autoregulation, analysis of flanking sequences and identification of a complementary system in *Salmonella typnimurium*. *Res. Microbiol.*, **142**, 169.
118. Hill, T. M., Tecklenburg, M. L., Pelletier, A. J., and Kuempel, P. L. (1989) *tus*, the trans-acting gene required for termination of DNA replication in *Escherichia coli*, encodes a DNA-binding protein. *Proc. Natl. Acad. Sci. USA*, **86**, 1593.
119. Kadner, R. J. (1995) Expression of the Uhp sugar-phosphate transport system of *Escherichia coli*. In *Two-component signal transduction* (ed. J. A. Hoch and T. J. Silhavy), p. 263. ASM Press, Washington, DC.
120. Island, M. D., Wei, B.-Y., and Kadner, R. J. (1992) Structure and function of the *uhp* genes for the sugar phosphate transport system in *Escherichia coli* and *Salmonella typhimurium*. *J. Bacteriol.*, **174**, 2754.
121. Arthur, M., Depardieu, F., Holman, T., Wu, Z., Wright, G., Walsh, C. T., and Courvalin, P. (1995) Regulation of glycopeptide resistance genes of enterococcal transposon Tn1546 by the VanR–VanS two-component regulatory system. In *Two-component signal transduction* (ed. J. A. Hoch and T. J. Silhavy), p. 387. ASM Press, Washington, DC.
122. Lyristis, M., Bryant, A. E., Sloan, J., Awad, M. M., Nisbet, I. T., Stevens, D. L., and Rood, J. I. (1994) Identification and molecular analysis of a locus that regulates extracellular toxin prodution in *Clostridium perfringens*. *Mol. Microbiol.*, **12**, 761.
123. Schell, M. A., Denny, T. P., and Huang, J. (1993) Vsr A, a second two-component sensor regulating virulence genes of *Pseudomonas solanacearum*. *Mol Microbiol.*, **11**, 489.
124. Quirk, P. G., Guffanti, A. A., Clejan, S., Cheng, J., and Krulwich, T. A. (1994) Isolation of Tn917 insertional mutants of *Bacillus subtilis* that are resistant to the protonophore carbonyl cyanide *m*-chlorophenylhydrazone. *Biochim. Biophys. Acta.*, **1186**, 27.
125. Gaffney, T. D., Lam, S. T., Ligon, J., Gates, K., Frazelle, A., Di Maio, J., Hill, S., Goodwin, S., Torkewitz, N., Allshouse, A. M., Kempf, H.-J., and Becker, J. O. (1994) Global regulation of expression of antifungal factors by a *Pseudomonas gluorescens* biological control strain. *Mol. Plant-Microbe Int.*, **7**, 455.
126. Stock, A. M., Koshland, D. E., Jr, and Stock, J. B. (1985) Homologies between the *Salmonella typhimurium* CheY protein and proteins involved in the regulation of chemotaxis, membrane protein synthesis, and sporulation. *Proc. Natl. Acad. Sci. USA*, **82**, 7989.
127. Stock, J. B., Ninfa, A. J., and Stock, A. M. (1989) Protein phosphorylation and the regulation of adaptive responses in bacteria. *Microbiol. Rev.*, **53**, 450.
128. Stock, J. B., Surrette, M. G., Levit, M., and Park, P. (1995) Two-component signal trans-

duction systems: structure–function relationships and mechanism of catalysis. In *Two-component signal transduction* (ed. J. A. Hoch and T. J. Silhavy), p. 25. ASM Press, Washington, DC.

129. Volz, K. (1995) Structural and functional conservation in response regulators. In *Two-component signal transduction* (ed. J. A. Hoch and T. J. Silhavy), p. 53. ASM Press, Washington, DC.
130. Ninfa, A. J. (1991) Protein phosphorylation and the regulation of cellular processes by the homologous two-component regulatory systems of bacteria. *Genet. Eng.*, **13**, 39.
131. Albright, L. M., Huala, E., and Ausubel, F. M. (1989) Procaryotic signal transduction mediated by sensor and regulator protein paris. *Annu. Rev. Genet.*, **23**, 311.
132. Kofoid, E. C. and Parkinson, J. S. (1988) Transmitter and receiver modules in bacterial signaling proteins. *Proc. Natl. Acad. Sci. USA*, **85**, 4981.
133. Alex, L. A. and Simon, M. I. (1994) Protein histidine kinases and signal transduction in prokaryotes and eukaryotes. *Trends. Genet.*, **10**, 133.
134. Bourret, R. B., Borkovich, K. A., and Simon, M. I. (1991) Signal transduction pathways involving protein phosphorylation in prokaryotes. *Annu. Rev. Biochem.*, **60**, 401.
135. Swanson, R. V., Alex, L. A., and Simon, M. I. (1994) Histidine and asparate phosphorylation: two-component systems and the limits of homology. *Trends Biol. Sci.*, **19**, 485.
136. Gross, R., Arico, B., and Rappuoli, R. (1989) Families of bacterial signal-transducing proteins. *Mol. Microbiol.*, **3**, 1661.
137. Saier, M. H., Jr (1994) Bacterial sensor kinase/response regulator systems: an introduction. *Res. Microbiol.*, **145**, 349.
138. Pao, G. M., Tam, R., Lipschitz, L. S., and Saier, M. H., Jr (1994) Response regulators, structure, function, and evolution. *Res. Microbiol.*, **145**, 356.
139. Igo, M. M., Slauch, J. M., and Silhavy, T. J. (1990) Signal transduction in bacteria: kinases that control gene expression. *New Biol.*, **1**, 5.
140. Ronson, C. W., Nixon, B. T., and Ausubel, F. M. (1987) Conserved domains in bacterial regulatory proteins that respons to environmental stimuli. *Cell*, **49**, 579.
141. Hoch, J. A. and Silhavy, T. J. (eds) (1995) *Two-component signal transduction*. ASM Press, Washington, DC.
142. Maeda, T., Wurgler-Murphy, S. M., and Saito, H. (1994) A two-component system that regulates an osmosensing MAP kinase cascade in yeast. *Nature*, **369**, 242.
143. Ota, I. M. and Varshavsky, A. (1993) A yeast protein similar to bacterial two-component regulators. *Science*, **262**, 566.
144. Yu, G., Deschenes, R. J., and Fassler, J. S. (1995) The essential transcription factor, Mcm1, is a downstream target of Sln1, a yeast 'two-component' regulator. *J. Biol. Chem.*, **270**, 8739.
145. Chang, C., Kwok, S. F., Bleecker, A. B., and Meyerowitz, E. M. (1993) Arabidopsis ethylene-response gene ETR1: similarity of product to two-component regulators. *Science*, **262**, 539.
146. Yen, H.-C., Lee, S., Tanksley, S., Lanahan, M. B., Klee, H. J., and Giovannoni, J. J. (1995) The tomato *never-ripe* locus regulates ethylene inducible gene expression and is linked to a homologue of the *Arabidopsis ETR1* gene. *Plant Physiol.*, **107**, 1343.
147. Popov, K. M., Zhao, Y., Shimomura, Y., Kuntz, M. J., and Harris, R. A. (1992) Branched-chain a-ketoacid dehydrogenase kinase: molecular cloning, expression, and sequence similarity with histidine protein kinases. *J. Biol. Chem.*, **267**, 13127.
148. Thummler, F., Algarra, P., and Fobo, G. M. (1995) Sequence similarities of phytochrome to protein kinases: implications for the structure, function, and evolution of the phytochrome gene family. *FEBS Lett.*, **357**, 149.

149. Brown, J. L., Bussey, H., and Stewart, R. C. (1994) Yeast Skn7 functions in a eucaryotic two-component regulatory system. *EMBO J.*, **13**, 5186.
150. Hess, J. F., Oosawa, K., Matsumura, P., and Simon, M. I. (1987) Protein phosphorylation is involved in bacterial chemotaxis. *Proc. Natl. Acad. Sci. USA*, **84**, 7609.
151. Keener, J. and Kustu, S. G. (1988) Protein kinase and phosphoprotein phosphatase activities of nitrogen regulatory protein NTRB and NTRC of enteric bacteria: roles of the conserved amino terminal domain of NTRC. *Proc. Natl. Acad. Sci. USA*, **85**, 4976.
152. Weiss, V. and Magasanik, B. (1988) Phosphorylation of nitrogen regulator I (NRI) of *Escherichia coli*. *Proc. Natl. Acad. Sci. USA*, **85**, 8919.
153. Hess, J. F., Bourret, R. B., and Simon, M. I. (1988) Histidine phosphorylation and phosphoryl group transfer in bacterial chemotaxis. *Nature*, **336**, 139.
154. Ninfa, A. J. and Bennett, R. L. (1991) Identification of the site of autophosphorylation of the bacterial protein kinase/phosphatase NRII. *J. Biol. Chem.*, **266**, 6888.
155. Roberts, D. L., Bennett, D. W., and Forst, S. A. (1993) Identification of the site of phosphorylation on the osmosensor, EnvZ, of *Escherichia coli*. *J. Biol. Chem.*, **269**, 8728.
156. Igo, M. M. and Silhavy, T. J. (1988) EnvZ, a transmembrane environmental sensor of *Escherichia coli* K-12, is phosphorylated *in vitro*. *J. Bacteriol.*, **170**, 5971.
157. Jin, S., Prusti, R. K., Roitsch, T., Ankenbauer, R. G., and Nester, E. W. (1990) Phosphorylation of the VirG protein of *Agrobacterium tumefaciens* by the autophosphorylated VirA protein: essential role in biological activity of VirG. *J. Bacteriol.*, **172**, 4945.
158. Mukai, K., Kawata, M., and Tanaka, T. (1990) Isolation and phosphorylation of the *Bacillus subtilis degS* and *degU* gene products. *J. Biol. Chem.*, **265**, 20000.
159. Uhl, M. A. and Miller, J. F. (1994) Autophosphorylation and phosphotransfer in the *Bordatella pertussis* BvgAS signal transduction cascade. *Proc. Natl. Acad. Sci. USA*, **91**, 1163.
160. Ninfa, A. J. and Magasanik, B. (1986) Covalent modification of the *glnG* product, NRI, by the *glnL* product, NRII, regulates the transcription of the *glnALG* operon in *Escherichia coli*. *Proc. Natl. Acad. Sci. USA*, **83**, 5909.
161. Wylie, D., Stock, A., Wong, C.-Y., and Stock, J. (1988) Sensory transduction in bacterial chemotaxis involves phosphotransfer between Che proteins. *Biochem. Biophys. Res. Commun.*, **151**, 891.
162. Hess, J. F., Oosawa, K., Kaplan, N., and Simon, M. I. (1988) Phosphorylation of three proteins in the signalling pathway of bacterial chemotaxis. *Cell*, **53**, 79.
163. Makino, K., Shinegawa, H., Amemura, M., Kawamoto, T., Yamada, M., and Nakata, A. (1989) Signal transduction in the phosphate regulon of *Escherichia coli* involves phosphotransfer between PhoR and PhoB proteins. *J. Mol. Biol.*, **210**, 551.
164. Walker, M. S. and DeMoss, J. A. (1993) Phosphorylation and dephosphorylation catalyzed *in vitro* by purified components of the nitrate sensing system, NarX and NarL. *J. Biol. Chem.*, **268**, 8391.
165. Aiba, H. and Mizuno, T. (1990) Phosphorylation of a bacterial activator protein, OmpR, by a protein kinase, EnvZ, stimulates the transcription of the *ompF* and *ompC* genes in *Escherichia coli*. *FEBS Lett.*, **261**, 19.
166. Aiba, H., Mizuno, T., and Mizushima, S. (1989) Transfer of phosphoryl group between two regulatory proteins involved in osmoregulatory expression of the *ompF* and *ompC* genes in *Escherichia coli*. *J. Biol. Chem.*, **264**, 8563.
167. Amemura, M., Makino, K., Shinegawa, H., and Nakata, A. (1990) Cross talk to the phosphate regulon of *Escherichia coli* by PhoM protein: PhoM is a histidine protein kinase and catalyzes phosphorylation of PhoB and PhoM-open reading frame 2. *J. Bacteriol*, **172**, 6300.

168. Burbulys, D., Trach, K. A., and Hoch, J. A. (1991) Initiation of sporulation in *B. subtilis* is controlled by a multicomponent phosphorelay. *Cell*, **64**, 545.
169. Deretic, V., Leveau, J. H. J., Mohr, C. D., and Hibler, N. S. (1992) *In vitro* phosphorylation of AlgR, a regulator of mucoidy in *Pseudomonas aeruginosa*, by a histidine protein kinase and effects of small phospho-donor molecules. *Mol. Microbiol.*, **6**, 2761.
170. Gilles-Gonzalez, M. A., Ditta, G. S., and Helinski, D. R. (1991) A haemoprotein with kinase activity encoded by the oxygen sensor of *Rhizobium meliloti*. *Nature*, **350**, 170.
171. Igo, M. M., Ninfa, A. J., Stock, J. B., and Silhavy, T. J. (1989) Phosphorylation and dephosphorylation of a bacterial transcriptional activator by a transmembrane receptor. *Genes and Dev.*, **3**, 1725.
172. Igo, M. M., Ninfa, A. J., and Silhavy, T. J. (1989) A bacterial environmental sensor that functions as a protein kinase and stimulates rtranscriptional activation. *Genes and Dev.*, **3**, 598.
173. Sanders, D. A., Gillece-Castro, B. L., Burlingame, A. L., and Koshland, D. E. Jr (1992) Phosphorylation site of NtrC, a protein phosphatase whose covalent intermediate activities transcription. *J. Bacteriol.*, **174**, 5117.
174. Sanders, D. A., Gilleco-Castro, B. L., Stock, A. M., Burlingame, A. L., and Koshland, D. E. Jr (1989) Identification of the site of phosphorylation of the chemotaxis response regulator protein, CheY. *J. Biol. Chem.*, **264**, 21770.
175. Aiba, H., Nakasaki, F., Mizushima, S., and Mizuno, T. (1989) Evidence for physiological importance of the phosphotransfer between the two regulatory components, EnvZ and OmpR in osmoregulation of *Escherichia coli*. *J. Biol. Chem.*, **264**, 14090.
176. Chiang, R. C., Cavicchioli, R., and Gunsalus, R. P. (1992) Identification and characterization of *narQ*, a second nitrate sensor for nitrate-dependent gene regulation in *Escherichia coli*. *Mol. Microbiol.*, **6**, 1913.
177. Kamberov, E. S., Atkinson, M. R., and Ninfa, A. J. (1994) Effect of mutations in *Escherichia coli glnL* (*ntrB*), encoding nitrogen regulator II (NRII or NtrB), on the regulated phosphatase activity involved in bacterial nitrogen assimilation. *J. Biol. Chem.*, **269**, 28294.
178. Russo, F. D., Slauch, J. M., and Silhavy, T. J. (1993) Mutations that affect the separate functions of OmpR, the phosphorylated regulator of porin transcription in *Escherichia coli*. *J. Mol. Biol.*, **231**, 261.
179. Kamberov, E. S., Atkinson, M. R., Feng, J., Chandran, P., and Ninfa, A. J. (1994) Sensory components controlling bacterial nitrogen assimilation. *Cell. Mol. Biol. Res.*, **40**, 175.
180. Atkinson, M. R. and Ninfa, A. J. (1992) Characterization of *Escherichia coli glnL* mutations affecting nitrogen regulation. *J. Bacteriol.*, **174**, 4538.
181. Atkinson, M. R. and Ninfa, A. J. (1993) Mutational analysis of the signal-transducing kinase/phosphatase Nitrogen Regulator II (NRII or NtrB). *J. Bacteriol.*, **175**, 7016.
182. Lois, A. F., Weinstein, M., Ditta, G. S., and Helinski, D. R. (1993) Autophosphorylation and phosphatase activities of the oxygen-sensing protein FixL of *Rhizobium meliloti* are coordinately regulated by oxygen. *J. Biol. Chem.*, **268**, 4370.
183. Ohlsen, K. L., Grimsley, J. K., and Hoch, J. A. (1994) Deactivation of the sporulation transcription factor Spo0A by the Spo0E protein phosphatase. *Proc. Natl. Acad. Sci. USA*, **91**, 1756.
184. Perego, M., Hanstein, C., Welsh, K. W., Djavakhisvili, T., Glaser, P., and Hoch, J. A. (1994) Multiple protein aspartate phosphatases provide a mechanism for the integration of diverse signals in th control of development in *Bacillus subtillis*. *Cell*, **79**, 1047.
185. Stock, J. B., Surette, M. G., McCleary, W. R., and Stock, A. M. (1992) Signal transduction in bacterial chemotaxis. *J. Biol. Chem.*, **267**, 19753.

186. Wanner, B. L. and Wilmes-Reisenberg, M. R. (1992) Involvement of phosphotransacetylase, acetate kinase, and acetyl phosphate synthesis in the control of the phosphate regulon in *Escherichia coli. J. Bacteriol.*, **174**, 2124.
187. Lukat, G. S., McCleary, W. R., Stock, A. M., and Stock, J. B. (1992) Phosphorylation of bacterial response regulator proteins by low molecular weight phospho-donors. *Proc. Natl. Acad. Sci. USA*, **89**, 718.
188. Feng, J., Atkinson, M. R., McCleary, W., Stock, J. B., Wanner, B. L., and Ninfa, A. J. (1992) Role of pjhosphorylated metabolic intermediates in the regulation of glutamine synthetase synthesis in *Escherichia coli. J. Bacteriol.*, **174**, 6061.
189. Pruss, B. and Wolfe, A. J. 1994. Regulation of acetyl phosphate synthesis and degredation and the control of flagellar expression in *Escherichia coli. Mol. Microbiol.*, **12**, 973.
190. McCleary, W. R. and Stock, J. B. (1994) Acetyl phosphate and the activation of two-component response regulators. *J. Biol. Chem.*, **269**, 31567.
191. Shin, S. and Park, C. (1995) Modulation of flagellar expression in *Escherichia coli* by acetyl phosphate and osmoregulator OmpR. *J. Bacteriol.*, **177**, 4696.
192. McCleary, W. R., Stock, J. B., and Ninfa, A. J. (1993) Is acetyl phosphate a global signal in *Escherichia coli*? *J. Bacteriol.*, **175**, 2793.
193. Wanner, B. L. (1992) Is cross-regulation by phosphorylation of two-component response regulator proteins important in bacteria? *J. Bacteriol.*, **174**, 2053.
194. Monson, E. K., Weinstein, M., Ditta, G. S., and Helinski, D. R. (1992) The FixL protein of *Rhizobium meliloti* can be separated into a haem-binding oxygen-sensing domain and a functional C-terminal kinase domain. *Proc. Natl. Acad. Sci. USA*, **89**, 4280.
195. Gilles-Gonzalez, M. A. and Gonzalez, G. (1993) Regulation of the kinase activity of haem protein FixL from the two-component system FixL/FixJ of *Rhizobium meliloti. J. Biol. Chem.*, **268**, 16293.
196. Atkinson, M. R., Kamberov, E. S., Weiss, R. L., and Ninfa, A. J. (1994) Reversible uridylylation of the *Escherichia coli* PII signal transduction protein regulates its ability to stimulate the dephosphorylation of the transcription factor Nitrogen Regulator I (NRI or NtrC). *J. Biol. Chem.*, **269**, 28288.
197. Hoch, J. A. (1993) Regulation of the phosphorelay and the initiation of sporulation in *B. subtilis. Annu. Rev. Microbiol.*, **47**, 441.
198. Rabin, R. S. and Stewart, V. (1992) Either of two functionally redundant sensor proteins, NarX and NarQ, is sufficient for nitrate regulation in *Escherichia coli* K-12. *Proc. Natl. Acad. Sci. USA*, **89**, 8419.
199. Yang, Y. and Inouye, M. (1991) Intermolecular complementation between two defective mutant signal-transducing receptors of *Escherichia coli. Prov. Natl. Acad. Sci. USA*, **88**, 11057.
200. Swanson, R. V., Bourret, R. B., and Simon, M. I. (1993) Intermolecular complementation of the kinase activity of CheA. *Mol. Microbiol.*, **8**, 435.
201. Wolfe, A. J. and Stewart, R. C. (1993) The short form of the CheA protein restores kinase activity and chemotactic ability to kinase-deficient mutants. *Proc. Natl. Acad. Sci. USA*, **90**, 1518.
202. Ninfa, E. G., Atrkinson, M. R., Kamberov, E. S., and Ninfa, A. J. (1993) Mechanism of autophosphorylation of *Escherichia coli* nitrogen regulator II (NRII or NtrB): transphosphorylation between subunits. *J. Bacteriol.*, **175**, 7024.
203. Min, K.-T., Hilditch, C. M., Diedrich, B., Errington, J., and Yudkin, M. D. (1993) σ^F, the first compartment-specific transcription factor of *B. subtilis*, is regulated by an anti-σ factor that is also a protein kinase. *Cell*, **74**, 735.

204. Duncan, L. and Losick, R. (1993) SpoIIAB is an anti-σ factor that binds to and inhibits transcription by regulatory protein σ^F from *B. subtilis*. *Proc. Natl. Acad. Sci. USA*, **90**, 2325.
205. Benson, A. K. and Haldenwang, W. G. (1993) *Bacillus subtilis* σ^B is regulated by a binding protein (RsbW) that blocks its association with core RNA polymerase. *Proc. Natl. Acad. Sci. USA*, **90**, 2320.
206. Swanson, R. V., Schuster, S. C., and Simon, M. I. (1993) Expression of CheA fragments which define domains encoding kinase, phosphotransfer, and CheY binding activities. *Biochemistry*, **32**, 7623.
207. Morrison, T. B. and Parkinson, J. S. (1994) Liberation of an interaction domain from the phosphotransfer region of CheA, a signalling kinase of *Escherichia coli*. *Proc. Natl. Acad. Sci. USA*, **91**, 5485.
208. Ninfa, E. G., Stock, A., Mowbray, S., and Stock, J. B. (1991) Reconstitution of the bacterial chemotaxis signal transduction system from purified components. *J. Biol. Chem.*, **266**, 9764.
209. Schuster, S. C., Swanson, R. V., Alex, L. A., Bourret, R. B., and Simon, M. I. (1993) Assembly and function of a quarternary signal transduction complex monitored by surface plasmon resonance. *Nature*, **265**, 343.
210. Iuchi, S. (1993) Phosphorylation/dephosphorylation of the receiver module at the conserved aspartate residue controls transphosphorylation activity of histidine kinase in sensor protein ArcB of *Escherichia coli*. *J. BIol. Chem.*, **268**, 23972.
211. Stock, A. M., Mottonen, J. M., Stock, J. B., and Schutt, C. E. (1989) Three-dimensional structure of CheY, the response regulator of bacterial chemotaxis. *Nature*, **337**, 745.
212. Volz, K. and Matsumura, P. (1991) Crystal structure of *Escherichia coli* CheY refined at 1.7-A resolution. *J. Biol. Chem.*, **266**, 15511.
213. Bruix, M., Pascual, J., Santoro, J., Prieto, J., Serrano, L., and Rico, M. (1993) 1H- and 15N-NMR assignment and solution structure of the chemotactic *Escherichia coli* CheY protein. *Eur. J. Biochem.*, **215**, 573.
214. Volkman, B. F., Nohaile, M. J., Amy, N. K., Kustu, S., and Wemmer, D. E. (1995) Three-dimensional solution structure of the N-terminal receiver domain of NTRC. *Biochemistry*, **34**, 1413.
215. Lupas, A. and Stock, J. B. (1989) Phosphorylation of an N-terminal regulatory domain activates the CheB methylesterase in bacterial chemotaxis. *J. Biol. Chem.*, **264**, 17337.
216. Gu, B., Lee, J. H., Hoover, T. R., Scholl, D., and Nixon, B. T. (1994) *Rhizobium meliloti* DctD, a σ-^{54}-dependent transcriptional activator, may be negatively controlled by a subdomain in the C-terminal end of its two-component receiver module. *Mol. Microbiol.*, **13**, 51.
217. Grimsley, J. K., Tjalkens, R. B., Strauch, M. A., Bird, T. H., Spiegelman, G. B., Hostomsky, Z., Whiteley, J. M., and Hoch, J. A. (1994) Subunit composition and domain structure of the Spo0A sporulation transcription factor of *Bacillus subtilis*. *J. Biol. Chem.*, **269**, 16977.
218. Da Re, S., Bertagnoli, S., Fourment, J., Reyrat, J.-M., and Kahn, D. (1994) Intramolecular signal transduction within the FixJ transcriptional activator: *in vitro* evidence for the inhibitory effect of the phosphorylatable regulatory domain. *Nucl. Acids Res.*, **22**, 1555.
219. Tawa, P. and Stewart, R. C. (1994) Kinetics of CheA autophosphorylation and dephosphorylation reactions. *Biochemistry*, **33**, 7917.
220. Ninfa, E. G. (1992) PhD thesis, Princeton University, Princeton, NJ.
221. Alper, H., Duncan, L., and Losick, R. (1994) An adenosine nucleotide switch controlling the activity of a cell type-specific transcription factor in *B. subtilis*. *Cell*, **77**, 195.

222. Borkovitch, K. A., Kaplan, N., Hess, J. F., and Simon, M. I. (1989) Transmembrane signal transduction in bacterial chemotaxis involves ligand-dependent activation of phosphate group transfer. *Proc. Natl. Acad. Sci. USA*, **86**, 1208.
223. Ninfa, A. J., Ninfa, E. G., Lupas, A., Stock, A., Magasanik, B., and Stock, J. (1988) Crosstalk between bacterial chemotaxis signal transduction proteins and the regulators of transcription of the Ntr regulon: evidence that nitrogen assimilation and chemotaaxis are controlled by a common phosphotransfer mechansim. *Proc. Natl. Acad. Sci. USA*, **85**, 5492.
224. Moore, J. B., Shiau, S.-P., and Reitzer, L. J. (1993) Alteration of highly conserved residues in the regulatory domain of Nitrogen Regulator I (NtrC) of *Escherichia coli. J. Bacteriol.*, **175**, 2692.
225. Bourret, R. B., Hess, J. F., and Simon, M. I. (1990) Conserved aspartate residues are phosphorylation in signal transduction by the chemotaxis protein CheY. *Proc. Natl. Acad. Sci. USA*, **87**, 41.
226. Atkinson, M. R., Kamberov, E. S., Weiss, R. L., and Ninfa, A. J. (1994) Reversible uridylylation of the *Escherichia coli* PII signal transduction protein regulates its ability to stimulate the dephosphorylation of the transcription factor Nitrogen Regulator I (NRI or NtrC). *J. Biol. Chem.*, **269**, 28288.

9 | Switch systems

J. R. SAUNDERS

1. Introduction

Bacteria of many species are frequently observed to exhibit profound phenotypic changes, typically affecting colonial appearance or behavior. Most often, such changes are evident in pathogenic bacteria, where genetic switches modulate expression of a variety of virulence determinants (1). Such variation is the result of a diverse range of genetic mechanisms that are manifested in what are somewhat imprecisely grouped together as either phase variation or antigenic variation. Phase variation results from quantitative changes in gene expression that may range from what are, in essence, simple on–off genetic switches, through to modulation of the volume of expression of functional gene product at either the transcriptional or translational level. In contrast, antigenic variation is evident as qualitative changes in the nature of one or more surface components. In the main, such variations are detected by sequence changes involving more complicated switches that lead to the expression of an antigenically altered protein. Antigenically variable components have been studied largely because of their importance in bacterial virulence. However, equivalent changes in surface-exposed components of non-pathogenic bacteria almost certainly occur and may have particular selective advantages in their particular environments. Moreover, qualitative changes in a variable gene product could take place without necessarily having effects on the immunological properties of the variable component concerned.

Phase and antigenically variable switch systems are normally attributed special status and distinguished from more conventional mutation events. In reality, at the rates at which the sequence changes involved occur, many of the molecular events implicated and the functional outcome may be indistinguishable from uncomplicated point or other types of mutation. Furthermore, the underlying genetic mechanisms involved may be the same in some cases for both phase and antigenic variation. In certain cases, the genetic regulation of such variable systems is programmed, at least partially, in response to environmental signals; in others, genetic variation is created randomly.

2. Strand-slippage mechanisms and variable gene expression

Phase variation may be achieved simply by altering numbers of reiterated single bases or short repeat sequences that lie either upstream of a gene modulating transcription, or within a gene and altering the translational reading frame. Regions containing reiterated bases may be hot spots for transient mispairing of double-stranded DNA, as the result of local denaturation during passage of the replication fork (2, 3). This process, which is known as slipped-strand mispairing, occurs independently of homologous recombination, and can create randomness in expression of a number of genes critical in pathogenesis in certain species. DNA-containing tracts of poly purines or pyrimidines are able to adopt a number of configurations that differ from the more normal B-form of DNA. Under conditions of strain imposed by superhelical coiling, B-form DNA may undergo transitions to H-DNA comprising a triple-stranded and a single-stranded region (4). H-DNA is known to stimulate gene expression when positioned on the 5′ side of a promoter (5), and the presence of single-stranded DNA, even transiently, may promote slipped-strand mispairing (6). Where the repeat sequence is AT-rich, and hence prone to local denaturation, another potential mechanism for gaining or losing repeats would be the formation of intrastrand pairing of repeats as loops that would become targets for excision repair processes (7). Tracts of repeated sequence may also provide targets for unequal crossing-over between the two copies of a gene generated post-replication.

2.1 Transcriptional modulation of phase variation

The simplest and most economical target for switches that modulate gene expression is likely to be transcription. Alteration of the structure of promoter regions, notably by reducing or increasing the efficiency of binding by RNA polymerase, may be used to modulate levels of mRNA produced, and hence control amounts of product.

2.1.1 Opc and PorA proteins of *Neisseria meningitidis*

Opc is an outer membrane protein unique to *Neisseria meningitidis*, the causative agent of meningococcal meningitis, and is associated with adhesion and invasion in non-capsulate meningococci (8, 9). The gene encoding Opc is subject to a form of phase variation that not only effects an on–off switch, but also modulates expression quantitatively. Off phase Opc^- cells are unable to adhere, but Opc^{++} or Opc^+ meningococci, expressing the polypeptide at high or intermediate levels, respectively, are adherent. Only Opc^{++} meningococci can invade human cells (9). Opc is highly immunogenic, but Opc^{++} meningococci alone are susceptible to bactericidal antibodies specific for the polypeptide. Variants down-regulated for Opc may therefore be at a selective advantage during the later stages of invasive meningococcal disease. This is

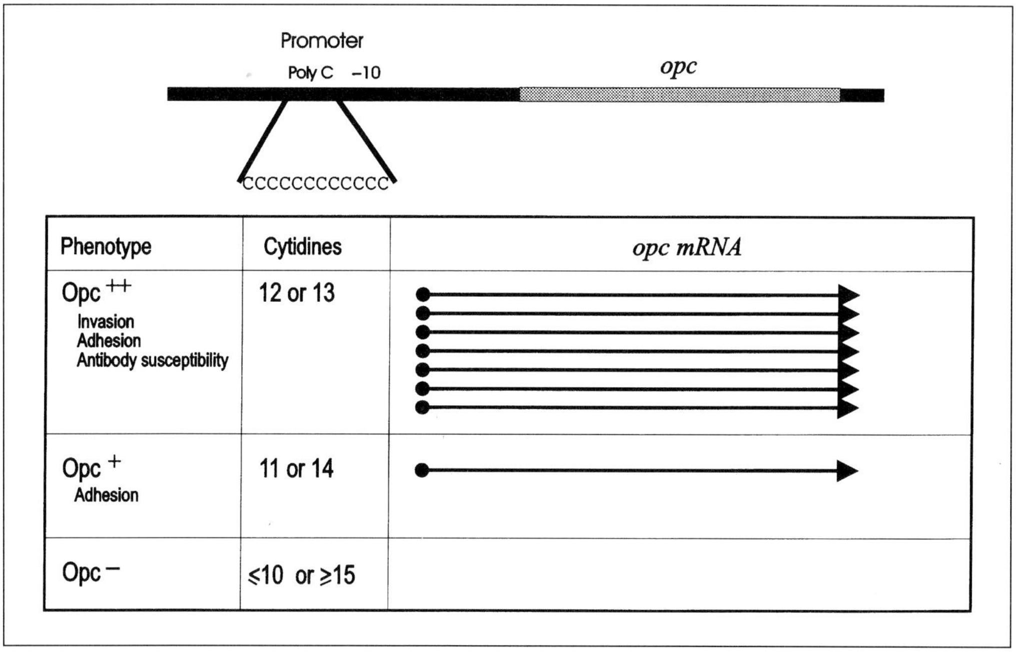

Fig. 1 Transcriptional regulation of the *Neisseria meningitidis opc* gene.

apparent since Opc^{++} bacteria are isolated more frequently from the nasopharynx than the bloodstream or cerebrospinal fluid (8, 10).

Transcription of *opc* starts 13 nucleotides downstream of the –10 region of an unusual promoter sequence lacking a –35 region, but containing a variable number of cytidine residues in the equivalent position. Efficient expression of Opc protein only occurs in variant *opc* genes containing 12 to 13 cytidine residues, intermediate expression requires 11 or 14 Cs, but with $\leq$10 or $\geq$15 residues, there is no expression (Fig. 1).

The class 1 outer membrane protein encoded by the *porA* gene of *N. meningitidis* exhibits a similar type of phase variation involving three different transcriptional levels (Table 1). The promoter region of this gene contains a run of contiguous guanidine residues between the –10 and –35 domains, with the presence of 11, 10, or 9 Gs resulting in high, medium, or no expression of *porA* mRNA, respectively, due to slipped-strand mispairing (11).

Similar situations are encountered in other bacteria. For example, a tract of cytidines located upstream of the pilin (*fim*) gene in *Bordetella pertussis* regulates its transcription, probably by altering interaction of the promoter with DNA-dependent RNA polymerase and regulatory proteins by a process dependent on the number of Cs (12). Similarly, promoters of the variable lipoprotein (*vlp*) genes of *Mycoplasma hominis* and *Mycoplasma hyorhinis* include a tract of A residues where strand slippage may alter efficiency of transcription. Variable lipoproteins (Vlp) are the major coat

Table 1 Some genes regulated by repeated sequence motifs

Gene regulated	Repeat	Repeat number	Phenotype	Reference
Transcriptional control (promoter strength)				
Bordetella pertussis fim	C	14	ON	(12)
		≤13	OFF	
Neisseria meningitidis opc	C	12, 13	ON++	(72)
		11, 14	ON+	
		10, 15, etc.	OFF	
Neisseria meningitidis porA	G	11	ON++	(11)
		10	ON+	
		9	OFF	
Mycoplasma hominis vlp	A	17	ON	(73)
		18, 20	OFF	
Haemophilus influenzae hifA,B	TA	10	ON	(15)
			OFF	
Translational control (frameshift)				
Yersinia pestis yopA	A	9	ON	(34)
		8	OFF	
Neisseria pilC	G	13 =	ON	(27)
		12 =	OFF	
Haemophilus influenzae lic	CAAT	16 =	ON	(24)
		15, 17	OFF	
Neisseria lgt	G	14 =	ON	(7)
		15 =	OFF	
Neisseria opa	CTCTT	9, 12, 15, etc	ON	(74)
		7, 8, 10, 11, etc	OFF	

proteins in such mycoplasma and are products of multiple but distinct, single-copy genes organized in a chromosomal cluster. Each *vlp* gene is linked to a promoter region containing a homopolymeric tract of adenine residues that controls phase variation and leads to combinations of distinct Vlp products as mosaics on the cell surface (Table 1) (13).

2.1.2 Hif fimbrial regulation in *Haemophilus influenzae*

LKP-fimbriae (pili) of *Haemophilus influenzae* assist nasopharyngeal colonization prior to localized infections or meningitis, but are not required for, or may impede, mucosal invasion (14). LKP fimbrial genes comprise *hifA* (the major fimbrial subunit), *hifB* (a periplasmic chaperon), *hifC* (an outer membrane usher), and *hif D* and *hifE* (minor subunits required for fimbrial biogenesis) (15). *hifB*, *hifC*, *hifD*, and *hifE* are co-regulated with *hifA* by overlapping divergently oriented promoters (15). Transcription in either direction is modulated by varying the number of TA residues in a region of poly-TA that lies across the –10 and –35 promoter sequences (16) (Fig. 2). This alters the spacing between the –10 and –35 regions, thus modulating the efficiency of RNA polymerase interaction with the promoters.

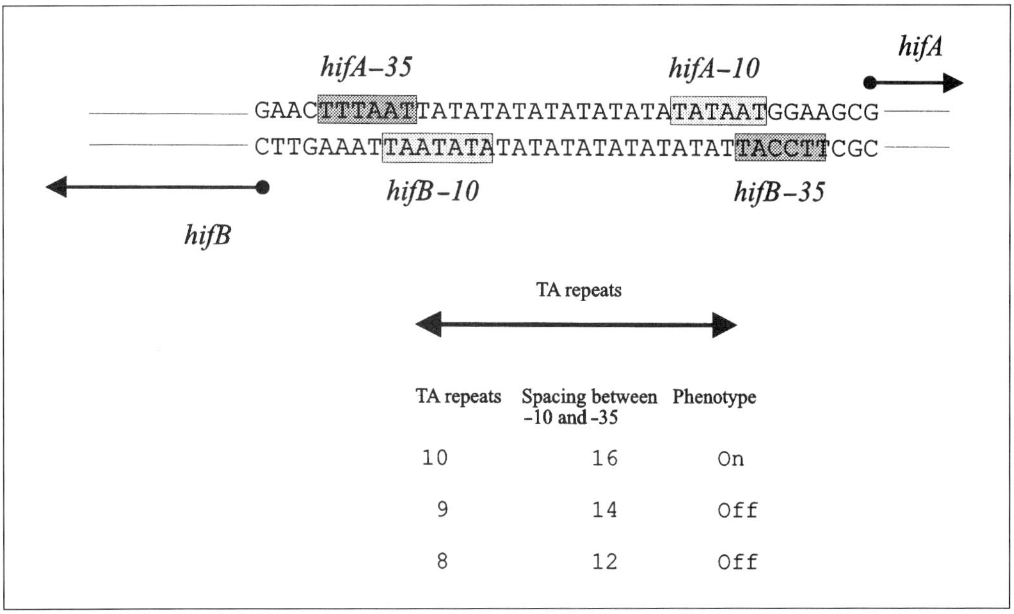

Fig. 2 Transcriptional regulation of *Haemophilus influenzae* LKP fimbriae by slipped-strand mispairing of poly-TA repeats.

2.2 Translational modulation of variation

Modulation of translation by slipped-strand mispairing provides a means of switching gene expression by altering the reading frame of genes, causing key domains within expressed gene products to be functional or otherwise. *recA*-independent translational control of phase variation achieved by slippage at repeated sequences is a strategy adopted by a number of bacterial pathogens (Table 1). For gene products encoded by multiple variant alleles, the superimposition of multiple on–off phase changes may also effect antigenic variation as different combinations of the variable locus are switched into and out of expression.

2.2.1 Opa protein expression in pathogenic *Neisseria* species

Probably the best studied example of translational slipped-strand mispairing control of gene expression occurs in the opacity (Opa) polypeptides of *Neisseria gonorrhoeae* and *N. meningitidis* (17). These are a family of antigenically variable outer-membrane proteins essential for invasion of epithelial cells by gonococci, and for the interaction of both gonococci and meningococci with human neutrophils. Opa proteins are encoded by a family of *opa* genes located at widely separated sites on the neisserial genome, with 10–11 *opa* loci in *N. gonorrhoeae* and 3–4 in *N. meningitidis*, the expression of which can be independently turned on or off (17). One variant, called Opa50, enables gonococci of strain MS11 to invade epithelial cells, but remaining variant Opa proteins show little or no specificity for epithelial cells, instead conferring interaction with human polymorphonuclear neutrophils (9, 18). Therefore, depending on which *opa* alleles are expressed, gonococci are capable of invading human epithelial cells or interacting with leukocytes (19). Mature Opa proteins are largely homologous, except for surface-exposed hypervariable regions (HV1 and HV2) which appear to mediate the above functional differences. Antigenic variation is accomplished by interaction of multiple phase variations to generate different combinations of expressed *opa* genes. One or more different *opa* alleles, or none at all can be expressed at different times and in different strains. Additional variation is possible by intragenomic recombination or horizontal gene transfer by transformation to create mosaic *opa* gene sequences (20, 21).

Reversible on–off switching occurs at each *opa* locus by alteration of the translational reading frame of the signal peptide coding region of each *opa* gene, which includes a tandem series of the pentameric repeat 5′-CTCTT-3′ (CR) (Fig. 3). A slipped-strand mispairing mechanism alters the numbers of repeats and hence shifts the reading frame of the signal region of the Opa polypeptide. Translation from an in-frame configuration, which requires 9 ±3 repeats, creates an Opa prepolypeptide that is correctly processed on export across the cytoplasmic membrane into a mature functional Opa polypeptide (17). The alternative repeat configurations of 8 (±3) and 10 (±3) result in translated polypeptides that are nonsense and consequently non-functional. Addition and subtraction of repeats in this region of *opa* loci has been attributed to slipped-strand mispairing between the CRs during replication, as a consequence of transient formation of single-stranded DNA during formation of H-DNA within the CR region (6,22).

2.2.2 Phase-variable expression of lipopolysaccharide in *H. influenzae* and the pathogenic *Neisseria*

Slipped-strand mispairing mechanisms operating at the translational level have been implicated in the phase variable expression of lipopolysaccharide (LPS) in *H. influenzae*, which is a significant process in modulating virulence in this organism (23, 24). The tetrameric motif CAAT is present in tandem direct repeat in the 5′ ends of the coding regions for three LPS biosynthetic (*lic*) genes required for expression of phase-variable LPS epitopes. Variation in numbers of repeats modulates translation of the downstream gene sequence in and out of frame with respect to the open reading frame of the functional gene product (Table 1). The *lic2A* locus requires 16 repeats for the on phenotype, and other numbers create the off switch. The AT-rich nature of the CAAT repeat may promote strand-slippage during replication, leaving unpaired repeats as the target of excision repair processes (7).

A similar translational control mechanism exists in *lgt* genes which are homologous to *lic2A* and are present in *N. meningitidis* (25) and *N. gonorrhoeae* (26), where they control expression of phase variable parts of the LPS. The first open reading frame of the expressed *N. meningitidis lgt* operon contains a homopolymeric tract of 14 guanosines towards its 5′ end (25). Changes in the numbers of Gs resulting from slipped-strand mispairing generate frameshift mutations in the operon causing a phase change (Table 1).

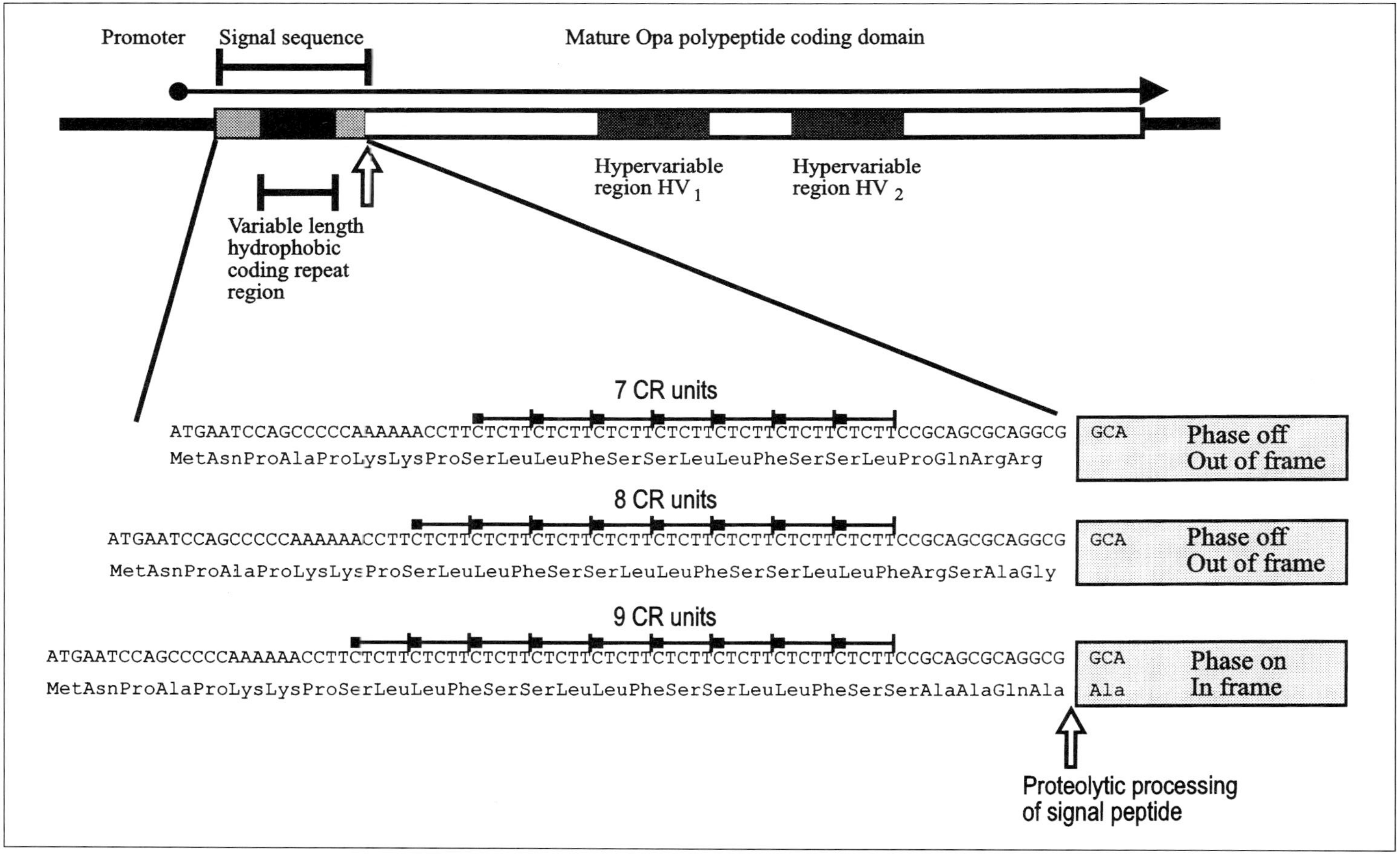

Fig. 3 Translational regulation of *Neisseria gonorrhoeae opa* genes by slipped-strand mispairing of CTCTT repeats in the signal peptide region.

2.2.3 PilC expression in pathogenic *Neisseria*

On–off–on phase variation by translational frameshifting also occurs at a tract of repetitious G residues that lie in the coding region for the N-terminal leader domain of the expressed polypeptide of the neisserial *pilC* gene. In addition to the principal pilus subunit (pilin or PilE) (see below), both gonococci and meningococci produce low quantities of the related phase-variable PilC1 and PilC2 proteins. PilC proteins have been implicated in pilus biogenesis and pilus-mediated epithelial cell adherence (27, 28). When PilC is not expressed, gonococci and meningococci produce few or no pili (28, 29). Functional activities of PilC proteins may be affected by presentation on PilE subunits of differing primary amino acid sequence (30, 31). PilC has been localized to the tip of the neisserial pilus and may be the ligand for adhering these bacteria to some human cell surfaces (32). Presence or absence of PilC modulated by phase variation may allow *N. gonorrhoeae* and *N. meningitidis* to produce variants capable of adsorbing or desorbing from particular sites during the course of infections. Most strains of *N. gonorrhoeae* and *N. meningitidis* carry two copies of *pilC*, which are largely homologous, but not identical, and are part of a duplication of a large genomic region that extends more than 2 kb downstream of the *pilC* coding region (33). *pilC1* and *pilC2* encode two related 110-kDa polypeptides, and are turned on and off at high frequency by frameshift mutations in a run of G residues located in the encoding region for the signal peptide (Table 1).

2.2.4 Other translationally modulated systems

Translational control of gene expression by strand-slippage appears to be a relatively common mechanism for producing phase variation. Where variation occurs at the amino-terminal end of the gene product, as discussed above, the effects of variation are apparent as on and off switching. However, variation may not be readily apparent where single base changes alter the translational reading frame at a position well into the open reading frame for the mature protein. Nevertheless, frameshifting at central or more C-terminal regions of genes is known to effect important phase changes in some pathogens. For example, the *yopA* gene of *Yersinia pestis* contains a tract of As and removal of one of these results in a frameshift that truncates the gene product (34) (Table 1).

A more complex example is provided by BvgS, an environmental sensor which together with BvgA, a transcriptional activator, provide a two-component global regulator of virulence in *Bordetella pertussis*. The *bvgS* gene contains a tract of C residues within its open reading frame, such that a change from six (on) to seven (off) Cs shifts the downstream coding region out of frame (35). The products of the *bvgAS* locus activate expression of most known *Bordetella* virulence factors, but also exert negative control over other genes, for example, the production of flagella and motility (36). Thus, simple frameshifting in this crucial gene may indirectly effect multiple changes in phenotype.

3. Phase switching by DNA inversion

Bacterial, plasmid, and phage genomes may contain invertible regions of DNA ranging from 100 bp to 40 kb and inverting at frequencies of up to about 10^{-3} to 10^{-4} per generation (37). Site-specific inversion of such elements can provide relatively simple on/off genetic switches for genes located within, or adjacent to, an invertible region. This is generally achieved by altering the spatial relationship between the region and a promoter or other controlling element.

3.1 Type 1 fimbriae in *Escherichia coli*

Perhaps the simplest model for invertible elements is provided by Type 1 fimbriae, which are found in both commensal and pathogenic *Escherichia coli*. These surface structures are phase variable and subject to selection during infections (38). Type 1 fimbriae promote adherence to mannose-containing receptors present on macrophages and certain epithelial cell types, and may protect bacteria from ingestion by macrophages. These fimbriae are, however, potent immunogens and may also block attachment of bacteria to epithelial cell surfaces.

Phase variation of Type 1 fimbriae involves the inversion of a 314-bp chromosomal DNA sequence located upstream of *fimA*, the gene encoding the major fimbrial structural subunit (Fig. 4). In the on orientation, *fimA* is transcribed from a promoter within the invertible element, but in the opposite (off) orientation, the structural gene is separated from its promoter. *Fim* inversion requires histone-like protein (H-NS), integration host factor (IHF), and leucine-responsive protein (Lrp) (see Chapter 6). *Fim* inversion is carried out by two proteins called FimB and FimE, which share 48 per cent sequence homology and belong to the lambda integrase family of site-specific recombinases. The *fim* switch is locked in either the on or off position in the

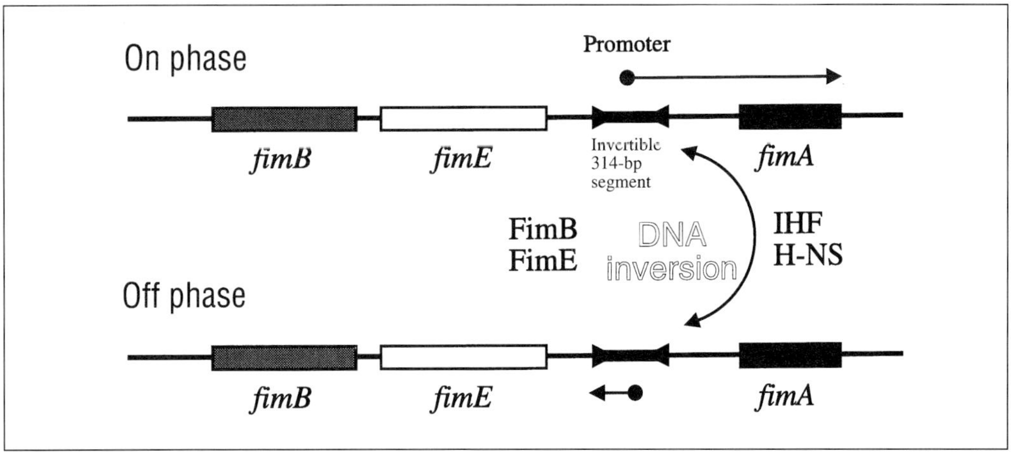

Fig. 4 Invertible control system for *Escherichia coli* Type 1 fimbriae.

absence of these two gene products. These polypeptides operate differently: FimB promoting both on–off and off–on inversions, and FimE only on–off inversions (39, 40). Expression of *fimE* is about 300 times greater than *fimB* in rich media at 37 °C, suggesting environmental regulation of the switch. In *fimBE*$^{+}$ cells, switching rate can be as high as 0.75 per generation. Inversion mediated by both *fimB* and *fimE* is stimulated by alanine and the branched-chain amino acids leucine and isoleucine/valine acting together with Lrp (41). Lrp has also been implicated in the regulation of other fimbrial antigens, such as pyleonephritis-associated pili (Pap) and K99. Lrp may be involved in bending DNA and together with IHF may align the terminal inverted repeats of the *fim* element before recombination (41). The *lrp* gene is not only feedback regulated by Lrp itself, but is also controlled by H-NS protein (42). Lrp binds in and adjacent to the *fim* switch (40).

3.2 *Salmonella* Hin-mediated inversion

Flagellar phase variation in *Salmonella typhimurium* involves the alternating expression of two possible antigenically distinct types of flagella, designated H1 and H2 (43). This constitutes a genetic switch of the form $H1_{ON}/H2_{OFF} \longleftrightarrow H1_{OFF}/H2_{ON}$ that is mediated by an invertible 993-bp DNA segment containing an outward reading promoter (Fig. 5). Control of expression of the genes encoding the protein subunits for either H1 (*fliC*) or H2 (*fljB*) is effected by inversion of the controlling segment by site-specific recombination between two flanking inverted repeat sequences of 14 bp

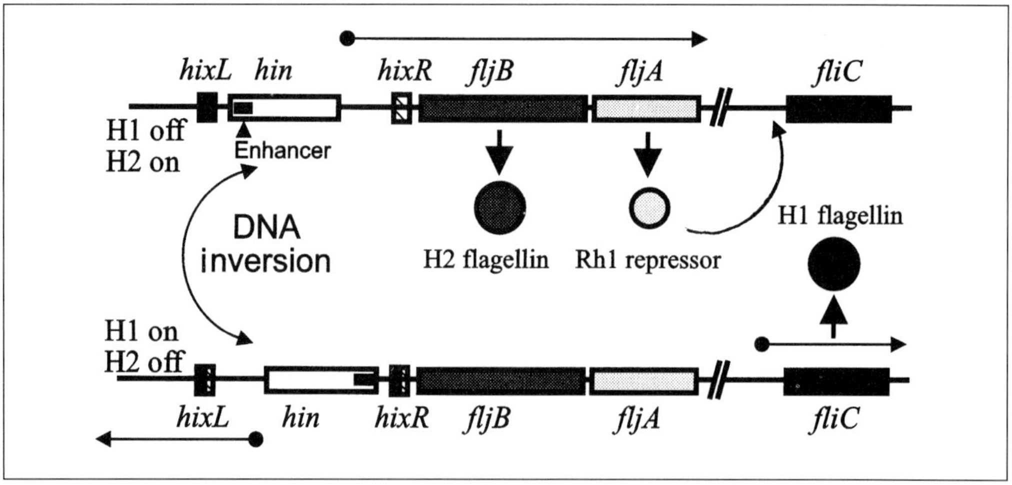

Fig. 5 Hin-mediated flagellar phase variation in *Salmonella typhimurium*. The promoter for *fliC* (H1 flagellin) is at a site distant from the main *hin* operon and is expressed constitutively, unless repressed by FljA (Rh1). Expression of *fljB* (H2 flagellin) is dependent on a promoter lying inside the invertible *hin* segment which contains the *hin* invertase gene itself, and an enhancer sequence. Hin-mediated inversion at *hixL* and *hixR* connects or disconnects the promoter from the *flj* operon. In one orientation, the promoter reads through the *flj* operon, leading to expression of *fljB* and *fljA* (repressor of *fliC*). In the opposite orientation the promoter reads non-productively outwards, leaving the *flj* operon unexpressed.

(*hixL* and *hixR*). This process is mediated by the Hin recombinase (*hin* gene product), an enzyme encoded within the invertible segment itself. In the $H1_{ON}/H2_{OFF}$ configuration, the promoter reads non-productively towards *hixL* (Fig. 5). Since *fliC* is located some distance away on the *Salmonella* genome and is constitutively expressed, H1 flagellin is produced rather than H2 flagellin. When the *hin* segment is inverted however, the promoter directs transcription through *hixR*.

The Hin recombinase requires the formation of a nucleoprotein complex called an invertasome (44). Here, the two *hix* recombination sites bound by Hin are assembled together at a Fis-bound recombinational enhancer and the intervening DNA segments are looped around (Fig. 5). HU promotes invertasome formation by binding non-specifically to the DNA between enhancer and *hix* site and facilitating DNA looping (45, 46). The looping brings the two termini of the invertible region close together in three-dimensional space. The enhancer region would then act as a scaffolding for subsequent breakage and reunion of the *hin* segment, leading to inversion.

3.3 Gin and other Hin family invertible systems

Hin is an example of a number of DNA invertases (Din) found in bacteria, many of which may be utilized in the determination of the host range of various extra-chromosomal elements. Perhaps the most noteworthy of these is the Gin (G inversion) system of bacteriophage Mu, where inversion of the G region of the genome leads to the expression of two different sets of proteins that determine phage host range. Gin exhibits significant homologies at DNA and protein levels with Hin and related systems such as Cin, a recombinase that mediates DNA inversion between two *cix* sites flanking genetic determinants for the host range of the *E. coli* bacteriophage P1 (47).

In one orientation of the Mu G region two proteins are expressed that allow the phage to adsorb to *E. coli*, and in the opposite orientation it produces different proteins for adsorbing to *Citrobacter freundii*. G inversion is mediated by the Gin invertase (48). Gin selectively mediates DNA inversion between two inversely oriented recombination sites (*gix*), requiring the assistance of negative supercoiling, an enhancer sequence, and Fis protein (Fig. 6). Deletion and fusion reactions are prevented because recombination only occurs through a particular synaptic complex that is specifically tailored for inversion. However, this specificity is determined by Gin, since a single amino acid change allows the protein to carry out deletions, fusions, and inversions without the accessory factors such as Fis. Processive recombination by wild-type Gin is not restricted by the number of base pairs separating the *gix* sites from each other and from the enhancer, suggesting that strand exchange proceeds through alternative pathways determined by the energetics of DNA coiling.

3.4 Methylation as a controlling switch

The expression of the *pap* fimbrial (pyelonephritis-associated pili) operon of *E. coli* is under control of a phase-variation mechanism whereby cells undergo a reversible

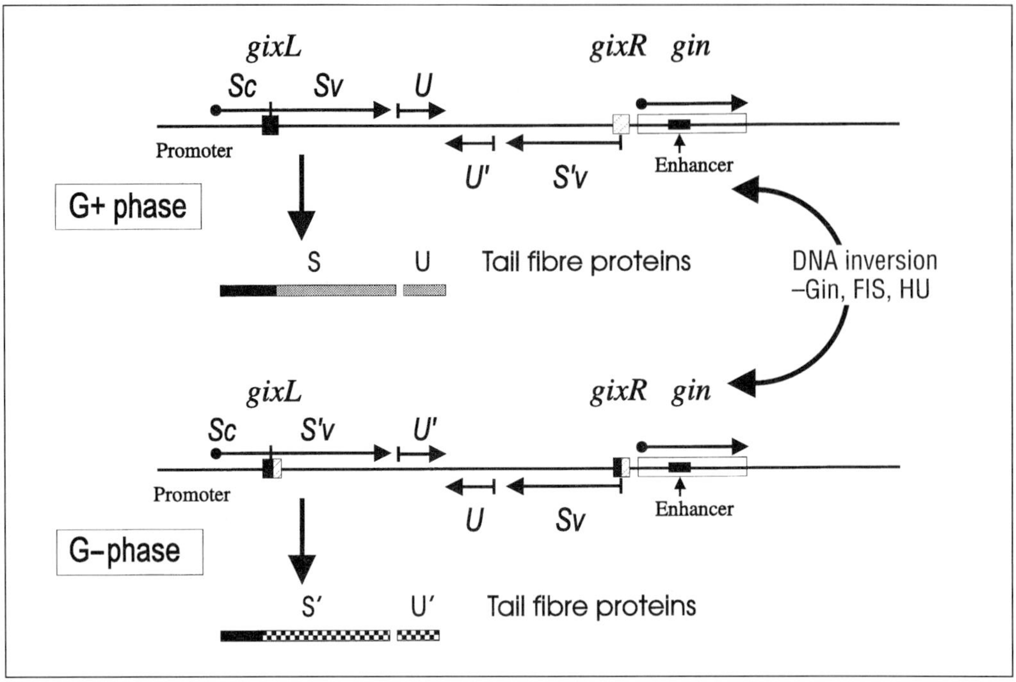

Fig. 6 Gin-mediated inversion of the G segment in bacteriophage Mu. Expression of two variants of the two tail proteins *S/S'* and *U/U'* is achieved by inversion between *gixL* and *gixR*. This fuses the conserved coding region S_C either translationally with S_V and transcriptionally with *U* (G^+ phase, plaquing on *E. coli*), or translationally with S_V and transcriptionally with *U'* (G^- phase, plaquing on *C. freundii*).

transition between transcriptionally active (on) and inactive (off) states. The phase off state results from repression of the *papBA* promoter by Lrp and H-NS, each of which can act independently as repressors of transcription (Fig. 7). Lrp, but not H-NS, requires DNA sequences upstream of the *papBA* promoter for repressor activity. In contrast, in the on state, Lrp, in conjunction with PapI, activates *pap* transcription to a level that is eightfold higher than basal. Thus, Lrp functions as a transcriptional activator in phase-on cells but as a repressor of basal transcription in phase-off cells (49).

The process involves differential DNA methylation by deoxyadenosine methylase (Dam) (50). Methylation of two GATC sites ($GATC_{1028}$ and $GATC_{1130}$) within the *pap* regulatory region is differentially inhibited in on and off phase cells. $GATC_{1028}$ in phase-on cells is non-methylated, whereas the $GATC_{1130}$ site is fully methylated (Fig. 7). Conversely, in phase-off cells, $GATC_{1028}$ is fully methylated, but $GATC_{1130}$ is not. Specific methylation inhibition requires PapI and Lrp as transcriptional activators. Lrp binds to a region surrounding $GATC_{1130}$, whereas PapI does not appear to bind to *pap* regulatory DNA on its own. However, Lrp and PapI seem to bind around the $GATC_{1028}$ site, and Dam methylation inhibits binding of Lrp/PapI near this site, as well as altering binding of Lrp at $GATC_{1130}$. Methylation of $GATC_{1028}$ blocks forma-

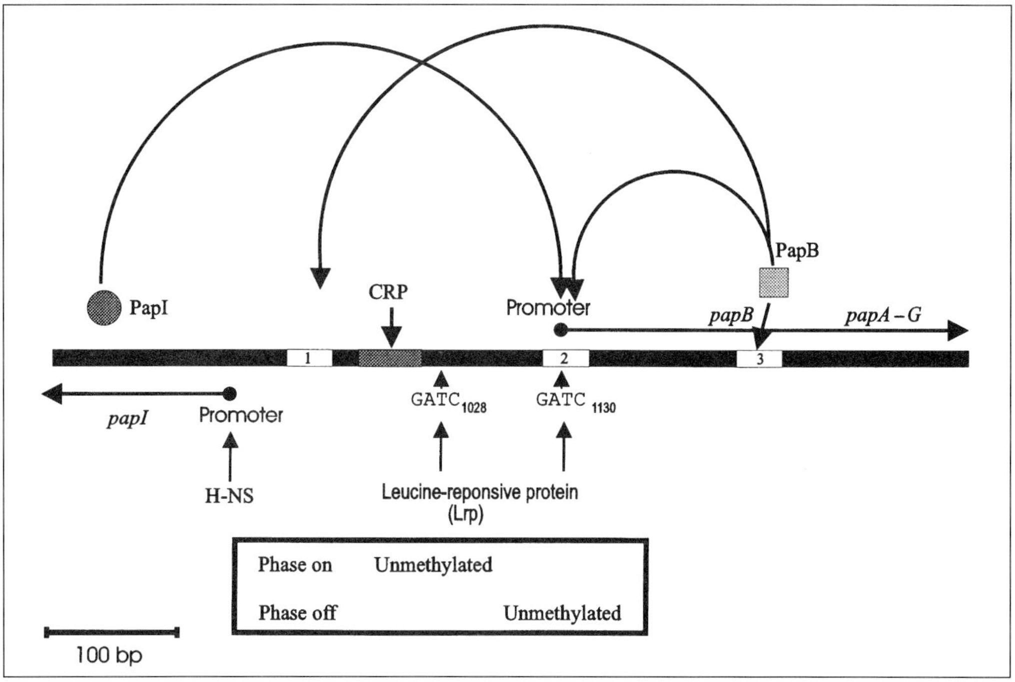

Fig. 7 Control of expression in the *Escherichia coli pap* operon by Dam methylation.

tion of the on state by inhibiting binding of Lrp and PapI regulatory proteins to this site. Conversely, methylation of $GATC_{1130}$ is required for the on state. This occurs by inhibition of binding of Lrp to sites overlapping the pilin promoter.

E. coli F1845 and S pili are also under phase variable control at the transcriptional level mediated by Lrp and Dam (51). Lrp is required for methylation protection of two differentially methylated GATC sites located within conserved DNA sequences in the regulatory regions of the *daa* and *sfa* operons for F1845 and S pili, respectively. Phase variation of these operons is regulated by a mechanism involving differential binding of Lrp similar to that controlling phase variation of the *pap* operon.

3.5 Shuffling systems—multiple inversions

Increased variation may be achieved by combining the effects of multiple inversions of gene sequence. The shufflon is a control switch that is found in R64 and related ColI1 plasmids, and determines the recipient specificity in liquid mating during conjugative transfer (52, 53). This genetic element consists of four DNA segments that are flanked and separated by seven 19-bp repeats (Fig. 8). Site-specific recombination between any pair of inverted repeats results in a complex series of DNA rearrangements where the four DNA segments invert independently or in blocks. This acts to select one of seven C-terminal segments of the *pilV* gene, which encodes the structural subunit of the thin sex pilus type encoded by R64. Conjugation transfer

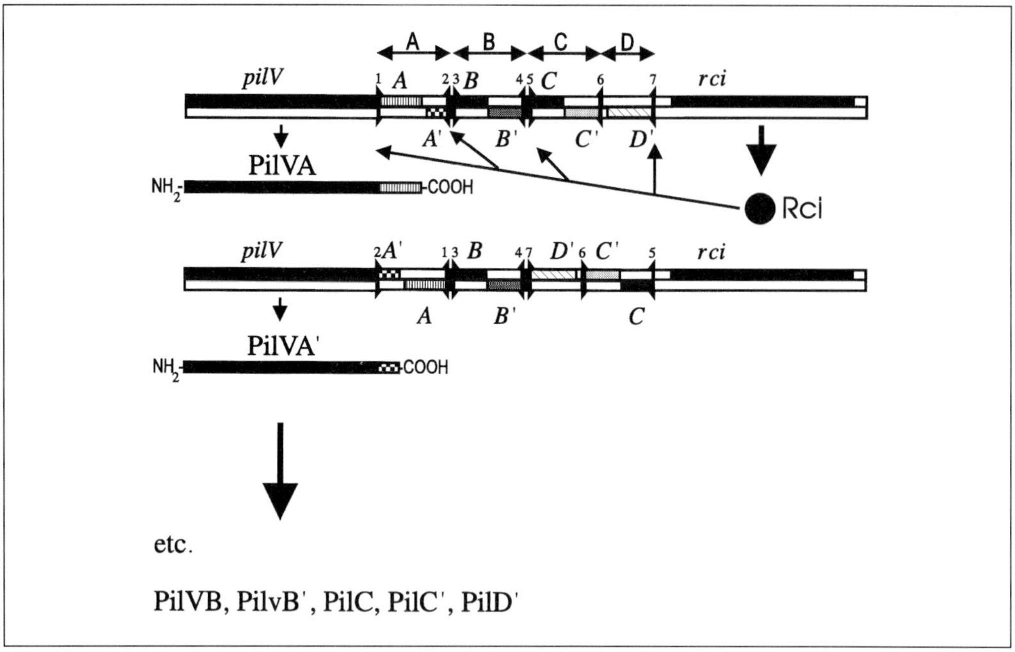

Fig. 8 The *pilV* shufflon of plasmid R64. The mature PilV protein is encoded by a constant N-terminal sequence (*pilV*$_c$) which may be linked to any one of seven possible C-terminal coding sequences (A, A', B, B', C, C', and D'). C-terminal coding sequences are contained on the four invertible segments A, B, C, D. Apart from segment D, these have an appropriate open reading frame in both possible orientations. Large black arrows indicate 19-bp repeat. The *rci* gene encodes a site-specific recombinase of the phage λ integrase family capable of shuffling and/or inverting each segment independently, as combinations, or *en bloc*.

frequencies depend on a combination of the nature of recipient bacteria and the C-terminal sequence of PilV in donor cells. Recombination between repeats is mediated by the product of a site-specific recombinase gene, *rci*, which encodes a basic protein of about 375 amino acids. In this case, site-specific recombination events invert and reorder the four different segments to produce seven possible C-terminal sequence variants (Fig. 8).

A similar mechanism appears to operate in determining which one of four variant linked genes *omp1A, B, C, D* encoding the major outer membrane protein of *Dichelobacter nodosus* is expressed (54, 55). The 5′ regions of all four genes for this surface-exposed protein of ~75 kdal possess short *nix* sequences, similar to the cross-over site sequences associated with DNA invertases such as Hin. The four variant *omp1* genes can be assembled from one of four structurally variant silent C-terminal coding regions and a conserved promoter and N-terminal coding region located on a 497-bp segment that is itself invertible (Fig. 9).

Similarly, interwoven, site-specific DNA inversions in the chromosome of the murine pathogen *Mycoplasma pulmonis* drive phase-variable expression of a *vsa* gene cluster which encodes the variable V-1 surface antigens (55).Only one *vsa* gene is associated with an expression site; the others are transcriptionally silent, since they

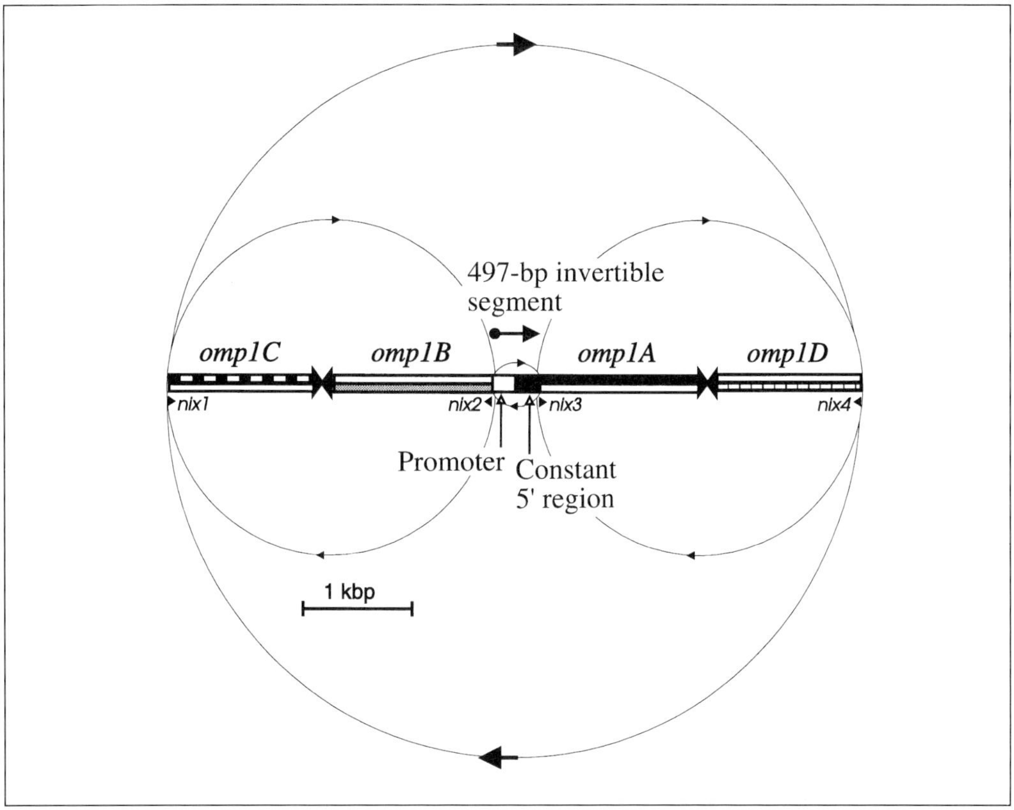

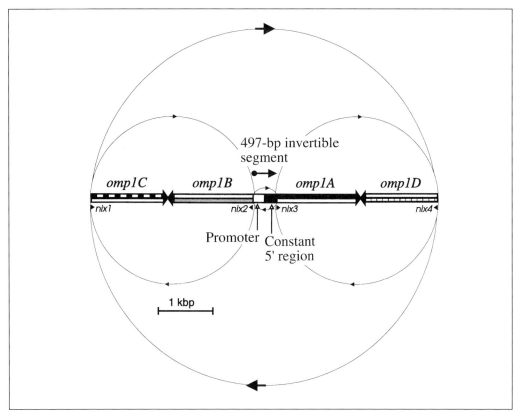

Fig. 9 Multiple invertible segment control of *omp1* expression in *Dichelobacter nodosus*. The central invertible region of 497 bp comprises the promoter and a constant 5' coding region for *omp1*. Variant coding regions for the remaining 3' end of *omp1* are contained in two pairs of opposing inverted copies that can invert independently. The entire segment of ~8 kbp can also invert. Depending on the orientation of the central promoter segment one of the four possible variant 3' segments of *omp1* will be expressed as an intact *omp1* reading frame.

lack the 5′ promoter and ribosome-binding site region of the expressed gene. DNA rearrangements reassort the 5′ region from an expressed gene with the 3′ region from a previously silent gene. The *vsa* rearrangements involve site-specific DNA inversions between copies of a specific 34-bp sequence conserved in each *vsa* gene. Interestingly, DNA inversions within the *vsa* locus occur in concert with inversion of the *hsd1* element, which regulates restriction and modification activity in *M. pulmonis* (56).

4. Variation by homologous recombination

A number of bacteria exploit homologous recombination, dependent on RecA or its analogues, between two or more complete or partial copies of genes encoding variable components. In some cases, for example *opa* genes of the pathiogenic *Neisseria* (see above), this may be an incidental process that generates variation at low frequency. In others, notably pilus variation in the same bacteria, homologous recombination is the driving force for creating variation at high rates.

4.1 Recombinational phase and antigenic variation in pilins of pathogenic *Neisseria*

N. gonorrhoeae is highly variable in a number of its surface components (see above) which allows the organism to cause repeated infections in humans. Pili are the most marked variable component of gonococci and meningococci. Phase variation in *Neisseria* may allow non-piliated, non-attaching variants to desorb from initial sites of infection and permit transport to other locations, where reversion to piliated forms would allow a new site of infection to be established. Non-piliated derivatives of gonococci and meningococci fail to adhere to tissue culture cells, whilst non-piliated *N. gonorrhoeae* cells may also show reduced or non-existent virulence (57).

Antigenic variation in gonococcal and meningococcal pili is accompanied by changes in apparent molecular weight of the pilin subunit which forms the major structural component of each pilus. Adherence phenotype is therefore affected radically by changes in pilin primary amino acid sequence. The mature gonococcal pilin, and its homologue in meningococci (Class I pilin), is divided into three distinct regions: N-terminal amino acids 1–53 constitute the conserved (C) domain; amino acids 54 to the first of two cysteine residues (cys1) at about 120 amino acids from the N-terminus constitute the semivariable (SV) region, which typically exhibits amino acid substitutions; and the region from ~121 to ~160 at the C-terminus forms the hypervariable (HV) region, which is characterized by deletions, insertions, and substitutions of amino acid sequence (58). Meningococcal Class 1 pilin and gonococcal pilin are unusual among bacterial proteins in being post-translationally modified glycoproteins (31, 59) containing the unusual trisaccharide digalactosyl 2,4-diacetamido-2,4,6-trideoxyhexose O-linked to Ser_{63} (60). The role of this modification is not yet clear. However, the presence of potentially variable carbohydrate moieties on pilin would provide an additional, untemplated source of variation in addition to changes in amino acid sequence.

Relatively subtle changes in primary amino acid sequence of Class I meningococcal pilin are sufficient to modulate adhesion of *N. meningitidis* to human epithelial cells. The asparagine at residue 60 of meningococcal Class I pilin is critical either directly for adhesion to a receptor on the epithelial surface or for interaction with an additional component which in turn is involved in receptor binding. In the related pilin produced by *N. gonorrhoeae* this residue is believed to stabilize the O-linked carbohydrate moiety attached to serine residue 63 (59–60). Intrastrain variability in adhesion phenotype of piliated meningococci has also been associated with changes in the hypervariable domain of pilin (61). These lead to the formation of highly adhesive pilins that are assembled into large bundles of pili which bind bacteria tightly to human cells and cause them to grow as colonies on infected monolayers.

Antigenic and phase variation in neisserial pili is effected by a varied and complex series of DNA transactions that may result in previously silent incomplete pilin

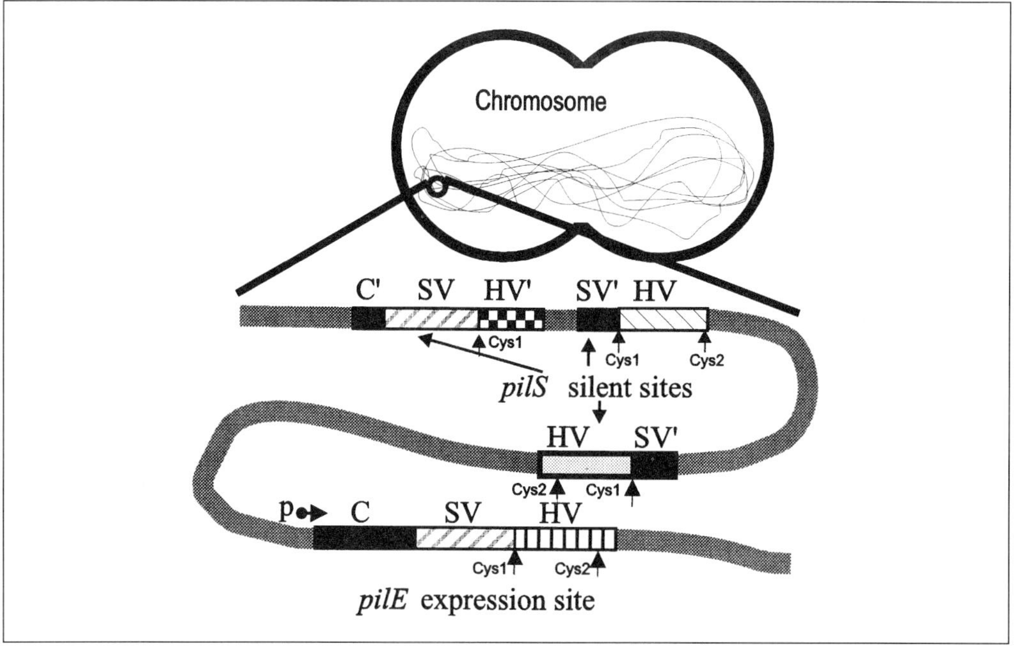

Fig. 10 Silent and expressed pilin sequences in the neisserial genome. The *pilE* locus contains a promoter and ribosome-binding site, together with an entire copy of the pilin-coding sequence. Silent (*pilS*) loci may contain copies of the *pilE* sequence that are truncated to greater or lesser extents from the 5' or 3' direction, and all lack promoter-proximal sequences. C, conserved; SV, semivariable; HV, hypervariable; Cys1 and Cys2 indicate the codons indicating the cysteine residues, Cys1 at the C-proximal end of SV and Cys2 towards the C-terminus of HV forming the disulfide loop; ' indicates partial deletion of complete C, SV, or HV regions. Different shading indicates that SV and HV regions may be variant. Not all possible *pilS* loci are shown.

sequences (*pilS* loci) replacing existing sequences in a single pilin expression locus (*pilE*) (62–64) (Fig. 10). Gonococci may contain up to 20 silent *pil* sequences, and meningococci somewhat fewer. Such sequences may be present as single copies, or frequently clustered in tandem arrays of up to five or six *pilS* copies separated by families of repeat sequences that appear to play a role in promoting recombination. Each *pilS* locus represents varying degrees of truncation when compared with the complete pilin coding region located at *pilE*: all or part of the C region is missing and deleted regions usually extend into the SV region at the 5′ end and into the HV region at the 3′ end of a complete pilin coding region. There are also sequence variations within each copy. Furthermore, some particular *pilS* copies appear to be associated with the production of off phase variants due to the generation of non-functional pilins.

Much phase (on–off) and all antigenic variation is dependent on variation at *pilE*, and is largely abolished in *recA* mutants: all that remains are frameshift mutations that result in non-functional pilin production (64). This suggests that homologous recombination is the primary driving force in pilus variation. The role of recombination in pilin variation is controversial. It has been proposed that small blocs of conserved gene sequence (mini-cassettes) interspersed within the SV region and surrounding the two cysteine codons of *pilE*, together with larger tracts of homology represented by the C region and sequences up and downstream of the expression locus, would provide sufficient homology for recombination between resident and incoming *pil* sequences (58, 63). Thus it would be possible for multiple and random rounds of intragenomic recombination to produce dramatic sequence variation within *pilE*. On the other hand, the tracts of homology present in *pilE* are shorter than would be ideal for, for example, *E. coli* RecA protein to function in promoting recombination. An alternative model (deletion repair) has been proposed that invokes use of incoming *pil* sequences to act as templates for the repair of deletions (65). The *pilE* locus appears to be prone to deletion, and laboratory isolates of gonococci frequently exhibit deletions extending across all or part of the C region which remove the promoter, leading to the formation of irreversible pilus phase-off (P^-) variants. It is clear also that there are two routes by which the genetic information stored in *pilS* loci is translocated to *pilE* (Fig. 11): one by intragenomic DNA transactions involving recombination between *pilE* and one or more *pilS* sequences, and the other involving transformation, where exogenous DNA bearing *pilS* (or *pilE*) sequences released from dead bacterial cells is taken up and incorporated into the resident *pilE* copy (66–67).

N. gonorrhoeae and *N. meningitidis* are naturally highly competent for transformation with species-related DNA, requiring both PilC and PilE for conversion of DNA

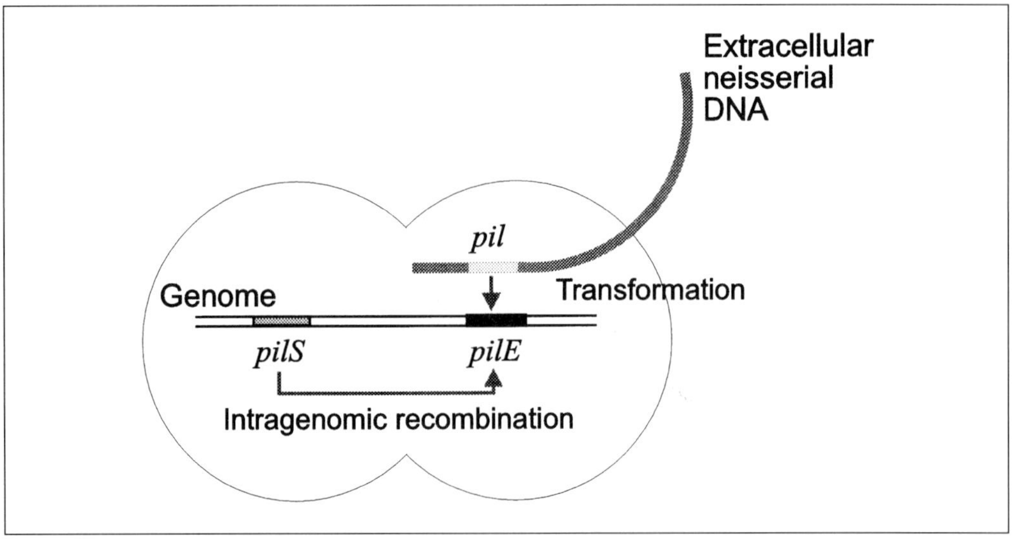

Fig. 11 Two routes for mediating sequence variation in *Neisseria pilE* genes. Sequence information stored in *pilS* loci may be transferred to *pilE* by homologous recombination, occurring either within the chromosome (intragenomic recombination), or between *pilE* and DNA fragments containing *pilS* or *pilE* sequences that have been taken up by transformation. In either case, homology for RecA-mediated recombination is provided by the C region and conserved flanking sequences and/or short internal homologies within SV and in the two Cys-coding regions.

carrying the *Neisseria*-specific DNA uptake sequence (5′-GCCGTCTGAA-3′) into a DNase-resistant form (68). Biogenesis of pili *per se* is neither essential nor sufficient for DNA uptake, but these two phase-variable *pil* proteins appear to cooperate in the recognition of specific DNA and/or translocation across the outer membrane. As a consequence of a highly efficient transformation system, the pathogenic *Neisseria* are able to acquire *pil* and other gene sequences from fellow bacteria that have released their DNA into the surrounding medium. Analysis of variant *pilE* sequences indicates that transformation events tend to be conservative, leading to the spread of particularly successful pilus types within populations. For example, if the incoming DNA contains entire *pilE* loci and flanking DNA, then the tendency is for total replacement of the resident *pilE* locus. In contrast, intragenomic recombination appears to be responsible for generating most of the variation within *pilE*.

Whatever the precise mechanism(s) involved, pilus antigenic variation is prodigious in the pathogenic *Neisseria*, leading to profound variation in antigenic properties and adhesion function, both within and between strains of both *N. meningitidis* and *N. gonorrhoeae*. Indeed, it has been estimated that over 10^6 potential combinations of functional pilin sequence are possible in *N. gonorrhoeae* (58).

4.2 Antigenic variation in *Borrelia*

The utilization of intragenomic recombination involving chromosomally located genes is a potentially risky strategy, since unwanted genomic rearrangements are an ever-present possibility where there is extensive duplication of homologous sequence. In the case of *Neisseria spp* this is witnessed by duplication of *pilE* in some strains and evidence for deletions of *pil* loci in most strains observed in laboratory culture. One solution to this problem is to locate the recombinant loci on plasmids. This occurs in the spirochaetes *Borrelia hermsii* (the cause of relapsing fever in mice) and *Borrelia burgdorferi* (the causative agent of Lyme disease) (69). In *B. hermsii* the active expression site for the variable major protein (Vmp) on the bacterial surface is carried in the teolmeric region of a linear plasmid. Antigenic variation is effected by the recombination between this active *vmp* site and distinct, but related, silent *vmp* gene sequences located on separate linear storage plasmids. This switches the nature of the expressed *vmp* gene, but a copy of the original displaced *vmp* sequence is retained on a storage plasmid. Additional variation is achieved by intragenic recombination within *vmp* copies (70).

In *B. burdorferi*, two similar genes, *ospA* and *ospB*, encode variable surface proteins and are located as an operon on a linear plasmid (71). Antigenic variation in this case occurs by homologous recombination between these two genes, or between *osp* loci on different copies of the plasmid.

5. Concluding remarks

It is clear that bacteria have exploited a variety of mechanisms for effecting the genetic switches that mediate phase and antigenic variation phenomena. Some pathogenic bacteria, such as *Haemophilus* and *Neisseria spp* have multiple switch systems controlling a diverse range of components. This may reflect the complex phenotypic changes necessary at various stages for such microorganisms to colonize and infect their hosts. All such systems seem to exploit normal DNA transactions to produce random quantitative or qualitative changes in gene expression. Highly variable species, such as *N. meningitidis,* must be able to tolerate a high rate of production of potentially unfit cells. Selective advantage gained or lost may then dictate which variants survive and/or cause disease.

References

1. Saunders, J. R. (1995) Population genetics of phase variable antigens. In *Population genetics of bacteria* (ed. S. Baumberg, J. P. W. Young, E. M. H. Wellington, and J. R. Saunders), p. 247. Cambridge University Press, Cambridge.
2. Streisinger, G. and Owen, J. E. (1985) Mechanisms of spontaneous and induced frameshift mutation in bacteriophage-T4. *Genetics*, **109**, 633.
3. Levinson, G. and Gutman, G. A. (1987) Slipped-strand mispairing—a major mechanism for DNA-sequence evolution. *Mol. Biol .Evol.*, **4**, 203.
4. Htun, H. and Dahlberg, J. E. (1989) Topology and formation of triple- stranded H-DNA. *Science*, **243**, 1571.
5. Kohwi, Y. and Kohwi-Shigematsu, T. (1991) Altered gene-expression correlates with DNA-structure. *Genes & Development*, **5**, 2547.
6. Belland, R. J. (1991) H-DNA formation by the coding repeat elements of neisserial *opa* genes. *Mol. Microbiol.*, **5**, 2351.
7. Moxon, E. R., Rainey, P. B., Nowak, M. A., and Lenski, R. E. (1994) Adaptive evolution of highly mutable loci in pathogenic bacteria. *Curr. Biol.*, **4**, 24.
8. Virji, M., Makepeace, K., Ferguson, D. J. P., Achtman, M., Sarkari, J., and Moxon, E. R. (1992) Expression of the Opc protein correlates with invasion of epithelial and endothelial-cells by *Neisseria meningitidis*. *Mol. Microbiol.*, **6**, 2785.
9. Virji, M., Makepeace, K., Ferguson, D. J. P., Achtman, M., and Moxon, E. R. (1993) Meningococcal *opa* and *opc* proteins—their role in colonization and invasion of human epithelial and endothelial-cells. *Mol. Microbiol.*, **10**, 499.
10. Olyhoek, A. J. M., Sarkari, J., Bopp, M., Morelli, G., and Achtman, M. (1991) Cloning and expression in *Escherichia coli* of *opc*, the gene for an unusual class-5 outer-membrane protein from *Neisseria meningitidis* (meningococci surface-antigen). *Microb. Pathogen.*, **11**, 249.
11. Van der Ende, A., Hopman, C. T. P., Zaat, S., Essink, B. B. O., Berkhout, B., and Dankert, J. (1995) Variable expression of class-1 outer-membrane protein in *Neisseria meningitidis* is caused by variation in the spacing between the –10- region and –35-region of the promoter. *J. Bacteriol.*, **177**, 2475.
12. Willems, R., Paul, A., Van der Heide, H. G. J., Teravest, A. R., and Mooi, F. R. (1990) Fimbrial phase variation in *Bordetella pertussis*—a novel mechanism for transcriptional regulation. *EMBO J.*, **9**, 2803.
13. Yogev, D., Watsonmckown, R., Rosengarten, R., Im, J., and Wise, K. S. (1995) Increased structural and combinatorial diversity in an extended family of genes encoding Vlp surface-proteins of *Mycoplasma hyorhinis*. *J. Bacteriol.*, **177**, 5636.
14. Farley, M. M., Whitney, A. M., Spellman, P., Quinn, F. D., Weyant, R. S., Mayer, L. and

Stephens, D. S. (1992) Analysis of the attachment and invasion of human epithelial-cells by *Haemophilus influenzae* biogroup *aegyptius*. *J. Infect. Dis.*, **165**, 111.
15. van Ham, S. M., van Alphen, L., Mooi, F. R., and van Putten, J. P. M. (1993) Phase variation of *Haemophilus influenzae* fimbriae -transcriptional control of 2 divergent genes through a variable combined promoter region. *Cell*, **73**, 1187.
16. van Ham, S. M., van Alphen, L., Mooi, F. R., and van Putten, J. P. M. (1994) The fimbrial gene cluster of *Haemophilus influenzae* type b. *Mol. Microbiol.*, **13**, 673.
17. Stern, A. and Meyer, T. F. (1987) Common mechanism controlling phase and antigenic variation in pathogenic *Neisseriae*. *Mol. Microbiol.*, **1**, 5.
18. Simon, D. and Rest, R. F. (1992) *Escherichia coli* expressing a *Neisseria gonorrhoeae* opacity-associated outer-membrane protein invade human cervical and endometrial epithelial cell lines. *Proc. Natl. Acad. Sci. USA*, **89**, 5512.
19. Kupsch, E. M., Knepper, B., Kuroki, T., Heuer, I., and Meyer, T. F. (1993) Variable opacity (*opa*) outer-membrane proteins account for the cell tropisms displayed by *Neisseria gonorrhoeae* for human-leukocytes and epithelial-cells. *EMBO J.*, **12**, 641.
20. Connell, T. D., Black, W. J., Kawula, T. H., Barritt, D. S., Dempsey, J. A., Kverneland, K., Stephenson, A., Schepart, B. S., Murphy, G. L., and Cannon J. G. (1988) Recombination among protein-II genes of *Neisseria gonorrhoeae* generates new coding sequences and increases structural variability in the protein-II family. *Mol. Microbiol.*, **2**, 227.
21. Schwalbe, R. S. and Cannon, J. G. (1986) Genetic-transformation of genes for protein-II in *Neisseria gonorrhoeae*. *J. Bacteriol.*, **167**, 186.
22. Belland, R. J., Morrison. S. G., Van der Ley, P., and Swanson, J. (1989) Expression and phase variation of gonococcal P-II genes in *Escherichia coli* involves ribosomal frame-shifting and slipped-strand mispairing. *Mol. Microbiol.*, **3**, 777.
23. Maskell, D. J., Szabo, M. J., Deadman, M. E., and Moxon, E. R. (1992) The *gal* locus from *Haemophilus influenzae*—cloning, sequencing and the use of *gal* mutants to study lipo-polysaccharide. *Mol. Microbiol.*, **6**, 3051.
24. High, N. J., Deadman, M. E., and Moxon, E. R. (1993) The role of a repetitive DNA motif (5′-CAAT-3′) in the variable expression of the *Haemophilus influenzae* lipopolysaccharide epitope alpha-gal(1–4)beta-gal. *Mol. Microbiol.*, **9**, 1275.
25. Jennings, M. P., Hood, D. W., Peak, I. R. A., Virji, M., and Moxon, E. R. (1995) Molecular analysis of a locus for the biosynthesis and phase variable expression of the lacto-*N*-neotetraose terminal lipopolysaccharide structure in *Neisseria meningitidis*. *Mol. Microbiol.*, **8**, 729.
26. Yang, Q. L. and Gotschlich E. C. (1995) Genetic locus for the biosynthesis of the variable portion of *Neiserria gonorrhoeae* lipooligosaccharide. *J. Exp. Med.*, **183**, 323
27. Jonsson, A. B., Nyberg, G., and Normark, S. (1991) Phase variation of gonococcal pili by frameshift mutation in *pilC*, a novel gene for pilus assembly. *EMBO J.*, **10**, 477.
28. Jonsson, A. B., Pfeifer, J., and Normark, S. (1992) *Neisseria gonorrhoeae pilC* expression provides a selective mechanism for structural diversity of pili. *Proc. Natl. Acad. Sci. USA*, **89**, 3204.
29. Rudel, T., Van Putten, J. P. M., Gibbs, C. P., Haas, R., and Meyer, T. F. (1992) Interaction of 2 variable proteins (*pilE* and *pilC*) required for pilus- mediated adherence of *Neisseria gonorrhoeae* to human epithelial-cells. *Mol. Microbiol.*, **6**, 3439.
30. Nassif, X., Beretti, J., Lowy, J., Stenberg, P., O'Gaora, P., Pfeifer, J., Normark, S., and So, M. (1994) Roles of pilin and PilC in adhesion of *Neisseria meningitidis* to human epithelial and endothelial cells. *Proc. Natl. Acad. Sci. USA*, **91**, 3769.
31. Virji, M., Saunders, J. R., Sims, G., Makepeace, K., Maskell, D., and Ferguson, D. J. P.

(1993) Pilus-facilitated adherence of *Neisseria meningitidis* to human epithelial and endothelial-cells—modulation of adherence phenotype occurs concurrently with changes in primary amino-acid-sequence and the glycosylation status of pilin. *Mol. Microbiol.*, **10**, 1013.

32. Rudel, T., Scheuerpflug, I., and Meyer, T. F. (1995) *Neisseria* PilC protein identified as type-4 pilus tip located adhesin. *Nature*, **373**, 357.
33. Jonsson, A. B., Rahman, M., and Normark, S. (1995) Pilus biogenesis gene, *pilC*, of *Neisseria gonorrhoeae*—*pilC1* and *pilC2* are each part of a larger duplication of the gonococcal genome and share upstream and downstream homologous sequences with *opa* and *pil* loci. *Microbiology*, **141**, 2367.
34. Rosqvist, R., Skurnik, M., and Wolfwatz, H. (1988) Increased virulence of *Yersinia pseudotuberculosis* by 2 independent mutations. *Nature*, **334**, 522.
35. Stibitz, S., Aaronson, W., Monack, D., and Falkow, S. (1989) Phase variation in *Bordetella pertussis* by frameshift mutation in a gene for a novel 2-component system. *Nature*, **338**, 266.
36. Akerley, B. J. and Miller, J. F. (1993) Flagellin gene-transcription in *Bordetella bronchiseptica* is regulated by the *bvgAS* virulence control- system. *J. Bacteriol.*, **175**, 3468.
37. Dybvig, K. (1993) DNA rearrangements and phenotypic switching in prokaryotes. *Mol. Microbiol.*, **10**, 465.
38. McClain, M. S., Blomfield, I. C., Eberhardt, K. J., and Eisenstein, B. I. (1993) Inversion-independent phase variation of type-1 fimbriae in *Escherichia coli*. *J. Bacteriol.*, **175**, 4335.
39. Gally, D. L., Bogan, J. A., Eisenstein, B. I., and Blomfield, I. C. (1993) Environmental-regulation of the *fim* switch controlling type-1 fimbrial phase variation in *Escherichia coli* K-12—effects of temperature and media. *J. Bacteriol.*, **175**, 6186.
40. McClain, M. S., Blomfield, I. C., and Eisenstein, B. I. (1991) Roles of *fimB* and *fimE* in site-specific DNA inversion associated with phase variation of type-1 fimbriae in *Escherichia coli*. *J. Bacteriol.*, **173**, 5308.
41. Blomfield, I. C., Calie, P. J., Eberhardt, K. J., McClain, M. S., and Eisenstein, B. I. (1993) Lrp stimulates phase variation of type-1 fimbriation in *Escherichia coli* K-12. *J. Bacteriol.*, **175**, 27.
42. Oshima, T., Ito, K., Kabayama, H., and Nakamura, Y. (1995) Regulation of *lrp* gene-expression by H-NS and Lrp proteins in *Escherichia coli* dominant negative mutations in *lrp*. *Mol. Gen. Genet.*, **247**, 521.
43. Hughes, K. T., Youderian. P., and Simon, M. I. (1988) Phase variation in salmonella—analysis of *hin* recombinase and *hix* recombination site interaction *in vivo*. *Genes Develop.*, **2**, 937.
44. Haykinson, M. J. and Johnson, R. C. (1993) DNA looping and the helical repeat *in vitro* and *in vivo*—effect of HU protein and enhancer location on Hin invertasome assembly. *EMBO J.*, **12**, 2503.
45. Johnson, R. C., Glasgow, A. C., and Simon, M. I. Spatial relationship of the Fis binding-sites for *hin* recombinational enhancer activity. *Nature*, **329**, 462.
46. Goshima, N., Kano, Y., Tanaka, H., Kohno, K., Iwaki, T., and Imamoto, F. (1994) IHF suppresses the inhibitory effect of H-NS on Hu function in the Hin inversion system. *Gene*, **141**, 17.
47. Rozsa, F. W., Viollier, P., Fussenegger, M., Hiestandnauer, R., and Arber, W. (1995) Gin-mediated recombination at secondary crossover sites on the *Escherichia coli* chromosome. *J. Bacteriol.*, **177**, 1159.
48. Crisona, N. J., Kanaar. R., Gonzalez. T. N., Zechiedrich. E. L., Klippel. A., and Cozzarelli,

N. R. (1994) Processive recombination by wild-type Gin and an enhancer- independent mutant—insight into the mechanisms of recombination selectivity and strand exchange. *J. Mol. Biol.*, **243**, 437.

49. Gally, D. L., Rucker, T. J., and Blomfield, I. C. (1994) The leucine-responsive regulatory protein binds to the *fim* switch to control phase variation of type-1 fimbrial expression in *Escherichia coli* K-12. *J. Bacteriol.*, **176**, 5665.
50. Nou, X. W., Skinner, B., Braaten, B., Blyn, L., Hirsch, D., and Low, D. (1993) Regulation of pyelonephritis-associated pili phase-variation in *Escherichia coli*—binding of the *papL* and the *lrp* regulatory proteins is controlled by DNA methylation. *Mol. Microbiol.*, **7**, 545.
51. Van der Woude, M. W. and Low, D. A. (1994) Leucine-responsive regulatory protein and deoxyadenosine methylase control the phase variation and expression of the *sfa* and *daa* pili operons in *Escherichia coli*. *Mol. Microbiol.*, **11**, 605.
52. Kim, S. R. and Komano, T. (1992) Nucleotide-sequence of the R721 shufflon. *J. Bacteriol.*, **174**, 7053.
53. Komano, T., Funayama, N., Kim, S. R., and Nisioka, T. (1990) Transfer region of IncI1 plasmid R64 and role of shufflon in R64 transfer. *J. Bacteriol.*, **172**, 2230.
54. Moses, E. K., Good, R. T., Sinistaj, M., Billington, S. J., Langford, C. J., and Rood, J. I. (1995) A multiple site-specific DNA-inversion model for the control of *omp1* phase and antigenic variation in *Dichelobacter nodosus*. *Mol. Microbiol.*, **17**, 183.
55. Bhugra, B., Voelker, L. L., Zou, N. X., Yu, H. L., and Dybvig, K. (1995) Mechanism of antigenic variation in *Mycoplasma pulmonis* -interwoven, site-specific DNA inversions. *Mol. Microbiol.*, **18**, 703.
56. Bhugra, B. and Dybvig, K. (1992) High-frequency rearrangements in the chromosome of *Mycoplasma pulmonis* correlate with phenotypic switching. *Mol. Microbiol.*, **6**, 1149.
57. Seifert, H. S., Wright. C. J., Jerse. A. E., Cohen, M. S., and Cannon, J. G. (1994) Multiple gonococcal pilin antigenic variants are produced during experimental human infections. *J. Clin. Invest.*, **93**, 2744.
58. Robertson, B. D. and Meyer, T. F. (1992) Genetic-variation in pathogenic bacteria. *Trend Genet.*, **8**, 422.
59. Parge, H. E., Forest, K. T., Hickey, M. J., Chrisensen, D. A., Getzoff, E. D., and Tainer, J. A. (1995) Structure of the fibre-forming protein pilin at 2.6A resolution. *Nature*, **378**, 32.
60. Stimson, E., Virji, M., Makepeace. K., Dell, A., Morris, H. R., Payne. G., Saunders, J. R., Jennings. M. P., Barker, S., Panico, M., and Moxon, E. R. (1995) Meningococcal pilin—a glycoprotein substituted with digalactosyl 2,4-diacetamido-2,4,6-trideoxyhexose. *Mol. Microbiol.*, **17**, 1201.
61. Marceau, M., Beretti, J. L., and Nassif, X. (1995) High adhesiveness of encapsulated *Neisseria meningitidis* to epithelial-cells is associated with the formation of bundles of pili. *Mol. Microbiol.*, **17**, 855.
62. Meyer, T. F., Billyard, E., Haas, R., Storzbach, S., and So, M. (1984) Pilus genes of *Neisseria gonorrhoeae*—chromosomal organization and DNA-sequence. *Proc. Natl. Acad. Sci. USA*, **81**, 6110.
63. Haas, R. and Meyer, T. F. (1986) The repertoire of silent pilus genes in *Neisseria gonorrhoeae* - evidence for gene conversion. *Cell*, **44**, 107.
64. Koomey, M., Gotschlich, E. C., Robbins, K., Bergstrom, S., and Swanson, J. (1987) Effects of *recA* mutations on pilus antigenic variation and phase- transitions in *Neisseria gonorrhoeae*. *Genetics*, **117**, 391.
65. Hill, S. A., Morrison, S. G., and Swanson, J. (1990) The role of direct oligonucleotide repeats in gonococcal pilin gene variation. *Mol. Microbiol.*, **4**, 1341.

66. Seifert, H. S., Ajioka. R. S., Marchal, C., Sparling, P. F., and So, M. (1988) DNA transformation leads to pilin antigenic variation in *Neisseria gonorrhoeae*. *Nature*, **336**, 392.
67. Gibbs, C. P., Reimann, B. Y., Schultz, E., Kaufmann, A., Haas, R., and Meyer, T. F. (1989) Reassortment of pilin genes in *Neisseria gonorrhoeae* occurs by 2 distinct mechanisms. *Nature*, **338**, 651.
68. Rudel, T., Facius, D., Barten, R., Scheuerpflug, I., Nonnenmacher, E., and Meyer, T. F. (1995) Role of pili and the phase-variable PilC protein in natural competence for transformation of *Neisseria gonorrhoeae*. *Proc. Natl. Acad. Sci. USA*, **92**, 7986.
69. Barbour, A. G. (1990) Antigenic variation of a relapsing fever *Borrelia* species. *Ann. Rev. Microbiol.*, **44**, 155.
70. Kitten, T., Barrera, A. V., and Barbour, A. G. (1993) Intragenic recombination and a chimeric outer-membrane protein in the relapsing fever agent *Borrelia hermsii*. *J. Bacteriol.*, **175**, 2516.
71. Wilske, B., Barbour, A. G., Bergstrom, S., Burman, N., Restrepo, B. I., Rosa, P. A., Schwan, T., Soutschek, E., and Wallich, R. (1992) Antigenic variation and strain heterogeneity in *Borrelia spp*. *Res. Microbiol.*, **143**, 583.
72. Sarkari, J., Pandit, N., Moxon, E. R., and Achtman, M. (1994) Variable expression of the Opc outer-membrane protein in *Neisseria meningitidis* is caused by size variation of a promoter containing poly-cytidine. *Mol. Microbiol.*, **13**, 207.
73. Yogev, D., Rosengarten, R., Watsonmckown, R., and Wise, K. S. (1991) Molecular-basis of mycoplasma surface antigenic variation—a novel set of divergent genes undergo spontaneous mutation of periodic coding regions and 5′ regulatory sequences. *EMBO J.*, **10**, 4069.
74. Bhat, K. S., Gibbs, C. P., Barrera, O., Morrison, S. G., Jahnig, F., Stern, A., Kupsch, E. M., Meyer, T. F., and Swanson, J. (1991) The opacity proteins of *Neisseria gonorrhoeae* strain MS11 are encoded by a family of 11 complete genes. *Mol. Microbiol.*, **5**, 1889.

10 Integration of control devices. I. Pathogenicity

CHARLES J. DORMAN

1. Introduction

One of the driving forces behind microbiological research has been the desire to understand how and why some bacteria cause disease in humans, in animals, or in plants. The employment of molecular genetic and cell biological techniques has led to the identification of factors which make key contributions to the ability of bacteria to be pathogenic. These may be classified as dedicated virulence factors (for example, toxins) which one might expect to find exclusively in pathogenic bacteria and, accessory virulence factors (for example, nutrient uptake systems) which might be equally important in the commensal and the pathogenic situation (1).

A successful life in association with a host requires the bacterium to have the ability to meet its nutritional requirements in the host environment and the ability to survive or overcome the host defences. Association with a particular niche on or within the host imposes selective pressures on bacteria and influences the composition of their genome and the manner in which the information encoded in the genome is expressed. Many dedicated virulence factors are associated with mobile genetic elements such as plasmids or transposable elements, indicating that they have been acquired relatively recently and that their maintenance may be selected for by the composition of the host environment (2).

It has been hypothesized that habitual association with different environments can result in differential expression of the same gene in related bacteria (3). This may reflect 'fine-tuning' of gene expression to suit the detail of each environment. Variations in gene expression occur within apparently homogenous bacterial populations as well as between related populations. In many cases, bacterial cell surface components such as proteins and lipopolysaccharide are subject to variability in their expression. The molecular detail of variation has been worked out in many cases and usually involves changes in genome structure. These can be as small as a change affecting one base pair (deletion, substitution, or modification) or can involve much larger rearrangements. Some small sequence variations arise by processes involving errors and repair during DNA replication; the larger rearrangements can be catalysed by site-specific recombination mechanisms in which dedicated recombinases vary the expression of a particular cell surface component. Examples of these processes have been reviewed in other chapters in this volume.

An important development in studies of bacterial pathogenicity has been the discovery that expression of virulence factors is not usually constitutive but is subject to regulation. Frequently, regulation is at the level of transcription and it occurs in response to specific environmental signals. The reception of these signals by the bacterium and their transduction to the genome results in an alteration in the transcriptional profile of the cell. This produces a change (subtle or dramatic) in cellular composition which renders the bacterium more fit for survival on or within the host. Depending on many other factors, including the health of the host, the bacterium may engage in activities that produce a pathological condition (i.e. a disease). The ease with which a particular strain of bacterium produces disease symptoms is a measure of its virulence. The expression of a virulent phenotype by a bacterium is, therefore, a manifestation of gene regulatory processes which that bacterium shares with its siblings in the population, and of any variations in gene expression arising from changes in genome structure which are peculiar to that bacterium. In other words, the composition of a particular bacterium in a particular environment is determined by a combination of stereotypic processes (classical gene regulation) and stochastic processes (variations in expression of specific genes).

2. The host as an environment

Association with the host can bring many benefits to the bacterium, such as an improved nutrient supply and freedom from competition (especially if the bacterium can enter a privileged niche such as the internal compartments of host cells). This association places adaptive demands on the bacterium, just as the colonization of any new environment does. The microbe will encounter variations in temperature, pH, oxygen supply, osmolarity, iron availability, and many other parameters. It will also encounter specific host defences. Furthermore, the host is not a single, homogeneous environment but a series of microenvironments or ecological hurdles which the bacterium must negotiate as it infects. Thus, the environment represented by the host is dynamic and the bacterium must mount a dynamic response (1, 2).

Other chapters in this volume deal with the molecular basis of environmental signal transduction to the bacterial genome. This chapter will review the contributions of these systems to the control of bacterial virulence gene expression and show how multiple systems collaborate to regulate coordinately the expression of genetic information which assists the bacterium during infection. This will be done by reviewing knowledge of specific pathogens of humans, animals, and plants. It should become apparent to the reader that, diverse as the environments associated with these hosts are, there is an impressive degree of convergence among the strategies used by the infecting bacteria to regulate expression of their virulence genes.

3. Thermo-osmotic control of virulence gene expression in *Shigella flexneri*

3.1 *Shigella flexneri* pathogenicity

Shigella flexneri is a facultative intracellular pathogen and the causative agent of bacillary dysentery. It is closely related to *Escherichia coli*, and particularly to enteroinvasive *E. coli* (or EIEC). Invasiveness is a sophisticated pathogenic phenotype and it depends on the expression of genetic information carried within a 230-kb virulence plasmid (4). Here, genes code for the ability to enter and spread within human epithelial cells. Typically, invasion occurs in the epithelia lining the colon and in the phagocytes within connective tissue making up the lamina propria of intestinal vili. The resulting disease, dysentery, is characterized by a bloody diarrhoea, severe abdominal cramps, and fever (5). The destructive effects of *S. flexneri* on the cells of the colon and the associated fluid loss contribute directly to these symptoms. Expression of the plasmid-encoded invasion genes is under environmental control, being regulated in response to changes in temperature and osmolarity (see below).

3.2 Virulence plasmid-encoded regulators

Two plasmid-encoded transcription factors, VirF and VirB, are required to activate the structural gene promoters. VirF is a member of the AraC family of transcription factors (6) and is required for transcription initiation at the *virB* promoter (7); the VirB protein activates the structural genes (8). The *virB* promoter is a major site of environmental regulatory input. If transcription of the *virB* gene is driven from a heterologous promoter, invasion gene expression can be divorced from thermal control (8). The binding site for VirF has been determined at the *virB* promoter, and this protein is central to the transmission of the thermal signal (9). At present it seems likely that VirF is bound to the *virB* promoter at both the permissive and non-permissive temperatures, but requires a thermal signal before interacting with RNA polymerase to facilitate initiation of transcription. This signal may operate through a change in the supercoiling of the DNA at the *virB* promoter (see below).

3.3 Negative regulation by H-NS

In addition to positive control by VirF, the *virB* promoter is subject to negative control by the nucleoid-associated protein, H-NS (9). H-NS is an abundant protein of 15.6 kDa which binds preferentially to curved DNA and regulates transcription of many unrelated genes; it is also an important structural component of bacterial chromatin (10; see also Chapter 6). Its role in *S. flexneri* virulence gene expression was first detected when a mutation in its chromosomally located structural gene, *hns*, was found to derepress invasion gene transcription at low temperatures; normally, transcription of these genes is repressed at 30 °C but in an *hns* mutant transcription approaches levels normally seen at 37 °C (11–13).

Subsequently, it was found that transcriptional activation of virulence gene expression was impossible at the inducing temperature if the bacteria were grown in a low osmolarity medium. Inactivation of the *hns* gene relieved this low-osmolarity-repression, suggesting that H-NS has a role in osmotic control of *virB* transcription (14).

H-NS has a negative effect on the *virB* promoter and its binding site has been determined by DNase I protection studies *in vitro* (9). The binding site intrudes into the pribnow box at which RNA polymerase holoenzyme would form an open promoter complex. This may indicate that H-NS simply excludes polymerase by steric hindrance. The signal that removes H-NS from the DNA or modifies its interaction with DNA so that it no longer blocks transcription is unknown. Presumably it is associated with an increase in osmotic pressure, in temperature, or both.

3.4 Sensitivity to variations in DNA supercoiling

The involvement of H-NS in *virB* regulation is suggestive of a role for DNA topology in controlling this promoter. Mutations in *hns* result in alterations in reporter plasmid linking number, and the H-NS protein has been shown to compact DNA strongly *in vivo* and *in vitro* and has the ability to constrain supercoils in DNA (10, 11). The *virB* promoter is sensitive to changes in DNA supercoiling and appears to require a critical level of supercoiling for optimal expression (9). The topology of bacterial DNA is also altered by changes in the growth temperature of the culture and by variations in osmotic pressure (15, 16). These observations suggest that supercoiling variations at *virB* produced by fluctuations in environmental parameters could contribute to *virB* gene regulation.

3.5 Regulatory model

These findings permit a model of *virB* transcriptional control to be proposed which takes all of these diverse regulatory factors into account. In this model an interaction is required between operator-bound VirF protein and RNA polymerase bound to the *virB* promoter in a closed complex. This interaction is modulated by variations in the supercoiling of the DNA in the vicinity of the promoter. These variations originate in signals from the environment, specifically temperature and osmotic pressure. The positive influence of VirF on RNA polymerase activity is antagonized by H-NS which is bound in the region of the pribnow box. The interaction of H-NS with DNA is also environmentally influenced, perhaps especially by variations in osmotic pressure. The departure of H-NS from the promoter DNA or a change in the quality of its interaction with that DNA may also modulate local DNA topology, promoting a better interaction between VirF and polymerase. Equally, an increase in growth temperature could alter *virB* promoter supercoiling such that a productive interaction between VirF and RNA polymerase is promoted. Evidence in support of this model comes from an experiment in which the level of negative supercoiling at the *virB* promoter was increased by a method which did not involve an increase in temperature. The *virB* promoter became active (albeit only in the presence of VirF), even at temperatures which were normally non-permissive (17). This strongly suggests that the thermal signal functions by altering the level of supercoiling at the *virB* promoter.

3.6 Other regulatory factors

The *S. flexneri* virulence genes are also influenced by the chromosomally encoded regulator EnvZ (18). This is a member of the histidine protein kinase family of environmental signal transducing proteins (see Chapter 8), and *S. flexneri* mutants failing to express it cannot express their invasion genes. In other enteric bacteria such as *E. coli,* EnvZ has been shown primarily to transduce osmotic pressure signals to the genome via a DNA binding protein OmpR (19). The EnvZ protein is inner-membrane-associated and undergoes autophosphorylation in response to increased osmotic pressure. Phosphotransfer to OmpR primes the latter for transcription activation of its subservient promoters. In *E. coli,* the best-characterized OmpR-dependent promoters are those of the porin genes *ompC* and *ompF.* Significantly, the product of the *S. flexneri ompC* gene is required for invasion of host cells by that bacterium (20).

The chromosomally encoded Rho factor has also been shown to contribute to the control of invasion gene expression in *S. flexneri* (21). In *E. coli,* Rho provides an essential function which catalyses transcription termination at certain sites (22). A *rho* mutation in *S. flexneri* results in transcription of the *virB* gene at 30 °C and a consequent transcription of the structural virulence genes (21). Mutations in *E. coli rho* have been shown previously to alter the superhelicity of DNA (23, 24) and it has been suggested that the supercoiling-sensitive expression of *virB* is altered in the presence of a *S. flexneri rho* lesion (21).

3.7 Gene regulation and the ecology of the disease

Why this level of gene regulatory complexity? To express constitutively the proteins encoded by the large invasion gene operons carried on the *S. flexneri* virulence plasmid would represent a very large energetic drain on the cell. Furthermore, as these include several surface-expressed proteins, the cell may needlessly prime the host defences by their premature exposure. To ensure that the invasion proteins are expressed in the niche where they are required (in the lower intestine of the human host) their expression is controlled by two environmental signals characteristic of that niche: a temperature of 37 °C and an osmotic pressure equivalent to that of physiological saline.

The *S. flexneri* virulence system serves to illustrate another feature common to many environmentally responsive genetic systems. This is the possession of specific (or 'private') regulators, here VirF and VirB, and a sensitivity to control by regulators shared with many other systems ('public' regulators) such as H-NS and environmentally determined levels of DNA supercoiling. The involvement of the more pleiotropic control elements ensures that the expression of this specific virulence system is coordinated with that of other systems (including house-keeping systems) which enable the organism to adapt to its host (reviewed in Reference 25).

4. Pleiotropic regulators of gene expression and virulence in *Salmonella typhimurium*

4.1 *Salmonella typhimurium* is a facultative intracellular pathogen

The Gram-negative enteric bacterium *Salmonella typhimurium* shares with *Shigella flexneri* the ability to invade and survive within mammalian cells. While shigellosis involves a condition in which the bacteria invade and destroy cells lining the colon, salmonellosis is a disease in which bacteria simply cross the epithelial layer in order to gain access to deeper tissues. *S. typhimurium* causes a typhoid-like disease in mice and this pathogenic interaction has been studied extensively and intensively by many groups as a model for human typhoid (26). Like *Sh. flexneri*, *S. typhimurium* possesses a high molecular mass plasmid which codes for virulence functions. However, the invasion genes of *Salmonella* are found on the chromosome rather than on this plasmid. As in *Shigella*, the number of invasion genes in *Salmonella* is very large. Interestingly, 12 of the *Salmonella* genes have been found to have the same gene order and to share significant sequence homology with their plasmid-borne counterparts in *Shigella* (27). However, several key invasion functions from *Shigella* have no known counterparts in *Salmonella* (28), pointing to important differences in detail between the invasion systems of these organisms. The *Salmonella* invasion genes are found on a 40-kb region of chromosomal DNA which is absent from the corresponding location on the *E. coli* chromosome, and it has been suggested that these genes were acquired as a block during the evolution of *Salmonella* as a pathogen (29).

4.2 Gene regulation and *Salmonella* virulence

A characteristic feature of research with *S. typhimurium* has been attempts to disable it as a pathogen so that it can be exploited as a live vaccine or a live delivery system for heterologous (i.e. non-*Salmonella*) antigens (reviewed in Reference 26). These studies have involved more or less systematic approaches to finding and inactivating genes which may contribute to virulence in order to achieve different levels of virulence attenuation (30). Any promising mutants are then tested for virulence in a mouse model infection system. Several regulatory genes have been shown to contribute to *Salmonella* virulence using this approach (2).

4.3 cAMP–CRP mutants and virulence attenuation

An early example of evidence linking a pleiotropic regulator of transcription with virulence in *S. typhimurium* was the demonstration that strains deficient in the cAMP–CRP system failed to kill mice (31). (This system has been described in

Chapter 4.) Furthermore, mice infected with the mutant bacteria were found to be protected against challenge with the wild-type organism. Presumably, loss of cAMP–CRP also results in loss of expression of one or more key structural genes whose products contribute to virulence. It is also possible that a gene negatively controlled by this control system is derepressed in the mutant, although cAMP–CRP is only rarely a repressor of transcription (32).

4.4 Histidine protein kinase and response regulator 'two-component' systems as regulators of virulence gene expression in *Salmonella*

A null mutation in the gene coding for the DNA-binding-protein OmpR attenuates the virulent characteristics of *S. typhimurium* in a mouse infection system (33). Since OmpR is the response regulatory partner of the EnvZ histidine protein kinase and both are required for the osmotic regulation of porin gene expression, it is likely that *ompR* mutants are attenuated due to a lost ability to respond appropriately to environmental stimuli. The *ompC* and *ompF* genes, coding for the OmpC and OmpF porins, respectively, are OmpR-dependent for transcription (19), and when both genes are mutated the result is a level of attenuation of virulence which approaches that seen in the *ompR* null mutant (34). Inactivating the *ompC* or *ompF* genes individually does not produce this attenuating effect (34). Moreover, since the double porin mutant is not as attenuated as the *ompR* regulatory gene mutant, this implies that other, as yet unknown, OmpR-dependent genes contribute to mouse virulence.

The PhoP/PhoQ 'two-component' system (see Chapter 8) is also an important contributor to *S. typhimurium* in mice. PhoQ shows homology to the histidine protein kinases, which include EnvZ, while PhoP resembles the response regulator proteins, which include OmpR (35). Structural genes have been identified which are under PhoP/PhoQ control and whose products contribute to virulence. One of these is *pagC* whose product is an 18-kDa envelope protein which helps *S. typhimurium* to survive in macrophages (36). Genes subject to positive control by PhoP/PhoQ are termed *pag* (*p*hoP-*a*ctivated *g*ene). PhoP-repressible genes (*prg*) have also been described and found to contribute to virulence. For example, the *prgH* gene is involved in epithelial cell invasion by *S. typhimurium* (37). Thus, the PhoP/PhoQ system regulates differentially the transcription of *S. typhimurium* genes involved in invasion (*prgH*) and in macrophage survival (*pagC*). In addition, this two-component system controls expression of house-keeping genes, such as the structural gene coding for acid phosphatase, *phoN*, a gene with no known role in bacterial virulence (36, 38, 39).

Several groups have attempted to discover the nature of the environmental signal which is transduced by the PhoP/PhoQ system when *S. typhimurium* encounters the host. Work with the organism in the commensal and the pathogenic situation has shown that members of the PhoP regulon are under multifactorial control, being sensitive to changes in carbon, nitrogen, phosphorus (hence the name of the regulatory system), and pH (31, 38, 39). One proposal concerning host intracellular conditions is that the PhoP/PhoQ system transmits information about phagosomal pH. Miller and co-workers have found that PhoP-activated genes (*pag*) are induced by 50- to 77-fold following several hours spent inside murine macrophages (40). In contrast, the same *pag* genes are uninduced when the bacteria are harboured within epithelial cells. Treatments with weak bases which raise the pH of the acidic compartments within the macrophage abolish the macrophage-specific induction of PhoP-dependent genes. It has been suggested that phagosome acidification following bacterial phagocytosis is a key signal for induction of the PhoP regulon. Furthermore, survival of the bacteria seems to involve the expression of a defensive system which delays acidification (40).

PhoP/PhoQ is also interesting in that it regulates a second two-component system, PmrA/PrmB. This regulates resistance to cationic peptides and makes an indirect contribution to the control of lipopolysaccharide modification (41). The *pmrApmrB* genes are organized as an operon and are co-transcribed with the upstream *pagB* gene, which is under the control of PhoP. At least seven genes identified initially as being under PhoP control are now known to be regulated instead by PmrA, showing that these two systems work as a regulatory cascade (42).

4.5 The plasmid-encoded *spv* virulence genes: regulation by growth phase

The involvement of a high molecular mass plasmid in *S. typhimurium* virulence has been referred to above. Unlike the plasmid of *Shigella flexneri,* that of *S. typhimurium* plays a rather ill-defined role in determining virulence. The evidence available suggests that it enhances the ability of the bacterium to survive and proliferate in the reticuloendothelial system of infected mice (43). It is important to point out that many *Salmonella* serovars do not contain high-molecular-weight plasmids but are still virulent. In addition, the plasmid structure varies among those serovars which do contain them. When these plasmids are compared, a common region can be discerned which is crucial for the virulent properties of the bacteria harbouring the plasmid. This region consists of an operon of at least four structural genes *spvABCD* and a regulatory gene *spvR* (43). The product of the regulatory gene, SpvR, is a member of the LysR family of transcription regulatory DNA binding proteins (44). SpvR is required for the transcriptional activation of the structural genes; it also regulates its own gene positively (45–48). Activation of the *spv* genes occurs as the bacterium enters stationary phase and has been shown to depend on the stationary-phase-specific sigma factor RpoS (49, 50). This control appears to operate through a regulatory cascade in which both RpoS and SpvR are required for *spvR* activation, and then SpvR activates the structural gene promoter independently of RpoS (51). In at least some strains of *S. typhimurium, spv* activation is regulated negatively by cAMP–CRP (52). This may act via the known negative effect of cAMP–CRP on expression of *rpoS,* the gene coding for the stationary phase sigma factor (53). Presumably, *S. typhimurium* growing in particular niches of the host experiences growth conditions approximating to those defined as stationary phase in laboratory cultures.

Under these growth conditions, *spv* gene expression is activated. Significantly, expression of this operon has been shown to be induced in bacteria within macrophages (54, 55).

4.6 The H-NS protein as a regulator of *Salmonella* virulence

The role of the nucleoid-associated protein H-NS in controlling expression of specific virulence genes in *Shigella flexneri* is well documented (see above). When attempts were made to investigate the potential for H-NS to influence the virulence of *S. typhimurium,* the results were found to be quite complex. Insertion mutations at the *hns* locus result in attenuation of mouse virulence, but when the mutants are transduced to wild-type, not all transductants regain virulence (56). This suggests that *hns* mutations have an enabling function which allows other, attenuating mutations to occur in the cell. This is consistent with the well-documented effects of *hns* mutations on genome stability in *E. coli* (57, 58). Where these putative second-site mutations map is unknown.

H-NS levels do alter the expression of at least one specific virulence system in *S. typhimurium.* This is the plasmid-linked *spv* operon (see previous section). When the *hns* gene is over-expressed from a multicopy recombinant plasmid, it prevents completely the stationary phase induction of *spv* gene expression. Strains harbouring a transposon insertion mutation in *hns* express *spv* genes to higher levels than wild-type strains in stationary phase (but not in exponential phase) (59).

4.7 Regulation of *Salmonella* invasion gene expression

Using transcriptional and translational reporter gene fusions to the *S. typhimurium invA* invasion gene, Galán and Curtiss (60) investigated its regulatory characteristics. Expression of *invA* was found to be induced strongly by increased osmolarity, showing that this gene belongs to the osmotic stimulon of *S. typhimurium.* Mutations in the gene coding for the osmoregulatory DNA binding protein OmpR had no effect on *invA* expression, showing that it did not belong to the EnvZ/OmpR regulon. Given previous data showing that osmotic stress alters the supercoiling of DNA in *S. typhimurium* (16), the effects of DNA gyrase inhibition and topoisomerase I gene inactivation on *invA* expression were tested. Increasing concentrations of the gyrase inhibitors novobiocin and coumermycin repressed expression of *invA,* while loss of a functional *topA* gene (coding for DNA topoisomerase I) resulted in a strong reduction in *invA* expression (60). These data suggest that normal expression of *invA* depends on a critical level of DNA supercoiling and that manipulations which alter *in vivo* supercoiling levels result in aberrant expression of this gene.

The physiological relevance of these data was demonstrated by an experiment in which a wild-type strain and a mutant deleted for *topA* were compared for their ability to invade tissue culture cells. Here, the *topA* mutant was found to be strongly impaired in its invasive abilities (60). Presumably, the high osmolarity environment of the colon is a crucial signal for invasion gene expression in *Salmonella,* as it is in

Shigella. Other growth conditions have been found to influence expression of an invasive phenotype, these include anaerobiosis and growth phase (61, 62). Anaerobic growth conditions promote invasiveness, while stationary phase is inhibitory. These findings suggest that a combination of high osmolarity and anaerobic growth conditions will promote invasiveness in actively growing *S. typhimurium* cells. Significantly, the degree of supercoiling in bacterial DNA is influenced by anaerobiosis and growth phase (in addition to osmolarity) (63, 64). Thus, *S. typhimurium* appears to share with *Sh. flexneri* a sensitivity to DNA topology in the expression of invasiveness.

5. The virulence gene regulatory cascade of *Vibrio cholerae*

5.1 Cholera and cholera toxin

The Gram-negative bacterium *Vibrio cholerae* is the causative agent of cholera, a diarrhoeal disease, in humans. The disease is caused by the activity of the bacteria in the host intestine and does not involve entry of the pathogen into host cells. Instead, virulence factors are expressed in the intestinal lumen which disrupt host cellular function. A major virulence determinant is cholera toxin, an ADP-ribosylating enzyme with an A_1B_5 subunit structure (65). The A subunit contributes the enzymatic activity of the toxin, while the B subunit is required for host cell binding. Following B-subunit-mediated attachment to the ganglioside GM_1 receptor on the host cell surface, the A subunit crosses the membrane and catalyses the ADP-ribosylation of guanyl-nucleotide-binding regulatory proteins (66–69). This leads to activation of host cell adenylate cyclase producing an accumulation of cAMP in the intestinal mucosa resulting in the severe fluid loss which characterizes cholera.

5.2 Cholera toxin genes are subject to amplification

The genes coding for the subunits of cholera toxin, *ctxA* (A subunit) and *ctxB* (B subunit) are located on the chromosome and are arranged as an operon, *ctxAB* (70). The operon is associated with a repetitive DNA sequence called RS1 which can promote amplification of *ctxAB* copies on the chromosome (70). Growth of the bacteria in the intestines of rabbits has been shown to exert a selective pressure which favours amplification of the *ctxAB* genes. Presumably, *V. cholerae* strains which have increased levels of *ctxAB* expression due to gene amplification are better adapted to the intestinal environment than those without amplified *ctx* genes. This hypothesis is supported by the observation that recently isolated *V. cholerae* strains of clinical origin have a higher *ctxAB* copy number than strains which have been stored in the laboratory for prolonged periods (71).

5.3 Cholera toxin genes are encoded by a bacteriophage

The *ctxAB* genes and their associated RS1 sequences form a complex mobile genetic element known as CTX. This can undergo site-specific recombination with specific

17-bp sites on the chromosome called *att*RS1 (72). The CTX element can replicate as a plasmid and is transmissible in the form of a bacteriophage whose particles contain only single-stranded CTX DNA. This phage is called CTXϕ and is related to the coliphage M13. It uses the toxin co-regulated pilus (TCP) as a receptor and so can only enter *V. cholerae* cells grown under conditions which permit expression of this pilus. The phage infects *V. cholerae* cells growing in the gastrointestinal tract of mice much more efficiently than it infects bacteria grown under laboratory conditions, showing that horizontal transfer contributes to dissemination of *ctx* and may depend on *in vivo* gene expression (73).

5.4 ToxR, ToxS, and the control of *ctxAB* expression

Transcription of the *ctxAB* operon depends on a membrane-located DNA binding protein called ToxR. This protein is thought to bind to sequences upstream of the *ctx* promoter which include the tandemly repeated element TTTTGAT. Deletion of these reiterated sequences reduces dramatically ToxR-dependent *ctx* gene expression and prevents gel retardation of purified *ctx* DNA by a ToxR–PhoA fusion protein (74).

This regulatory system seems to permit the *ctx* promoter to respond to a wide range of environmental signals, including osmolarity, pH, temperature, and the presence of the L-amino acids asparagine, arginine, glutamate, and serine. The membrane location of ToxR is unusual for a DNA binding protein but lends itself well to signal transduction. The N-terminal domain of the protein is found in the cytoplasm and includes a region of approximately 100 amino acids showing homology to the C-terminal (DNA-binding) domain of OmpR, the response regulatory partner of the EnvZ/OmpR 'two-component' signal transduction system. This observation appeared at first to indicate that ToxR was a protein combining the functions of sensor and response regulator in a single polypeptide. Subsequently, a gene immediately downstream of *toxR* was found to code for another protein required for activation of the *ctx* promoter (75). This gene, *toxS*, codes for an inner-membrane-associated protein of 19 kDa which is mainly located in the periplasm. ToxS interacts directly with ToxR, probably in the periplasm, and the function of ToxS is thought to be to dimerize ToxR (76).

5.5 *Vibrio cholerae* has a ToxR-dependent virulence gene regulon

In addition to the *ctx* operon, *V. cholerae* possesses other genes whose expression depends on ToxR. These include the genes coding for the outer membrane proteins OmpT and OmpU, those coding for the toxin co-regulated pilus (TCP), and genes which specify the accessory colonization factor (ACF) (for review see Reference 77). Expression of the cytoplasmic enzyme aldehyde dehydrogenase (encoded by *aldA*) is also controlled via ToxR; this enzyme is not thought to contribute to the virulence phenotype of *V. cholerae* (78). The direct relationship between ToxR and *ctxAB* promoter (involving the TTTTGAT repeat sequences) has proved harder to establish for these other genes. It is now appreciated that ToxR probably regulates transcription of many of its subservient genes indirectly, acting through a regulatory cascade.

5.6 ToxT is an intermediate regulator of ToxR-dependent genes

DiRita *et al.* discovered that the *toxT* gene of *V. cholerae* can activate ToxR-dependent genes in *E. coli* in the absence of other *V. cholerae*-encoded factors (79). The ToxT protein has been discovered independently by Ogierman and Manning (80) and called TcpN. These investigators have shown TcpN/ToxT to be a putative regulatory protein with homology to the AraC family of transcription activators and to be encoded by a gene within the TCP pilin operon. Expression of *toxT*-encoded mRNA is dependent on ToxR and is influenced by the environmental factors which modulate transcription of the ToxR regulon members, supporting the hypothesis that ToxT lies below ToxR in a regulatory cascade. Interestingly, ToxT can also regulate *ctxAB* transcription, indicating that it may amplify the positive inputs of ToxR at this promoter (79).

5.7 The heat shock response and the expression of ToxR

The *toxR* gene occupies a busy location on the *V. cholerae* chromosome. In addition to the *toxS* gene located downstream, a gene coding for a possible heat shock protein lies immediately upstream and is transcribed on the opposite DNA strand. This upstream gene, *htpG*, codes for the bacterial homologue of heat shock protein Hsp90 and when cloned in *E. coli* depends on that organism's heat shock sigma factor σ^{32} for transcription (81). It has been proposed that in *V. cholerae*, heat shock activation of the *htpG* gene has a repressive effect on transcription of the downstream *toxR* gene, and this may explain why transcription of ToxR-dependent genes is downregulated during thermal shifts from 22 °C to 37 °C (81).

5.8 DNA topology and *Vibrio cholerae* virulence gene expression

The *acf* genes of *V. cholerae* encode the accessory colonization factor, ACF. As stated above, expression of these genes is controlled positively by ToxR. The *acfA* and *acfD* genes are transcribed in opposite directions from an intergenic region of 104 bp of AT-rich DNA (82). Cloning of these two genes and their intergenic region in low copy number plasmids led to a loss of normal regulation and this was interpreted as being due to possible differences in the topology of the DNA when carried on the plasmid rather than on the chromosome. This hypothesis was supported by the finding that the gyrase inhibiting antibiotics nalidixic acid and novobiocin inhibit strongly ToxR-dependent expression of these genes (82). The data indicate that the promoters of the *acfA* and *acfD* genes require a critical level of DNA supercoiling in addition to ToxR (and ToxT) for normal activation.

6. Control of virulence gene expression in *Bordetella pertussis*

6.1 The virulence determinants of *Bordetella pertussis*

Salmonella typhimurium, *Shigella flexneri*, and *Vibrio cholerae* all cause diseases of the human or animal intestine. In contrast, *Bordetella pertussis* is the causative agent of whooping cough in humans, a respiratory disease. Although this pathogen is adapted to a very different environment to that colonized by *S. typhimurium*, *Sh. flexneri*, and *V. cholerae*, it employs some of the same regulatory features to control expression of its virulence genes in response to external signals. *B. pertussis* adheres to the cilia of the upper respiratory tract, multiplies and releases several toxic substances. The known major virulence determinants are expressed on the surface of the bacterium or are secreted and released. They include fimbriae, pertactin, filamentous haemagglutinin, pertussis toxin, dermonectrotic toxin, tracheal toxin, and adenlyate cyclase haemolysin. Pertussis toxin may be responsible for the characteristic paroxysmal cough associated with the disease; it is also believed to be an adhesin. The dual functional adenylate cyclase haemolysin can penetrate host cells where it elevates cAMP levels; this protein may be an antiphagocytic factor. Dermonecrotic toxin and tracheal cytotoxin may induce ciliary paralysis and local tissue damage, impeding clearance of the organisms and contributing to the progress of the disease (83).

6.2 Coordinate control of virulence factor expression

Expression of *B. pertussis* virulence factors is influenced by specific environmental stimuli. Growth with NaCl promotes expression of a virulent phenotype whereas exposure to $MgSO_4$ or nicotinic acid represses virulence (83, 84). Temperature is also an important influence; virulence factors are expressed at 37 °C but not 25 °C (85). A regulatory locus required for the coordinate control of virulence gene expression has been detected genetically, and shown to code for a histidine protein kinase and response regulator partnership called BvgS/A (83, 85, 86). BvgS is inner membrane associated and is the putative kinase of the system; it also possesses a C-terminal domain with homology to the signal receiver element of response regulator proteins (87). This C-terminal domain may play a role in modulating the strength of the signal transmitted to the partner protein, BvgA, which is a 23-kDa DNA-binding protein (88, 89). Genes subject to Bvg control fall into two classes. The best-characterized are known as *vag* or '*vir*-*a*ctivated *g*enes' (*bvg* was known formerly as *vir*). The members of the other class are called *vrg* or '*vir*-*r*epressed *g*enes' (90, 91). The BvgAS system is conserved in the human pathogen *B. parapertussis* and in the animal pathogen *B. bronchiseptica* (92).

6.3 Transcriptional control of BvgAS expression

The genes coding for BvgA and BvgS are arranged in an operon, and in addition to controlling expression of the structural genes for *B. pertussis* virulence factors, BvgA regulates positively the transcription of *bvgAS*. Upstream from *bvgA* and transcribed from the opposite strand lies *fhaB,* the first of three genes required for expression of the filamentous haemagglutinin. There are five promoters in the intergenic region, four of which influence *bvg* gene expression (88, 93). Promoters P1 and P3 are regulated by BvgA and are subject to repression by $MgSO_4$ and nicotinic acid. Promoter P2 is constitutively active and may function to ensure a basal level of *bvgAS* expression in uninduced cells. The P4 promoter directs the synthesis of an RNA that is complementary to the untranslated 5′ leader sequence of the *bvgAS* mRNA. P4 is responsive to BvgA and external environmental factors. The *bvg* promoters do not appear to have –10 and –35 hexameric motifs typical of σ^{70}-driven promoters, and the issue of which sigma factors are responsible for their function has not been settled (85). The fifth promoter, P_{FHA}, directs transcription of the *fhaB* gene.

6.4 Temporal control of gene expression in the Bvg regulon

The elevation of virulence gene transcript levels does not occur at the same rate for all members of the Bvg regulon. Experiments with temperature shifts have shown that some promoters respond quickly to an increase in temperature while others display a prolonged lag (94). Promoters showing an early response to a shift from 25 °C to 37 °C include the BvgA-regulated P1 and P4 promoters at *bvg* and the P_{FHA} promoter at *fhaB*. Two hours later, the P3 promoter at *bvg* is activated, as are the promoters of the genes coding for pertussis toxin (*ptx*) and adenylate cyclase haemolysin (*cya*). Scarlato and co-workers have reasoned that induction of adhesin expression (filamentous haemagglutinin) before toxin expression might confer a selective advantage by allowing virulence factors to be supplied as they are needed by the pathogen (94). Consistent with this hypothesis, fimbriae are also induced transcriptionally immediately after a temperature upshift (85).

6.5 Direct and indirect control of Bvg regulon promoters

Interaction between BvgA and the *fhaB–bvgA* intergenic region has been demonstrated *in vitro* (89). Interactions with the promoters of the genes coding for pertussis toxin (*ptx*) and adenylate cyclase haemolysin (*cya*) have proven very difficult to demonstrate (89, 95). This has raised the possibility that BvgA may regulate these (the 'late' genes) indirectly or may require a co-factor for transcriptional activation at these promoters. A putative intermediate regulator has been described by Huh and Weiss (96), but the identity of this 23-kDa protein has not been established. At the time of writing the mechanism by which BvgA-dependent activation of the late genes is achieved remains uncertain, although the possibility that variations in local

DNA supercoiling may contribute to the control of *ptx* promoter function has been explored (see Section 10.6.7).

6.6 Multilevel control of *Bordetella pertussis* fimbrial gene expression

Bordetella pertussis produces two serologically distinct fimbriae, type 2 and type 3, encoded by the genes *fim2* and *fim3* (97). Transcription of fimbrial genes is regulated in response to environmental stimuli via the *bvg* locus, as outlined above. Any strain of *B. pertussis* may produce both types of fimbriae, one type, or no fimbriae at all. This phenomenon is known as fimbrial phase variation and is believed to assist the bacterium in avoiding the immune system of the host. Phase variation operates at the level of the individual gene to switch it between high and low expression states. This is in contrast to Bvg-mediated regulation which acts on all *fim* genes (and other regulon genes) simultaneously. The *fim2*, *fim3*, and a third gene called *fimX*, all share certain sequence motifs in their promoter regions. These include a run of approximately 15 C residues found immediately upstream of the pribnow box of the promoter. The length of this C tract can vary by small deletions or insertions and this correlates with variations in *fim* gene expression. It has been proposed that changes in the length of the C tract might modulate interactions between RNA polymerase and a transcription activator (probably BvgA) bound upstream of the promoter (97). This system illustrates the phenomenon of a stereotypic response (Bvg-mediated regulation) superimposed on an apparently stochastic process (phase variation). The result is a population-wide random variation in Fim expression which is integrated with classical control of *fim* gene transcription. A system of this type combines the potential of mutation and gene regulation, and allows many more possibilities for determining the expression of cell surface components (such as fimbriae) than one relying on transcriptional regulation alone. This is likely to afford selective advantages in dealing with a dynamic and hostile external environment.

6.7 DNA topology and pertussis toxin gene expression

The promoter of the pertussis gene, *ptx*, is classified as a 'late' virulence gene promoter because it requires 2 hours to become fully active following receipt of an inducing temperature signal (Section 10.6.4). The difficulty of demonstrating interaction of BvgA with the promoter *in vitro* and the fact that BvgAS does not appear to activate it in *E. coli* has been referred to above (Section 10.6.5). By placing the *ptx* promoter in different recombinant plasmids, Scarlato and co-workers (98) were able to demonstrate *trans* activation of *ptx* by BvgAS in *E. coli*. Moreover, the *ptx* promoter was found also to be environmentally regulated. This led them to hypothesize that the late promoters might require a critical level of DNA supercoiling in addition to BvgA for activation. They tested the sensitivity of the *ptx* promoter to changes in DNA supercoiling by treating *E. coli* cultures harbouring it and the *bvg* system with the DNA gyrase inhibitor novobiocin. Inhibition of gyrase enhanced expression of *ptx* mRNA, although it was not shown if this effect was direct or not (98). Experiments involving *in-vitro* transcription with total protein extract from *B. pertussis* showed that when purified DNA topoisomerase I was added to the reaction, *ptx* mRNA levels were unaffected, whereas addition of DNA gyrase inhibited expression of *ptx* mRNA (98). These data suggest strongly a role for DNA gyrase in influencing the expression of the *ptx* gene although the level at which this influence is exerted remains unclear.

7. Regulation of virulence gene expression in *Agrobacterium tumefaciens*, a plant pathogen

Agrobacterium tumefaciens has taken the process of adaptation to its host to an almost extreme level. It has the ability to transmit bacterial genes to the nucleus of its host's cells and to have them expressed there, altering the metabolic activity of the host cell in ways that benefit the bacterium. Unlike the bacteria considered in earlier sections of this chapter, *A. tumefaciens* is an obligately aerobic soil bacterium and a pathogen of plants. Despite this, many of the regulatory mechanisms which it uses to control virulence gene expression are similar to those found in pathogens of humans and animals.

7.1 The virulence system of *Agrobacterium tumefaciens*

A high molecular mass plasmid harbours the major virulence genes of *A. tumefaciens*. The plasmid is known as the Ti (or *T*umour-*i*nducing) plasmid because the result of successful infection by this bacterium is the production of a crown gall tumour on the root of the plant. The plasmid may be regarded as carrying the components of a binary virulence system; the *vir* genes provide the regulatory functions and the environmental sensing mechanism, and their activities are central to the control of expression of the second part of the virulence system. This comprises the T-DNA, which is that part of the plasmid which is transferred to and then expressed within the nucleus of the host plant cell (99).

7.2 The infection process and the plant host environment

Agrobacterium tumefaciens infects plants via wounds and the expression of virulence factors is triggered by the receipt of wound-released molecules. Chemotaxis towards wound exudates is an important early step in infection, although there is controversy concerning which exudates are involved as attractants (100, 101). Chromosomal genes appear to contribute to the next key step in infection; adherence. Although specific receptors and adhesins are believed to exist on the plant cell and on the bacterium, respectively, little is known of the molecular structure of either (102, 103).

Expression of Ti plasmid-encoded virulence genes is controlled primarily at the level of transcription in response to signal molecules released from plant wound sites. These signal molecules are phenolic compounds and sugars. Two particularly powerful inducers, isolated from tobacco plants, are acetosyringone and hydroxy-acetosyringone (104). There is some evidence of strain adaptation to certain plants; *A. tumefaciens* strains associated with the Douglas fir show strong induction of virulence gene expression by coniferin, whereas strains adapted to other plants are less well induced by this phenolic compound (105).

Among effective sugars, monomers of plant cell wall polysaccharides are often potent inducers, for example, D-glucuronic acid, D-galacturonic acid, arabinose, fucose, glucose, galactose, and xylose (106).

In addition to phenolics and carbohydrates, the pH of the external medium is of critical importance in determining the degree of virulence gene induction which is achieved. pH values in the range 5.0 to 5.5 are optimal, although this level of acidity is bacteriostatic (107).

7.3 T-DNA transfer and expression

The T-DNA sequence within the Ti plasmid is flanked by 25-bp imperfect direct repeats known as the 'border repeats' (108). These act in a polar manner, with the right repeat being essential for tumourigenesis (109). Near the right border is a region known as 'overdrive' which functions as an enhancer of DNA transfer (110). Following receipt of an inducing signal, the bottom strand of the T-DNA is nicked close to the border repeat and the strand becomes displaced (111, 112). Ti plasmid-encoded VirD1 and VirD2 proteins catalyse these events (113), with VirD1 being characterized as a potential DNA topoisomerase (114). Production of single-stranded DNA is enhanced by binding of the VirC1 and VirC2 proteins to the overdrive sequence (115). The VirD2 protein appears to guide the displaced single strand of T-DNA (T-strand) to the plant nucleus (116). Another protein, VirE2, binds the T-strand. This binding is not sequence specific and may serve to coat and so protect the single strand of bacterial DNA on its journey to the host (117). The mechanism by which the T-strand is passed from prokaryotic to eukaryotic cell has been compared to bacterial conjugation (118). Once the T-strand is integrated with the host genome, the genes carried within the T-DNA are expressed using the host gene expression machinery; these bacterial genes possess eukaryotic transcription initiation signals and are transcribed as mRNAs with polyadenylation signals (99). The products of the T-DNA genes alter the phytohormone balance of the host, resulting in tumourigenesis, or specify production of opines. Opines are derivatized amino acids or sugars which are released from the tumour and taken up and utilized by the bacteria (99).

7.4 Regulation of virulence gene expression in *Agrobacterium tumefaciens*

The expression of the complex virulence system described in the previous section represents a considerable metabolic demand for the bacterium. It was noted in

Section 10.7.2 that environmental signals of plant origin play a major role in controlling virulence gene expression. The signal transduction pathway through which this environmental information flows consists of a histidine protein kinase and response regulator 'two-component' system encoded by the *virA* and *virG* genes on the Ti plasmid (19). The VirA protein is the kinase and this 829 amino acid polypeptide is associated with the inner membrane. It possesses an N-terminal domain of approximately 270 amino acids within the periplasm, transmembrane segments, and a cytoplasmic C-terminal domain (19). VirA has been shown to undergo autophosphorylation on a histidine residue (number 474) within its C-terminal domain and to be able to transfer its phosphate to the response regulator protein VirG *in vitro* (119). Phosphorylation renders VirG proficient for activating transcription of virulence genes which carry a sequence known as the 'vir box' in their promoter regions; the *virG* gene also possesses a vir box at its promoter rendering it sensitive to autoregulation in a manner reminiscent of *bvgA* in *Bordetella pertussis* (Section 10.6.3; 120). In addition to control by VirG, the *virG* gene possesses a phosphate-starvation-controlled promoter (120). Another similarity between the VirA/G system of *A. tumefaciens* and the BvgAS system of *B. pertussis* concerns the existence of a region in the C terminus of the kinase (VirA and BvgS) which shows homology to the phosphate-receiving N-terminal domain of the response regulator (VirG and BvgA). The function of this domain is unknown but it might perform a signal attenuation role, requiring VirA to receive repeated or prolonged inputs from the environment before transmitting the signal to the response regulator.

8. Autoinducers and virulence gene expression

Recently, it has become appreciated that the phenomenon of quorum sensing contributes in important ways to coordinating the expression of virulence (and other) genes within bacterial populations. Low molecular mass, diffusible compounds, known as autoinducers, are synthesized by bacteria and interact with gene regulatory proteins both in the producing and in neighbouring bacteria. The autoinducers belong to the homoserine lactone family of compounds and they must reach a critical concentration before becoming effective, linking efficacy to the density of the bacterial population which synthesizes the inducer.

Originally identified in marine bacteria engaged in symbiotic relationships with fish, autoinducer control of transcription has now been detected in pathogenic species of bacteria. In the plant pathogen *Erwinia carotovora*, autoinducers control expression of genes encoding virulence factors and other genes which contribute to antibiotic production (121, 122). Homoserine lactones regulate conjugative transfer of the high molecular mass virulence plasmid in *Agrobacterium tumefaciens* (123, 124) and control expression of elastase in the opportunistic pathogen *Pseudomonas aeruginosa* (121, 125). In the case of elastase, multiple autoinducers appear to be involved, suggesting that regulation is complex (126). It is very likely that further examples of virulence gene regulation by autoinducers will be described in the near future.

9. Host cell contact and virulence gene expression

The previous section introduced the concept of communication between bacterial cells leading to a change in the gene expression profile of the population. It is now becoming apparent that host cells can participate in a form of communication which results in altered expression of virulence factors in bacteria. Research with uropathogenic *E. coli* has indicated that host cell contact via the P pilus adhesin results in increased expression of a sensor regulator gene, *barA* (also called *airS*) required for the bacterial iron starvation response (see below). The BarA protein is very similar (54 per cent) to the BvgS histidine protein kinase of *Bordetella pertussis* described earlier in this chapter. Mutants unable to express BarA fail to synthesize iron-chelating siderophores when grown in an iron depleted medium. The mechanism by which P pilus-mediated adhesion induces *barA* expression is not understood (127).

Yersinia species pathogenic for humans contain a high molecular mass virulence plasmid which encodes secreted proteins called Yops. When *Y. pseudotuberculosis* contacts eukaryotic cells, transcription of the *yop* genes is increased. Contact with the host cell triggers export of LcrQ, a repressor of *yop* gene expression. The resulting reduction of the intracellular concentration of repressor permits the *yop* genes to be transcribed. LcrQ export depends on a Yop-specific protein secretion system, and the activity of this system appears to provide the interface between host cell contact and *yop* gene activation (128).

10 Iron and bacterial virulence gene expression

Iron has long been recognized as a significant factor in controlling expression of certain bacterial virulence genes. Although rich in iron, the mammalian host stores the metal in forms which make it generally unavailable to bacteria, condemning these organisms to a state of iron starvation *in vivo*. The bacteria respond by producing siderophores with a high iron affinity to chelate the metal. The ability to produce such compounds is in itself a determinant of virulence and genes encoding iron chelation systems have been detected on the chromosomes and on plasmids in many pathogens.

The 17-kDa DNA-binding protein Fur regulates transcription of a wide range of genes involved in iron acquisition in Gram-negative bacteria. Among pathogenic bacteria, the *fur* gene has been isolated from *Legionella pneumophila* (129), *Neisseria gonorrhoeae* (130), *Neisseria meningitidis* (131), *Pseudomonas aeruginosa* (132), *Salmonella typhimurium* (133), *Serratia marcencens* (134), *Shigella* spp. (135), *Vibrio cholerae* (136), and *Yersinia* spp. (137). Fur acts as a repressor of transcription, binding to a 19-bp consensus sequence, known as the Fur box, which is located in the promoter regions of genes in the Fur regulon. When intracellular concentrations of Fe(II) are high, the metal binds to Fur and potentiates its transcription repressing activity (138, 139). In addition to controlling expression of genes contributing to iron uptake, Fur also influences the expression of genes involved in the response to oxidative stress, the acid tolerance response, and the synthesis of toxins (138, 140, 141). In this way it plays a coordinating role in iron metabolism, stress response, and virulence.

Related proteins have been found to regulate virulence gene expression in other bacteria where iron is known to be a determining factor in the elaboration of a virulent phenotype. For example, the DtxR protein of *Corynebacterium diphtheriae* regulates transcription of the diphtheria toxin genes in response to iron. When the intracellular iron concentration is high, DtxR represses transcription, by analogy with Fur (142, 143). A reduction in the iron concentration relieves repression of the toxin genes and the resulting expression of the toxin can lead to host tissue damage, possibly providing the bacterium with new sources of iron.

11. Concluding remarks

This chapter has attempted to illustrate how the control devices regulating expression of genetic information in bacteria contribute to pathogenicity. The examples chosen are from organisms adapted to different environmental niches, but despite the diversity of their ecology, these bacteria employ similar control devices. An obvious example is the use of histidine protein kinase and response regulator partnerships to control, in response to specific environmental signals, the transcription of genes contributing directly or indirectly to the expression of a virulent phenotype. Although the examples considered here are from Gram-negative organisms, 'two-component' regulators of this type have also been discovered in Gram-positive pathogens, such as *Staphylococcus aureus,* where they control virulence gene expression (19). Gene regulatory mechanisms involving an interaction between a regulatory protein and a specific DNA sequence found in association with one or more subservient genes or operons may be regarded as the classical mode of transcriptional control. The 'two-component' regulators just referred to are of this type, as are the so-called 'alternative' sigma factors.

A more recent advance in understanding concerns the abilities of DNA-binding proteins showing little or no sequence specificity to affect the transcription of very large groups of (often unrelated) genes. The H-NS protein discussed in this chapter is an example of such a control element. The involvement of this chromatin-organizing protein in transcriptional regulation points to the emerging awareness of the importance of DNA topology in modulating transcription of very many genes, including genes required for virulence. Future research aimed at understanding how this form of regulatory mechanism is integrated with more classical methods will greatly assist our understanding of the pathogenic bacterium and its behaviour during infection.

Acknowledgements

Research in the author's laboratory is supported by the Wellcome Trust (UK), the European Union (DG XII), the Health Research Board of Ireland, and Forbairt (Ireland).

References

1. Smith, H. (1995) The revival of interest in mechanisms of bacterial pathogenicity. *Biol. Rev.*, **70,** 277.
2. Dorman, C. J. (1994) *Genetics of bacterial virulence.* p. 204. Blackwell Scientific, Oxford.
3. Groisman, E. A. and Ochman, H. (1994) How to become a pathogen. *Trends Microbiol.*, **2,** 289.
4. Sansonetti, P. J., Kopecko, D. J., and Formal, S. B. (1982) Involvement of a plasmid in the invasive ability of *Shigella flexneri. Infect. Immun.*, **35,** 852.
5. Sansonetti, P. J. (1992) Molecular and cellular biology of epithelial invasion by *Shigella flexneri* and other enteroinvasive pathogens. In *Molecular biology of bacterial infection: Current status and future perspectives* (ed. C. E. Hormaeche, C. W. Penn, and C. J. Smyth). *Soc. Gen. Microbiol. Symp.*, **49,** 47.
6. Gallegos, M.-T., Michán, C., and Ramos, J. L. (1993) The XylS/AraC family of regulators. *Nucl. Acids Res.*, **21,** 807.
7. Adler, B., Sasakawa, C., Tobe, T., Makino, S., Komatsu, K., and Yoshikawa, M. (1989) A dual transcriptional activation system for the 230 kb plasmid genes for virulence-associated antigens of *Shigella flexneri. Mol. Microbiol.*, **3,** 627.
8. Tobe, T., Nagai, S., Adler, B., Yoshikawa, M., and Sasakawa, C. (1991) Temperature-regulated expression of invasion genes in *Shigella flexneri* is controlled through transcriptional activation of the *virB* gene on the large invasion plasmid. *Mol. Microbiol.*, **5,** 887.
9. Tobe, T., Yoshikawa, M., Mizuno, T., and Sasakawa, C. (1993) Transcriptional control of the invasion regulatory gene *virB* of *Shigella flexneri*: activation by VirF and repression by H-NS. *J. Bacteriol.*, **175,** 6142.
10. Ussery, D. W., Hinton, J. C. D., Jordi, B. J. A. M., Granum, P. E., Seirafi, A., Stephen, R. J., Tupper, A. E., Berridge, G., Sidebotham, J. M., and Higgins, C. F. (1994) The chromatin-associated protein H-NS. *Biochimie*, **76,** 968.
11. Dorman, C. J., Ní Bhriain, N., and Higgins, C. F. (1990) DNA supercoiling and the environmental regulation of virulence gene expression in *Shigella flexneri. Nature*, **344,** 789–792.
12. Hromockyj, A. E., Tucker, S. C., and Maurelli, A. T. (1992) Temperature regulation of *Shigella* virulence: identification of the repressor gene *virR*, an analogue of *hns*, and partial complementation by tyrosyl transfer RNA ($tRNA_1^{Tyr}$). *Mol. Microbiol.*, **6,** 2113.
13. Maurelli, A. T. and Sansonetti, P. J. (1988) Identification of a chromosomal gene controlling temperature-regulated expression of *Shigella* virulence. *Proc. Natl. Acad. Sci. USA*, **85,** 2820.
14. Porter, M. E. and Dorman, C. J. (1994) A role for H-NS in the thermo-osmotic regulation of virulence gene expression in *Shigella flexneri. J. Bacteriol.*, **176,** 4187.
15. Goldstein, E. and Drlica, K. (1984) Regulation of bacterial DNA supercoiling: plasmid linking numbers vary with growth temperature. *Proc. Natl. Acad. Sci. USA*, **81,** 4046.
16. Higgins, C. F., Dorman, C. J., Stirling, D. A., Waddell, L., Booth, I. R., May, G., and Bremer, E. (1988) A physiological role for DNA supercoiling in the osmotic regulation of gene expression in *S. typhimurium* and *E. coli. Cell*, **52,** 569.
17. Tobe, T., Yoshikawa, M., and Sasakawa, C. (1995) Thermoregulation of *virB* transcription in *Shigella flexneri* by sensing of changes in local DNA superhelicity. *J. Bacteriol.*, **177,** 1094.
18. Bernardini, M. L., Fontaine, A., and Sansonetti, P. J. (1990) The two-component regulatory system OmpR-EnvZ controls the virulence of *Shigella flexneri. J. Bacteriol.*, **172,** 6274.

19. Stock, J. B., Ninfa, A. J., and Stock, A. M., (1989) Protein phosphorylation and regulation of adaptive responses in bacteria. *Microbiol. Rev.*, **53,** 450.
20. Bernardini, M. L., Sanna, M. G., Fontaine, A., and Sansonetti, P. J. (1993) OmpC is involved in invasion of epithelial cells by *Shigella flexneri. Infect. Immun.*, **61,** 3625.
21. Tobe, T., Yoshikawa, M., and Sasakawa, C. (1994) Deregulation of temperature-dependent transcription of the invasion regulatory gene, *virB,* in *Shigella* by *rho* mutation. *Mol. Microbiol.*, **12,** 267.
22. Platt, T. (1986) Transcription termination and the regulation of gene expression. *Annu. Rev. Biochem.*, **55,** 339.
23. Arnold, G. F. and Tessman, I. (1988) Regulation of DNA superhelicity by *rpoB* mutations that suppress defective Rho-mediated transcription termination in *Escherichia coli. J. Bacteriol.*, **170,** 4266.
24. Fassler, J. S., Arnold, G. F., and Tessman, I. (1986) Reduced superhelicity of plasmid DNA produced by the *rho-15* mutation in *Escherichia coli. Mol. Gen. Genet.*, **204,** 424.
25. Dorman, C. J. (1995) DNA topology and the global control of bacterial gene expression: implications for the regulation of virulence gene expression. *Microbiology*, **141,** 1271.
26. Cabello, F., Hormaeche, C., Mastroeni, P., and Bonina, L. (1993) *Biology of Salmonella.* Plenum, New York.
27. Maurelli, A. T. (1994) Virulence protein export systems in *Salmonella* and *Shigella*: a new family or lost relatives? *Trends Cell Biol.*, **4,** 240.
28. Groisman, E. A. and Ochman, H. (1993) Cognate gene clusters govern invasion of host epithelial cells by *Salmonella typhimurium* and *Shigella flexneri. EMBO J.*, **12,** 3779.
29. Mills, D. M., Bajaj, V., and Lee, C. A. (1995) A 40 kb chromosomal fragment encoding *Salmonella typhimurium* invasion genes is absent from the corresponding region of the *Escherichia coli* K-12 chromosome. *Mol. Microbiol.*, **15,** 749.
30. Chatfield, S. N., Li, J. L., Sydenham, M., Douce, G., and Dougan, G. (1992) *Salmonella* genetics and vaccine development. In *Molecular biology of bacterial infection: Current status and future perspectives* (ed. C. E. Hormaeche, C. W. Penn, and C. J. Smyth). *Soc. Gen. Microbiol. Symp.*, **49,** 299.
31. Curtiss, R., III and Kelly, S. M. (1987) *Salmonella typhimurium* deletion mutants lacking adenylate cyclase and cyclic AMP receptor protein are avirulent and immunogenic. *Infect. Immun.*, **55,** 3035.
32. Botsford, J. L. and Harman, J. G. (1992) Cyclic AMP in prokaryotes. *Microbiol. Rev.*, **56,** 100.
33. Dorman, C. J., Chatfield, S., Higgins, C. F., Hayward, C., and Dougan, G. (1989) Characterization of porin and *ompR* mutants of a virulent strain of *Salmonella typhimurium*: *ompR* mutants are attenuated *in vivo. Infect. Immun.*, **57,** 2136.
34. Chatfield, S. N., Dorman, C. J., Hayward, C., and Dougan, G. (1991) Role of *ompR*-dependent genes in *Salmonella typhimurium* virulence: mutants deficient in both OmpC and OmpF are attenuated *in vivo. Infect. Immun.*, **59,** 449.
35. Miller, S. I., Kukral, A. M., and Mekalanos, J. J. (1989) A two-component regulatory system (*phoP phoQ*) controls *Salmonella typhimurium* virulence. *Proc. Natl. Acad. Sci. USA*, **86,** 5054.
36. Groisman, E. A. and Saier, M. H. (1990) *Salmonella* virulence: new clues to intra-macrophage survival. *Trends Biochem. Sci.*, **15,** 30.
37. Behlau, I. and Miller, S. I. (1993) A PhoP-repressed gene promotes *Salmonella typhimurium* invasion of epithelial cells. *J. Bacteriol.*, **175,** 4475.
38. Kier, L. D., Weppleman, R. M., and Ames, B. N. (1977) Regulation of nonspecific acid phosphatase in *Salmonella typhimurium*: *phoN* and *phoP* genes. *J. Bacteriol.*, **138,** 155.

39. Weppleman, R., Kier, L. D., and Ames, B. N. (1977) Properties of two phosphatases and a cyclic phosphodiesterase of *Salmonella typhimurium*. *J. Bacteriol.*, **130,** 411.
40. Aranda, C. M. A., Swanson, J. A., Loomis, W. P., and Miller, S. I. (1992) *Salmonella typhimurium* activates virulence gene transcription within acidified macrophage phagosomes. *Proc. Natl. Acad. Sci. USA*, **89,** 10079.
41. Gunn, J. S. and Miller, S. I. (1996) PhoP–PhoQ activates transcription of *pmrAB*, encoding a two-component regulatory system involved in *Salmonella typhimurium* antimicrobial peptide resistance. *J. Bacteriol.*, **178,** 6857.
42. Soncini, F. C. and Groisman, E. A. (1996) Two-component regulatory systems can interact to process multiple environmental signals. *J. Bacteriol.*, **178,** 6796.
43. Gulig, P. A., Danbara, H., Guiney, D. G., Lax, A. J., Norel, F., and Rhen, M. (1993) Molecular analysis of *spv* virulence genes of the salmonella virulence plasmids. *Mol. Microbiol.*, **7,** 825.
44. Pullinger, G. D., Baird, G. D., Williamson, C. M., and Lax, A. J. (1989) Nucleotide sequence of a plasmid gene involved in the virulence of salmonellas. *Nucl. Acids Res.*, **17,** 7983.
45. Caldwell, A. L. and Gulig, P. A. (1991) The *Salmonella typhimurium* virulence plasmid encodes a positive regulator of a plasmid-encoded virulence gene. *J. Bacteriol.*, **173,** 7176.
46. Fang, F. C., Krause, M., Roudier, C., Fierer, J., Guiney, D. (1991) Growth regulation of a plasmid gene essential for virulence. *J. Bacteriol.*, **173,** 6783.
47. Matsui, H., Abe, A., Kawahara, K., Terakado, N., and Danbara, H. (1991) Positive regulator for the expression of Mba protein of the virulence plasmid, pKDSC50, of *Salmonella choleraesuis*. *Microb. Pathog.*, **10,** 459.
48. Taira, S., Riikonen, P., Saarilahti, H., Sukupolvi, S., and Rhen, M. (1991) The *mkaC* virulence gene of the *Salmonella* serovar typhimurium 96 kb plasmid encodes a transcriptional activator. *Mol. Gen. Genet.*, **228,** 381.
49. Fang, F. C., Libby, S. J., Buchmeier, N. A., Loewen, P. C., Switala, J., Harwood, J., and Guiney, D. G. (1992) The alternative σ factor KatF (RpoS) regulates *Salmonella* virulence. *Proc. Natl. Acad. Sci. USA*, **89,** 11978.
50. Norel, F., Robbe-Saule, V., Popoff, M. Y., and Coynault, C. (1992) The putative sigma factor KatF (RpoS) is required for the transcription of the *Salmonella typhimurium* virulence gene *spvB* in *Escherichia coli*. *FEMS Microbiol. Lett.*, **99,** 271.
51. Kowarz, L., Coynault, C., Robbe-Saule, V., and Norel, F. (1994) The *Salmonella typhimurium katF* (*rpoS*) gene: cloning, nucleotide sequence, and regulation of *spvR* and *spvABCD* virulence plasmid genes. *J. Bacteriol.*, **176,** 6852.
52. O'Byrne, C. P. and Dorman, C. J. (1994) The *spv* virulence operon of *Salmonella typhimurium* LT2 is regulated negatively by the cyclic AMP (cAMP)–cAMP receptor protein system. *J. Bacteriol.*, **176,** 905.
53. Lange, R. and Hengge-Aronis, R. (1991) Identification of a central regulator of stationary-phase gene expression in *Escherichia coli*. *Mol. Microbiol.*, **5,** 49.
54. Fierer, J., Eckmann, L., Fang, F., Pfeifer, C., Finlay, B. B., and Guiney, D. (1993) Expression of the Salmonella virulence plasmid gene *spvB* in cultured macrophages and non-phagocytic cells. *Infect. Immun.*, **61,** 5231.
55. Rhen, M., Riikonen, P., and Taira, S. (1993) Transcriptional regulation of *Salmonella enterica* virulence plasmid genes in cultured macrophages. *Mol. Microbiol.*, **10,** 45.
56. Harrison, J. A., Pickard, D., Higgins, C. F., Khan, A., Chatfield, S. N., Ali, T., Dorman, C. J., Hormaeche, C. E., and Dougan, G. (1994) Role of *hns* in the virulence phenotype of pathogenic salmonellae. *Mol. Microbiol.*, **13,** 133.
57. Barr, G. C., Ní Bhriain, N., and Dorman, C. J. (1992) Identification of two new genetically

active regions associated with the *osmZ* locus of *Escherichia coli*: role in regulation of *proU* expression and mutagenic effect at *cya*, the structural gene for adenylate cyclase. *J. Bacteriol.*, **174,** 988.

58. Lejeune, P. and Danchin, A. (1990) Mutations in the *bglY* gene increase the frequency of spontaneous deletions in *Escherichia coli* K-12. *Proc. Natl. Acad. Sci. USA*, **87,** 360.
59. O'Byrne, C. P. and Dorman, C. J. (1994) Transcription of the *Salmonella typhimurium spv* virulence locus is regulated negatively by the nucleoid-associated protein H-NS. *FEMS Microbiol. Lett.*, **121,** 99.
60. Galán, J. E. and Curtiss, III, R. (1990) Expression of *Salmonella typhimurium* genes required for invasion is regulated by changes in DNA supercoiling. *Infect. Immun.*, **58,** 1879.
61. Ernst, R. K., Dombroski, D. M., and Merrick, J. M. (1990) Anaerobiosis, type 1 fimbriae, and growth phase are factors that affect invasion of HEp-2 cells by *Salmonella typhimurium*. *Infect. Immun.*, **58,** 2014.
62. Lee, C. A. and Falkow, S. (1990) The ability of *Salmonella* to enter mammalian cells is affected by bacterial growth state. *Proc. Natl. Acad. Sci. USA*, **87,** 4304.
63. Dorman, C. J., Barr, G. C., Ní Bhriain, N., and Higgins, C. F. (1988) DNA supercoiling and the anaerobic and growth phase regulation of *tonB* gene expression. *J. Bacteriol.*, **170,** 2816.
64. Yamamoto, N. and Droffner, M. L. (1986) Mechanisms determining aerobic or anaerobic growth in the facultative anaerobe *Salmonella typhimurium*. *Proc. Natl. Acad. Sci. USA*, **82,** 2077.
65. Wren, B. (1992) Bacterial enterotoxin interactions. In *Molecular biology of bacterial infection: Current status and future perspectives* (ed. C. E. Hormaeche, C. W. Penn, and C. J. Smyth). *Soc. Gen. Microbiol. Symp.*, **49,** 127.
66. Cassel, D. and Pfeuffer, T. (1978) Mechanism of cholera action: covalent modification of the guanyl-binding protein of the adenylate cyclase system. *Proc. Natl. Acad. Sci. USA*, **75,** 2269.
67. Gill, D. M. and Meren, R. (1978) ADP-ribosylation of membrane proteins catalysed by cholera toxin: basis of the activation of adenylate cyclase. *Proc. Natl. Acad. Sci. USA*, **75,** 3040.
68. Mekalanos, J. J., Collier, R. J., and Romig, W. R. (1979) Enzymic activity of cholera toxin. I. New method of assay and the mechanism of ADP-ribosyl transfer. *J. Biol. Chem.*, **254,** 5849.
69. Mekalanos, J. J., Collier, R. J., and Romig, W. R. (1979) Enzymic activity of cholera toxin. II. Relationships to proteolytic processing, disulphide bond reduction, and subunit composition. *J. Biol. Chem.*, **254,** 5855.
70. Mekalanos, J. J., Swartz, D. J., Pearson, G. D. N., Harford, N., Groyne, F., and de Wilde, M. (1983) Cholera toxin genes: nucleotide sequence, deletion analysis and vaccine development. *Nature*, **306,** 551.
71. Mekalanos, J. J. (1983) Duplication and amplification of toxin genes in *Vibrio cholerae*. *Cell*, **35,** 253.
72. Pearson, G. D. N., Woods, N. A., Chiang, S. L., and Mekalanos, J. J. (1993) CTX genetic element encodes a site-specific recombination system and an intestinal colonization factor. *Proc. Natl. Acad. Sci. USA*, **90,** 3750.
73. Waldor, M. K. and Mekalanos, J. J. (1996) Lysogenic conversion by a filamentous phage encoding cholera toxin. *Science*, **272,** 1910.
74. Miller, V. L., Taylor, R. K., and Mekalanos, J. J. (1987) Cholera toxin transcriptional activator ToxR is a transmembrane DNA binding protein. *Cell*, **48,** 271.
75. Miller, V. L., DiRita, V. J., and Mekalanos, J. J. (1989) Identification of *toxS*, a regulatory

gene whose product enhances ToxR-mediated activation of the cholera toxin promoter. *J. Bacteriol.*, **171,** 1288.
76. DiRita, V. J. and Mekalanos, J. J. (1991) Periplasmic interaction between two membrane regulatory proteins, ToxR and ToxS, results in signal transduction and transcriptional activation. *Cell*, **64,** 29.
77. DiRita, V. J. (1992) Coordinate expression of virulence genes by ToxR in *Vibrio cholerae. Mol. Microbiol.*, **6,** 451.
78. Parsot, C. and Mekalanos, J. J. (1991) Expression of the *Vibrio cholerae* gene encoding aldehyde dehydrogenase is under the control of ToxR, the cholera toxin transcriptional activator. *J. Bacteriol.*, **173,** 2842.
79. DiRita, V. J., Parsot, C., Jander, G., and Mekalanos, J. J. (1991) Regulatory cascade controls virulence in *Vibrio cholerae. Proc. Natl. Acad. Sci. USA*, **88,** 5403.
80. Ogierman, M. A. and Manning, P. A. (1992) Homology of TcpN, a putative regulatory protein of *Vibrio cholerae*, to the AraC family of transcriptional activators. *Gene*, **116,** 93.
81. Parsot, C. and Mekalanos, J. J. (1990) Expression of ToxR, the transcriptional activator of the virulence factors in *Vibrio cholerae*, is modulated by the heat shock response. *Proc. Natl. Acad. Sci. USA*, **87,** 9898.
82. Parsot, C. and Mekalanos, J. J. (1992) Structural analysis of the *acfA* and *acfD* genes of *Vibrio cholerae*: effects of DNA topology and transcriptional activators on expression. *J. Bacteriol.*, **174,** 5211.
83. Coote, J. G. (1991) Antigenic switching and pathogenicity: environmental effects on virulence gene expression in *Bordetella pertussis. J. Gen. Microbiol.*, **137,** 2493.
84. Lacey, B. W. (1960) Antigenic modulation of *Bordetella pertussis. J. Hyg.*, **58,** 57.
85. Scarlato, V., Aricò, B., Domenighini, M., and Rappuoli, R. (1993) Environmental regulation of virulence factors in *Bordetella* species. *BioEssays*, **15,** 99.
86. Aricò, B., Miller, J. F., Roy, C., Stibitz, S., Monack, D. M., Falkow, S., Gross, R., and Rappuoli, R. (1989) Sequences required for expression of *Bordetella pertussis* virulence factors share homology with prokaryotic signal transduction proteins. *Proc. Natl. Acad. Sci. USA*, **86,** 6671.
87. Stibitz, S. and Yang, M.-S. (1991) Subcellular localization and immunological detection of proteins encoded by the *vir* locus of *Bordetella pertussis. J. Bacteriol.*, **173,** 4288.
88. Scarlato, V., Prugnola, A., Aricò, B., and Rappouli, R. (1990) Positive transcriptional feedback at the *bvg* locus controls expression of virulence factors in *Bordetella pertussis. Proc. Natl. Acad. Sci. USA*, **87,** 6753.
89. Roy, C. R. and Falkow, S. (1991) Identification of *Bordetella pertussis* regulatory sequences required for transcriptional activation of the *fhaB* gene and autoregulation of the *bvgSA* operon. *J. Bacteriol.*, **173,** 2385.
90. Beattie, D. T., Knapp, S., and Mekalanos, J. J. (1990) Evidence that modulation requires sequences downstream of the promoters of two *vir*-repressed genes of *Bordetella pertussis. J. Bacteriol.*, **172,** 6997.
91. Knapp, S. and Mekalanos, J. J. (1988) Two *trans*-acting regulatory genes (*vir* and *mod*) control antigenic modulation in *Bordetella pertussis. J. Bacteriol.*, **170,** 5059.
92. Aricò, B., Scarlato, V., Monack, D. M., Falkow, S., and Rappouli, R. (1991) Structural and genetic analysis of the *bvg* locus in *Bordetella* species. *Mol. Microbiol.*, **5,** 2481.
93. Roy, C. R., Miller, J. F., and Falkow, S. (1990) Autogenous regulation of the *Bordetella pertussis bvgABC* operon. *Proc. Natl. Acad. Sci. USA*, **87,** 3763.
94. Scarlato, V., Aricò, B., Prugnola, A., and Rappouli, R. (1991) Sequential activation and environmental regulation of virulence genes in *Bordetella pertussis. EMBO J.*, **10,** 3971.

95. Miller, J. F., Roy, C. R., and Falkow, S. (1989) Analysis of *Bordetella pertussis* virulence gene regulation by use of transcriptional fusions in *Escherichia coli. J. Bacteriol.*, **171,** 6345.
96. Huh, Y. J. and Weiss, A. A. (1991) A 23-kilodalton protein, distinct from BvgA, expressed by virulent *Bordetella pertussis* binds to the promoter region of *vir*-regulated toxin genes. *Infect. Immun.*, **59,** 2389.
97. Willems, R., Paul, A., van der Heide, H. G. J., ter Avest, A. R., and Mooi, F. R. (1990) Fimbrial phase variation in *Bordetella pertussis*: a novel mechanism for transcriptional regulation. *EMBO J.*, **9,** 2803.
98. Scarlato, V., Aricò, B., and Rappouli, R. (1993) DNA topology affects transcriptional regulation of the pertussis toxin gene of *Bordetella pertussis* in *Escherichia coli* and *in vitro. J. Bacteriol.*, **175,** 4764.
99. Winans, S. C. (1992) Two-way chemical signalling in *Agrobacterium*–plant interactions. *Microbiol. Rev.*, **56,** 12.
100. Ashby, A. M., Watson, M. D., Loake, G. J., and Shaw, C. H. (1988) Ti plasmid-specified chemotaxis of *Agrobacterium tumefaciens* C58C[1] towards *vir*-inducing phenolic compounds and soluble factors from monocotyledonous and dicotyledonous plants. *J. Bacteriol.*, **170,** 4181.
101. Hawes, M. C. and Smith, L. Y. (1989) Requirement for chemotaxis in pathogenicity of *Agrobacterium tumefaciens* on roots of soil grown pea plants. *J. Bacteriol.*, **171,** 5668.
102. Krens, F. A., Molendijk, L., Wullems, G. J., and Schliperoort, R. A. (1985) The role of bacterial attachment in the transformation of cell-wall-regenerating tobacco protoplasts by *Agrobacterium tumefaciens. Planta*, **166,** 300.
103. Lippincott, B. B. and Lippincott, J. A. (1969) Bacterial attachment to a specific wound site as an essential stage in tumor initiation by *Agrobacterium tumefaciens. J. Bacteriol.*, **97,** 620.
104. Stachel, S. E., Messens, E., Van Montague, M., and Zambryski, P. (1985) Identification of the signal molecules produced by wounded plant cells that activate T- DNA transfer in *Agrobacterium tumefaciens. Nature*, **318,** 624.
105. Morris, J. W. and Morris, R. O. (1990) Identification of an *Agrobacterium tumefaciens* virulence gene inducer from the pinaceous gymnosperm *Pseudotsuga mensiesii. Proc. Natl. Acad. Sci. USA*, **87,** 3614.
106. Cangelosi, G. A., Ankenbauer, R. G., and Nester, E. W. (1990) Sugars induce *Agrobacterium* virulence genes through a periplasmic binding protein and a transmembrane signal protein. *Proc. Natl. Acad. Sci. USA*, **87,** 6708.
107. Stachel, S. E., Nester, E. W., and Zymbryski, P. (1986) A plant cell factor induces *Agrobacterium tumefaciens vir* gene expression. *Proc. Natl. Acad. Sci. USA*, **83,** 379.
108. Yadav, N. S., Vanderlayden, J., Bennett, D. R., Barnes, W. M., and Chilton, M.-D. (1982) Short direct repeats flank the T-DNA on a nopaline Ti plasmid. *Proc. Natl. Acad. Sci. USA*, **79,** 6322.
109. Wang, K., Herrera-Estrella, L., Van Montague, M., and Zambryski, P. (1984) Right 25 bp terminus sequences of the nopaline T-DNA is essential for and determines direction of DNA transfer from *Agrobacterium* to the plant genome. *Cell*, **38,** 35.
110. Peralta, E. G., Hellmiss, R., and Ream, L. W. (1986) Overdrive, a T-DNA transmission enhancer on the *A. tumefaciens* tumor-inducing plasmid. *EMBO J.*, **5,** 1137.
111. Albright, L. M., Yanofsky, M. F., Leroux, B., Ma, D., and Nester, E. W. (1987) Processing of the T-DNA of *Agrobacterium tumefaciens* generates border nicks and linear, single-stranded T-DNA. *J. Bacteriol.*, **169,** 1046.
112. Wang, K., Stachel, S. E., Timmerman, B., Van Montague, M., and Zambryski, P. (1987)

Site-specific nick in the T-DNA border sequence as a result of *Agrobacterium vir* gene expression. *Science,* **235,** 587.
113. Wang, K., Herrera-Estrella, A., and Van Montague, M. (1990) Overexpression of the *virD1* and *virD2* genes in *Agrobacterium tumefaciens* enhances T-complex formation and plant transformation. *J. Bacteriol.,* **172,** 4432.
114. Ghai, J. and Das, A. (1989) The *virD* operon of *Agrobacterium tumefaciens* Ti plasmid encodes a DNA relaxing enzyme. *Proc. Natl. Acad. Sci. USA,* **86,** 3109.
115. Toro, N., Datta, A., Carmi, O. A., Young, C., Prusti, R. K., and Nester, E. W. (1989) The *Agrobacterium tumefaciens virC1* gene product binds to overdrive, a T-DNA transfer enhancer. *J. Bacteriol.,* **171,** 6845.
116. Howard, E. A., Winsor, B. A., De Vos, G., and Zambryski, P. (1989) Activation of the T-DNA transfer process in *Agrobacterium tumefaciens* results in the generation of a T-strand-protein complex: tight association of VirD2 with the 5′ ends of T-strands. *Proc. Natl. Acad. Sci. USA,* **86,** 4017.
117. Citovsky, V., De Vos, G., and Zambryski, P. (1988) Single-stranded DNA binding protein encoded by the *virE* locus of *Agrobacterium tumefaciens. Science,* **240,** 501.
118. Zambryski, P. (1988) Basic processes underlying *Agrobacterium*-mediated DNA transfer to plant cells. *Annu. Rev. Genet.,* **22,** 1.
119. Jin, S., Prusti, R. K., Roitsch, T., Ankenbauer, R. G., and Nester, E. W. (1990) Phosphorylation of the VirG protein of *Agrobacterium tumefaciens* by the autophosphorylated VirA protein: essential role in biological activity of VirG. *J. Bacteriol.,* **172,** 4945.
120. Winans, S. C. (1990) Transcriptional induction of an *Agrobacterium* regulatory gene at tandem promoters by plant-released phenolic compounds, phosphate starvation, and acidic growth media. *J. Bacteriol.,* **172,** 2433.
121. Jones, S., Yu, B., Bainton, N. J., Birdsall, M., Bycroft, B. W., Chhabra, S. R., Cox, A. J. R., Golby, P., Reeves, P. J., Stephens, S., Winson, M. K., Salmond, G. P. C., Stewart, G. S. A. B., and Williams, P. (1993) The *lux* autoinducer regulates the production of exoenzyme virulence determinants in *Erwinia carotovora* and *Pseudomonas aeruginosa. EMBO J.,* **12,** 2477.
122. Bainton, N. J., Bycroft, B. W., Chhabra, S. R., Stead, P., Gledhill, L., Hill, P. J., Rees, C. E. D., Winson, M. K., Salmond, G. P. C., Stewart, G. S. A. B., and Williams, P. (1992) A general role for the *lux* autoinducer in bacterial cell signalling: control of antibiotic biosynthesis in *Erwinia. Gene,* **116,** 87.
123. Piper, K. R., von Bodman, S. B., and Farrand, S. K. (1993) Conjugation factor of *Agrobacterium tumefaciens* regulates Ti plasmid transfer by autoregulation. *Nature,* **362,** 448.
124. Zhang, L., Murphy, P. J., Kerr, A., and Tate, M. E. (1993) *Agrobacterium* conjugation and gene regulation by *N*-acyl-L-homoserine lactone. *Nature,* **362,** 446.
125. Passador, L., Cook, J. M., Gambello, M. J., Rust, L., and Iglewski, B. H. (1993) Expression of *Pseudomonas aeruginosa* virulence genes requires cell-to-cell communication. *Science,* **260,** 1127.
126. Pearson, J. P., Passador, L., Iglewski, B. H., and Greenberg, E. P. (1995) A second *N*-acylhomoserine lactone signal produced by *Pseudomonas aeruginosa. Proc. Natl. Acad. Sci. USA,* **92,** 1490.
127. Zhang, J. P. and Normark, S. (1996) Induction of gene expression in *Escherichia coli* after pilus-mediated adherence. *Science,* **273,** 1234.
128. Pettersson, J., Nordfelth, R., Dubinina, E., Bergman, T., Gustafsson, M., Magnusson, K. E., and Wolf-Watz, H. (1996) Modulation of virulence factor expression by pathogen target cell contact. *Science,* **273,** 1231.

129. Hickey, E. K. and Cianciotto, N. P. (1994) Cloning and sequencing of the *Legionella pneumophila fur* gene. *Gene,* **143,** 117.
130. Berish, S. A., Subbarao, S., Chen, C.-Y., Trees, D. L., and Morse, S. A. (1993) Identification and cloning of a *fur* homolog from *Neisseria gonorrhoeae. Infect. Immun.,* **61,** 4599.
131. Thomas, C. E. and Sparling, P. F. (1994) Identification and cloning of a *fur* homolog from *Neisseria meningitidis. Mol. Microbiol.,* **11,** 725–737.
132. Prince, R. W., Storey, D. G., Vasil, A. I., and Vasil, M. L. (1991) Regulation of *toxA* and *regA* by the *Escherichia coli fur* gene and identification of a Fur homologue in *Pseudomonas aeruginosa* PA103 and PAO1. *Mol. Microbiol.,* **5,** 2823.
133. Ernst, J. F., Bennett, R. L., and Rothfield, L. I. (1978) Constitutive expression of the iron-enterochelin and ferrichrome uptake systems in a mutant strain of *Salmonella typhimurium. J. Bacteriol.,* **135,** 928.
134. Poole, K. and Braun, V. (1988) Iron regulation of *Serratia marcencens* hemolysin gene expression. *Infect. Immun.,* **56,** 5929.
135. Payne, S. M. (1989) Iron and virulence in *Shigella. Mol. Microbiol.,* **3,** 1301.
136. Goldberg, M. B., Boyko, S. A., and Calderwood, S. B. (1990) Transcriptional regulation by iron of a *Vibrio cholerae* virulence gene and homology of the gene to the *Escherichia coli* Fur system. *J. Bacteriol.,* **172,** 6863.
137. Staggs, T. M. and Perry, R. D. (1992) Fur regulation in *Yersinia* species. *Mol. Microbiol.,* **6,** 2507.
138. Calderwood, S. B. and Mekalanos, J. J. (1987) Iron-regulation of Shiga-like toxin expression in *Escherichia coli* is mediated by the *fur* locus. *J. Bacteriol.,* **169,** 4759.
139. DeLorenzo, V., Wee, S., Herrero, M., and Neilands, J. B. (1987) Operator sequences of the aerobactin operon of plasmid ColV-K30 binding the ferric-uptake regulation (Fur) repressor. *J. Bacteriol.,* **169,** 2624.
140. Niederhoffer, E. C., Naranjo, C. M., Bradley, K. L., and Fee, J. A. (1990) Control of *Escherichia coli* superoxide dismutase (*sodA* and *sodB*) genes by the ferric uptake regulation (*fur*) locus. *J. Bacteriol.,* **172,** 1930.
141. Hall, H. K. and Foster, J. W. (1996) The role of Fur in the acid tolerance response of *Salmonella typhimurium* is physiologically and genetically separable from its role in iron acquisition. *J. Bacteriol.,* **178,** 5683.
142. Boyd, J., Oza, M. N., and Murphy, J. R. (1990) Molecular cloning and DNA sequence analysis of a diphtheria *tox* iron-dependent regulatory element (*dtxR*) from *Corynebacterium diphtheriae. Proc. Natl. Acad. Sci. USA,* **87,** 5968.
143. Schmitt, M. P. and Holmes, R. K. (1991) Iron-dependent regulation of diphtheria toxin and siderophore expression by the cloned *Corynebacterium diphtheriae* repressor gene *dtxR* in *C. diphtheriae* C7 strains. *Infect. Immun.,* **59,** 1899.

11 Integration of control devices. II. Sporulation and antibiotic production

MICHAEL D. YUDKIN and KEITH F. CHATER

1. Introduction

An earlier chapter described the specialized characteristics that *E. coli* cells adopt during stationary phase, and the complex regulation that establishes this differentiated state. In equivalent situations, some other bacteria of diverse taxonomic groups show more obvious physiological and morphological differentiation, to give rise to antibiotic production and to resting cells called spores that are somewhat (or in some cases very much) tougher and longer-lived than stationary phase *E. coli* cells.

In this chapter we first consider the different kinds of sporulation exhibited by *Bacillus subtilis*—a simple rod-shaped organism—and *Streptomyces coelicolor*, a more complex, mycelial organism. More is known about the interplay of regulatory devices during sporulation by *B. subtilis*, so we give more attention to this system. Its study has told us a great deal about how a single bacterial cell can give rise to two cells with quite different fates—one to become a spore of quite astonishing resilience, the other to die after briefly cooperating in the development of that spore. In contrast, in the mycelial organism *S. coelicolor* sporulation is a multicellular process, giving rise to long chains of spores at the tips of aerial reproductive branches. Even at first sight these different kinds of sporulation raise general questions of obvious interest. What molecular mechanisms induce the bacteria to initiate sporulation? Which genes are needed for the formation of the spore? What keeps these genes inactive while the cells are growing, and how do they become activated when their expression is required? Do the sporulation genes need to be expressed in an ordered sequence, and if so what determines the order?

Bacillus and *Streptomyces* spp often produce antibiotics at the onset of stationary phase. In both kinds of bacteria the regulation of production shows some degree of interaction with the regulation of sporulation. Since *Streptomyces* is the most important genus for producing antibiotics for human and veterinary medicine and other food and agricultural uses, the analysis of antibiotic production by streptomycetes has attracted wide-ranging research attention. In the second half of this chapter, we describe current knowledge of the regulation of antibiotic production in both *B. subtilis* and *S. coelicolor*.

2. Spore formation in *B. subtilis*

A slightly closer look shows that the process of sporulation in *B. subtilis* is even more intriguing than it seems at first sight. The earliest morphological sign of commitment to sporulation is an *asymmetric division*, which divides the bacterium into two cells of unequal size, as opposed to the symmetrical division that is characteristic of growing cells (Fig. 1(a)). Thereafter, the smaller cell—the forespore—becomes the spore, but the larger—the mother cell—is also indispensable to the success of sporulation. We can therefore ask further questions. What is the division of labour between the two cells? How is the specialization between them initiated and maintained, that is, how is expression of sporulation-specific genes spatially determined? Do the two cells communicate during spore development, and if so by what means?

Such questions would be interesting even if this process were unique to *Bacillus* spp. and their relatives, and had no relevance to other biological phenomena; but in fact it is of far-reaching importance, because it exemplifies two fundamental biological processes—*development* and *differentiation*. Development, because during sporulation the characteristic morphological changes proceed in a precise sequence, just as in animal embryos, for instance, and involve the expression of *temporally regulated* genes. Differentiation, because the two cells formed by the original asymmetric division have independent, complementary tasks involving the expression of *cell-specific* genes. Thus, sporulation requires both *temporal* and *spatial* regulation, and in studying the system we may hope to address one of the problems that is central to developmental biology: how is *differential gene expression* established in the genetically identical sister cells resulting from asymmetric division? The past decade has seen remarkable progress in answering these questions (reviewed in Reference 1), thus vindicating the pioneer workers in the field who saw sporulation as a model for the widespread phenomena of differentiation and development (2). One reason for this outstanding success is that *B. subtilis* strain 168 is exceptionally amenable to analysis by biochemistry, classical genetics, and molecular genetics.

2.1 Morphological changes during sporulation of *B. subtilis*

The asymmetric septation that is the first visible sign that a cell is entering sporulation is followed by a curious event, the *engulfment* of the forespore within the mother cell. As a result, the two cell compartments are separated by two bilayer membranes, and an important part of subsequent development involves the laying down of material between these membranes: first the cortex, which is a modified form of the kind of cell wall found in growing *B. subtilis* cells, and then the spore coat, a tough

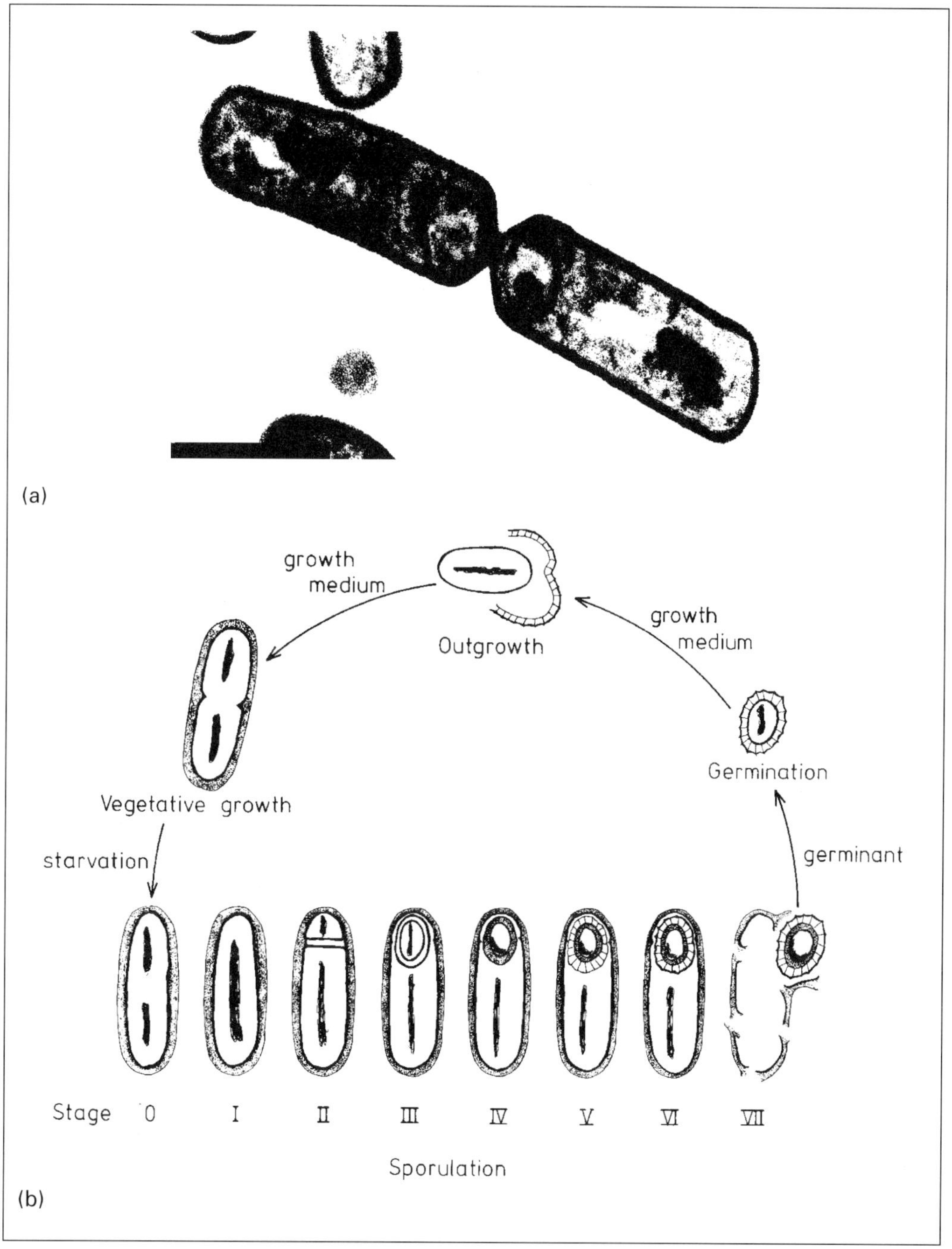

Fig. 1 Sporulation in *Bacillus subtilis*. (a) Asymmetric cell division is the first visible sign of sporulation. (b) Sketch of the stages of sporulation.

surface layer composed of an ordered array of different proteins, which confers on the spore several of its characteristic features. The final properties of the spore, its full resistance to chemical and physical insults and its ability to germinate so as to resume growth when favourable conditions return, then appear during a process called maturation; after this the mother cell lyses and the mature spore is released. Because of their development as an intracellular compartment of the mother cell, *B. subtilis* spores are more specifically termed *endospores*.

Particular stages that the cell reaches during development were recognized many years ago and given numbers for ease of reference (3) (Fig. 1(b)). Non-sporulating or growing cells are said to be in Stage 0. Asymmetric septation marks the end of Stage II (Stage I as originally defined is no longer recognized), and the completion of engulfment is called Stage III. Stage IV is the deposition of the cortex, Stage V the deposition of the spore coat, Stage VI maturation of the spore, and Stage VII the lysis of the mother cell. Although progress through these morphological changes is continuous, we shall see that the boundaries between Stages II and III, and III and IV, have turned out to mark important points in developmental regulation.

2.2 *spo* mutations and sporulation genes

It is easy to find *B. subtilis* mutants that cannot complete sporulation but are otherwise normal, because the mutant colonies remain colourless on prolonged incubation instead of developing the dark colour associated with one of the major spore coat proteins. Such strains carry mutations in any of perhaps 100 *spo* genes that are scattered round the circular chromosome (4).

Most *spo* mutants, when subjected to conditions that would induce sporulation in the wild type, proceed along the sequence of morphological changes up to a defined point and then become blocked. (This fact is presumptive evidence that each *spo* gene is expressed at a specific time during sporulation.) The stage of the block is characteristic of the mutation; thus, for example, we can say that a particular strain unable to proceed beyond Stage II is a Stage II mutant, and carries a mutation in a *spoII* gene. It turned out that several *spo* genes, which lie in different positions, or loci, on the chromosome are responsible for progress through each stage. Thus, several different *spo0* genes (*spo0A*, *spo0B* etc.) are required for cells even to enter the sporulation sequence, i.e. to advance from Stage 0.

Later it was discovered that some *spo* loci actually consist of more than one gene, with the result that the individual genes had to be named, for example, *spoIIAA* or *spoIVFB*. Furthermore, some genes which were originally given a non-descriptive name have more recently had their functions precisely identified, and often their names have then been changed to indicate the function (an example is the gene *spoIIAC*, which, as discussed below, is now called *sigF*). The protein product of each gene is given a name based on that of the gene that encodes it. Thus the gene *spoIIAA* encodes the protein product SpoIIAA. For the sake of clarity, we have tried to minimize the number of specific *spo* genes that we refer to in this chapter.

2.3 Different forms of RNA polymerase play a central role in sporulation

Bacillus subtilis possesses more than 10 different sigma (σ) factors, which are given the names σ^A, σ^B etc. (see Chapter 3). Each of them can, from time to time, become attached to the core RNA polymerase (abbreviated to the letter E) to make different species of RNA polymerase holoenzyme called $E\sigma^A$, $E\sigma^B$, etc. Most of the promoters that are expressed in growing cells are recognized by $E\sigma^A$. On the other hand, the promoters of most of the sporulation-specific genes are recognized by RNA polymerase containing other sigma factors, σ^E, σ^F, σ^G, and σ^K (5), which the bacteria make exclusively during sporulation. Active forms of these sigma factors appear at different times during sporulation, and two of them (σ^E and σ^K) are confined to the mother cell, and the other two (σ^F and σ^G) to the forespore. A further sigma factor, σ^H, which is important in the events immediately preceding asymmetric septation, is present in small quantities during vegetative growth, but its quantity increases as the culture starts to run out of nutrients and move towards sporulation (6). The specific temporal and spatial distribution of sigma factor activity ensures that, once asymmetric septation is complete, each gene that is expressed during sporulation is transcribed at a specific time and in just one of the two compartments (Table 1). Superimposed on this basic dependence of gene expression on the availability of a suitable $E\sigma$ form is the fine-tuning influence of repressor and activator proteins. In this chapter, we focus mainly on this beautifully orchestrated programme of gene expression, though we cannot altogether avoid (nor would we wish to!) reference to the physiological context that initiates, supports, and results from the underlying gene regulation.

Table 1 Timing and location of appearance of sporulation-specific sigma activities

	Forespore	Mother cell
Before asymmetric septation	σ^H	
Early post-septation	σ^F	σ^E
Late post-septation	σ^G	σ^K

2.4 Initiation of sporulation

To initiate sporulation, the *B. subtilis* cell has to transcribe several genes whose products are needed in the first stages of the process—the early sporulation genes. When the culture is well supplied with nutrients and is growing, these genes are kept transcriptionally inactive by a number of mechanisms. The most important of these involves a protein called AbrB, which is always present in growing cells (Fig. 2) (7). AbrB represses *sigH* and (indirectly) *sigE* and *sigF*, and also represses *sinI*, which encodes a negative regulator of another gene called *sinR*, whose product represses

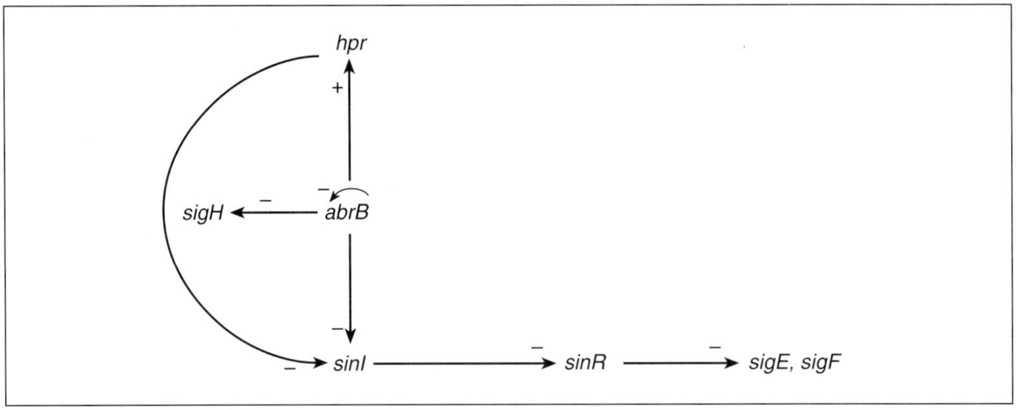

Fig. 2 The role of AbrB in preventing inappropriately timed sporulation. AbrB represses its own gene as well as the genes that encode σ^H and SinI. For a fuller explanation see text.

the major activator gene for sporulation (*spo0A* - see below) (8). (This kind of mind-torturing complexity is characteristic of the way in which sporulation is regulated; in order to keep track of repression cascades, it helps to remember that only *odd* numbers of repressing steps impose overall repression.) However, that is not all; AbrB is also an activator of some other genes, including another negative regulator of *sinI* (called *hpr*), thereby reinforcing the repression of *spo0A* by the *sinR* gene product.

To enter sporulation the cell must obviously prevent the synthesis of AbrB. The way in which it does so is remarkable, and depends on the incomplete repression of *spo0A* by AbrB and SinR. Thus, growing *B. subtilis* cells contain a small amount of Spo0A. As the cells sense the diminishing concentration of nutrients in their environment, they start to activate a so-called molecular phosphorelay (Fig. 3), which results in the *phosphorylation* of Spo0A (9). Spo0A-phosphate is a repressor of *abrB*. Since there is only just enough AbrB in the cell to prevent sporulation, continued growth of the culture (even at a diminishing rate as the nutrients become exhausted) soon causes its concentration to fall below the critical level needed to repress *sinI*, *spo0A*, *sigH*, *sigE*, and *sigF*. Since Eσ^H transcribes the *spo0A* promoter, the de-repression of *spo0A* is further enhanced. Moreover, Spo0A-phosphate is not only a repressor, it also directly activates several genes needed for Stage II of sporulation, including *sigE* and *sigF*, as well as *sinI* (hence leading to repression of *sinR* and consequent release of *spo0A* from repression by SinR—a further positive regulatory loop) (8, 10).

The molecular phosphorelay leading to Spo0A phosphorylation involves several proteins that act as intermediate carriers of the phosphate group as it passes from ATP (the usual donor of phosphate groups in the cell) to Spo0A. The cell can thus integrate a number of different signals indicating environmental and internal conditions. These provide information not only on the nutritional status of the culture but also on cell population density and, for each cell, its stage in the division cycle (11-13). All three of these parameters (and very likely others of which we are

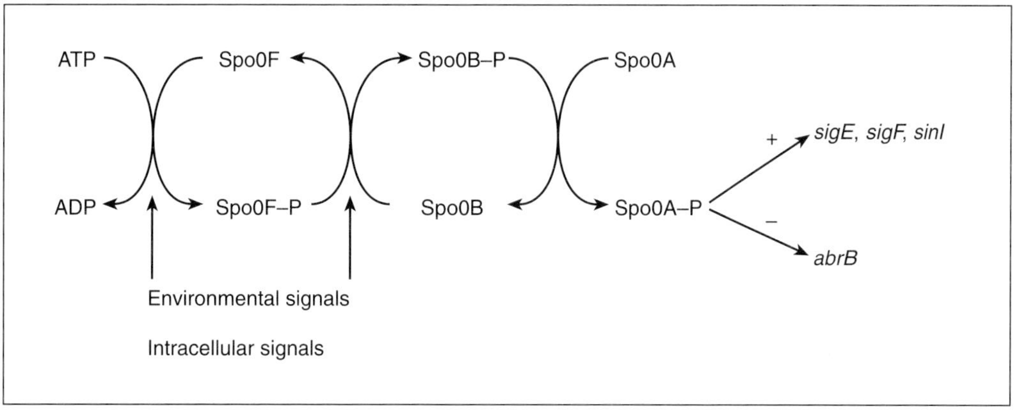

Fig. 3 A molecular phosphorelay that activates SpoOA. At each stage, organic phosphate is transferred from one component of the phosphorelay to the next. Several environmental and intracellular signals regulate the phosphorelay; further details are given in the text.

not yet aware) affect the operation of the phosphorelay (13). The best understood of the signals is that that indicates starvation. It appears that a lack of nutrients leads to a fall in the intracellular concentration of GTP (and possibly GDP) (14); by some mechanism whose details we do not understand, this reduction in the guanine nucleotide pool is important in inducing sporulation. Interestingly enough, cells can be induced to sporulate even in rich medium by treatment with decoyinine, a drug that inhibits the conversion of XMP to GMP (15).

During exponential growth, the cell needs to prevent Spo0A from being phosphorylated. To this end, it possesses several *protein phosphatases*, which hydrolyse Spo0F-phosphate and thus stop the phosphate ester group from reaching Spo0A (16). The phosphatases seem to be under the negative control of specific proteins, which are encoded by genes in the same transcription units as the phosphatases themselves. These proteins are secreted from the cell and cleaved, and the resulting peptides are then imported back into the cell and used to regulate the phosphatases. This complicated mechanism seems to function as a sensor of cell concentration: when the culture reaches a sufficiently high density, the concentration of the peptides will become high enough to downregulate the phosphatases, and only at that point can Spo0F-phosphate start to accumulate and set the phosphorelay in action (16).

We have mentioned that during growth AbrB repressor is maintained at a level only just high enough to maintain repression of the sporulation genes. In fact, the concentration of AbrB is poised on a knife-edge, because AbrB is a repressor of its own gene (Fig. 2) (17). If for any reason the AbrB concentration falls, the gene is expressed a little more strongly and more AbrB is synthesized; if the concentration becomes too high, the gene is more strongly repressed and continued growth of the cells dilutes the AbrB. Such autoregulatory negative feedback systems are seen for many—but not all—repressor genes in various bacteria.

2.5 Establishing compartment-specific gene expression

Asymmetric septation, to yield the small forespore and the larger mother cell, occurs about one hour after the onset of the starvation that induces sporulation. After septation, expression of each of the genes that are necessary for the completion of sporulation is confined to just one or the other compartment. Immediately after septation (Stage II), $E\sigma^F$ begins to transcribe genes in the forespore, and $E\sigma^E$ begins to transcribe other genes in the mother cell. Later in sporulation, after engulfment (Stage III), forespore-specific genes are transcribed by $E\sigma^G$ and mother-cell-specific genes by $E\sigma^K$ (Table 1). Within each temporal and spatial class, all the genes have promoters that are recognized by a given sigma factor.

In order to explain more fully how the spatial pattern of gene expression is determined, we should note that the different sporulation-specific sigma factors are arranged in a dependent sequence: $\sigma^F \rightarrow \sigma^E \rightarrow \sigma^G \rightarrow \sigma^K$ (18). That is to say, only if σ^F functions correctly will σ^E become active; only if σ^E functions correctly will σ^G become active; and only if σ^G functions correctly will σ^K become active (Fig. 4). So the establishment of compartment-specific gene expression depends absolutely on ensuring that σ^F becomes active, soon after the completion of asymmetric septation, in the forespore but not the mother cell.

The *sigF* gene is one of the set of three genes collectively called *spoIIA*. (The old name for *sigF* was *spoIIAC*.) The three genes are co-transcribed from a single promoter which is recognized by $E\sigma^H$, and consequently they are expressed much more strongly in cells that are beginning to sporulate (and therefore have increased σ^H levels) than in growing cells. Such dependence on σ^H might lead one to expect σ^F activity in both spore compartments, so how can we account for the fact that σ^F activity is confined to the forespore? This spatial specificity has turned out to involve the proteins that are encoded by the *spoIIAA* and *spoIIAB* genes, together with that encoded by *spoIIE*, a gene whose location on the chromosome is quite distant from that of *spoIIA*. SpoIIAB is capable of interacting with σ^F in such a way as to inhibit its activity (19); it has therefore been called an anti-sigma factor. Alternatively, SpoIIAB can interact with SpoIIAA (the anti-anti-sigma factor). The nature of this interaction seems to depend on the adenine nucleotide available (20, 21). When ATP is present, SpoIIAB acts as a protein kinase to transfer a phosphate group from ATP to SpoIIAA

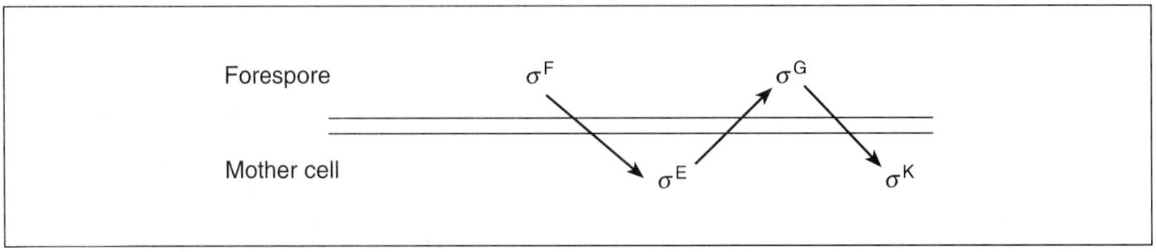

Fig. 4 A cascade of sigma factors controls the expression of sporulation genes both temporally and spatially. This figure stresses the interactions between the two compartments; in addition, the gene for σ^G is transcribed by RNA polymerase containing σ^F, and the gene for σ^K is transcribed by RNA polymerase containing σ^E.

to form SpoIIAA-phosphate (22); but in the presence of ADP SpoIIAB can (at least *in vitro*) form a complex with SpoIIAA, pre-empting the binding of SpoIIAB to σ^F. SpoIIE, the third regulatory protein of the system, is a specific protein phosphatase, which hydrolyses SpoIIAA-phosphate back to SpoIIAA (23).

How do these proteins interact *in vivo* to regulate σ^F? It has been suggested that SpoIIAB's preference for binding partner is sensitive to the concentrations of ATP and ADP, with ATP favouring the formation of a σ^F/SpoIIAB complex and ADP the formation of a SpoIIAA/SpoIIAB complex, and that the ATP:ADP ratio may be lower in the prespore compartment than in the mother cell (20, 21). However, measurements of the dissociation constants for these protein/protein complexes have cast doubt on this idea (24). Maybe a clue about the regulation can be found in the fact that, when the asymmetric division that partitions the prespore from the mother cell is complete, the SpoIIE phosphatase is localized to the asymmetric septum (25); there is some reason to believe that this location allows it to hydrolyse SpoIIAA-phosphate in the prespore but not in the mother cell (23, 26). In addition, the reaction in which SpoIIAB catalyses phosphorylation of SpoIIAA involves the formation of an extremely long-lived complex between the enzyme (SpoIIAB) and ADP (27). These observations suggest that, at about the time when σ^F becomes active, the simultaneous action of SpoIIAB and SpoIIE in the prespore could cause SpoIIAA to cycle repeatedly between the phosphorylated and the non-phosphorylated forms. Cycling of SpoIIAA in this way would sequester SpoIIAB into the long-lived complex; as a result SpoIIAB would be unable to bind to σ^F, and thus σ^F would be free to join with core RNA polymerase to initiate transcription (24) (Fig. 5).

Expression of *sigE*, like that of *sigF*, is activated by Spo0A-phosphate (see above) and therefore precedes asymmetric septation. Like σ^F, σ^E is controlled by a mechanism that allows it to become active in only one of the two compartments. This time, however, the mechanism does not involve an anti-σ factor. Instead, σ^E is

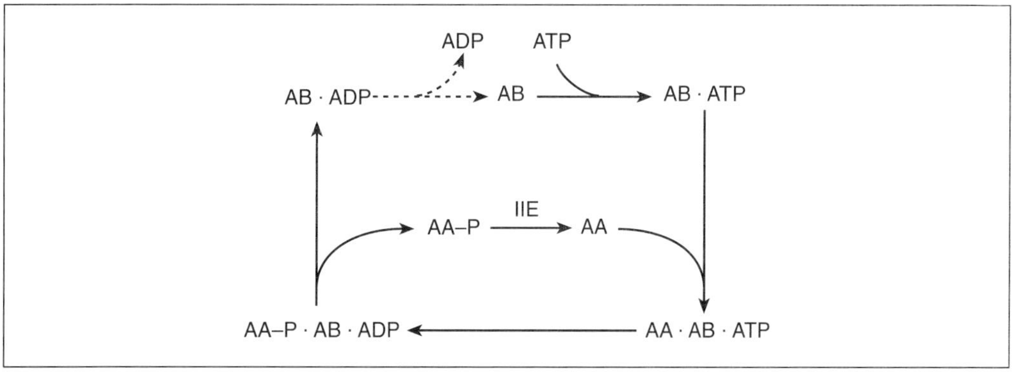

Fig. 5 A model for the activation of σ^F (24). The designation Spo is omitted for the proteins SpoIIAA, SpoIIAB and SpoIIE. The Figure shows the proposed cycling of SpoIIAA between the phosphorylated and the non-phosphorylated form. This cycling involves the sequestration of SpoIIAB in the form of a complex with ADP, from which the ADP is lost only very slowly (see dotted arrows). As a result, the concentration of SpoIIAB available to inhibit σ^F is diminished, and hence σ^F activity is liberated.

synthesized as a longer, inactive, precursor protein, and activation requires a specific protease that removes the surplus N-terminal part of the pro-σ^E (28). The activity of this protease is confined to the mother cell, and depends on Eσ^F-dependent transcription of a gene called *spoIIR* in the forespore (29). Expression of *spoIIR* apparently leads to the transmission to the mother cell of a signal, the directionality of which is crucial to its ability to activate cleavage of pro-σ^E (30). The result is that the activation of the two sigma factors, σ^F and σ^E, is roughly coordinated in time but separated in space.

2.6 Gene expression in the forespore after engulfment

About 40 minutes after asymmetric septation has been completed, the forespore compartment becomes engulfed by the mother cell (Stage III). Over the next two or three hours, more than a dozen genes are expressed specifically within the forespore, as a result of the activity first of σ^F and then of σ^G.

Only a few genes are known to be transcribed by Eσ^F; one is *spoIIR* (see above). A second encodes an enzyme called germination protease, which specifically hydrolyses storage proteins during spore germination (see below). A third encodes σ^G, which is responsible for the final wave of transcription in the forespore. Since the promoter of the *sigG* gene is recognized by Eσ^F but not by Eσ^E, σ^G appears only in the forespore and not in the mother cell. The *sigG* promoter is also recognized by Eσ^G itself (31). This arrangement ensures that σ^G continues to be synthesized for as long as it is needed, even though the activity of σ^F gradually declines after σ^G becomes active.

We noted before that the appearance of σ^G activity also depends on σ^E. An attractive but unproven mechanism for this dependence is that σ^G activity requires engulfment, which is known to be Eσ^E-dependent. This would ensure that the sequence of events in one compartment is not allowed to run ahead of the sequence in the other compartment (see also the activation of σ^K, described below).

σ^G is responsible for expressing not only its own gene but also many others whose products are needed for maturation of the forespore (Table 2). Some of these gene

Table 2 Examples of genes expressed in the forespore

Gene	Product
Genes transcribed by Eσ^F	
spoIIR	Inter-compartmental signalling protein
gpr	Germination protease
sigG	σ^G
Genes transcribed by Eσ^G	
sigG	σ^G
Several *ger* genes	Proteins needed for germination
Several *ssp* genes	Small acid-soluble proteins

products are eventually actively involved in spore germination. Others are small acid-soluble proteins, which either make complexes with the DNA inside the spore, and render it resistant to ultraviolet radiation, or act as storage proteins, to be broken down into amino acids by germination protease (see above) when favourable conditions return (32).

2.7 Gene expression in the mother cell

The activity of Eσ^E in the mother cell gives rise to numerous gene products that are essential to the developing spore. At least one is needed for engulfment. Others include some enzymes needed to synthesize the cortex, which is laid down during Stage IV between the two membranes that surround the forespore. In addition Eσ^E transcribes the *sigK* gene for σ^K, the final mother-cell-specific sigma factor (Table 3).

The σ^E-dependent genes are not all expressed simultaneously. The elegant mechanism for achieving this temporal variation (33) involves a protein, SpoIIID, that represses some of the genes but activates others. Genes belonging to the first of these classes are expressed as soon as Eσ^E becomes active, but their transcription comes to a halt as the concentration of SpoIIID builds up. Genes that require SpoIIID activation are transcribed only later, when sufficient SpoIIID has accumulated. The gene for σ^K is one of these. Some σ^E-dependent genes are unaffected by SpoIIID, and are expressed all the time σ^E is active. These cease to be expressed only when σ^E is replaced by σ^K.

σ^K is the last of the sporulation-specific sigma factors to appear (34). Like σ^G, it can bring about transcription of its own gene, allowing it to be made even after σ^E (which is responsible for its initial burst of synthesis) is no longer active. Like σ^E, it is synthesized as an inactive precursor protein which needs to be processed by a specific protease, whose activity depends on transcription of a gene (in this case Eσ^G-dependent) in the opposite compartment, ensuring that progress in both compartments is kept in step.

Many of the σ^K-dependent genes encode spore coat proteins, which appear to form the coat at least in part by a process of self-assembly. However, correct assembly also

Table 3 Examples of genes expressed in the mother cell

Gene	Product
Genes transcribed by Eσ^E	
spoVD, gerJ, gerM	Proteins needed for cortex synthesis
spoIIID	Temporal regulator of transcription by Eσ^E
sigK	σ^K
Genes transcribed by Eσ^K	
sigK	σ^K
Several *cot* genes	Coat proteins
dpa	Enzymes of dipicolinic acid synthesis
gerE	Temporal regulator of transcription by Eσ^K

relies on the fact that the different coat proteins are made at different times. A few are the products of σ^E-dependent genes, but there is also a variation in the time of expression of the σ^K-dependent coat protein genes. The mechanism of this variation is similar to that described above for SpoIIID (35): the σ^K-dependent *gerE* gene encodes a protein that represses certain genes and activates others (Table 3).

Dipicolinic acid, which (complexed with Ca^{2+}) is present in large quantities in mature spores and is believed to contribute to their characteristic heat-resistance, is synthesized by the products of two genes that are transcribed by $E\sigma^K$ and repressed by GerE; thus synthesis of dipicolinic acid, like the synthesis of most of the coat proteins, is a late function of the mother cell.

2.8 Understanding sporulation in *B. subtilis* is an unfinished story

We are now familiar with many of the molecular interactions that occur during sporulation in *B. subtilis* and in particular with the systems that control the spatial and temporal regulation of gene expression; but the striking changes in appearance that occur as the cells sporulate are still quite poorly understood. To relate them to the molecular events that have been investigated so fruitfully in the past 20 years or so will be a most important challenge for the future.

3. The development of spore chains in the aerial mycelium of *Streptomyces coelicolor*

When growing on agar media in the laboratory, *Streptomyces* spp. form colonies made up of a coherent mycelium of branched hyphal filaments. Like plants, they are 'rooted' to the spot, and they achieve dispersal through the formation of numerous spores at the tips of specialized aerial hyphae, which give the colonies a furry appearance (Fig. 6A and B). Although the spores are not much more heat resistant than vegetative hyphae, they are desiccation resistant and long-lived in soil.

Both vegetative and aerial hyphae of streptomycetes grow by extension at existing tips and by the formation of new branches, usually several tens of micrometres back from the tips. Likewise, cross-walls are not formed close to the tips, so the apical compartments (and, to a lesser extent, the older parts of hyphae) are multigenomic. When spores form in the aerial hyphae, tip growth ceases, and every copy of the genome in the apical compartment becomes separated from its neighbours by the formation of specialized sporulation septa. These are regularly spaced (1–2 μm apart) and—within any one hypha—synchronously formed (Fig. 6C). After sporulation septation, which once started takes only a few minutes, the individual prespore compartments change shape, from cylinders to ovoids, and they acquire thick spore walls over a period of a few hours. In most species, the spores become pigmented (grey in the case of *S. coelicolor* A3(2), the strain used in most of this work), providing a helpful visual sign that sporulation has taken place on the colony surface.

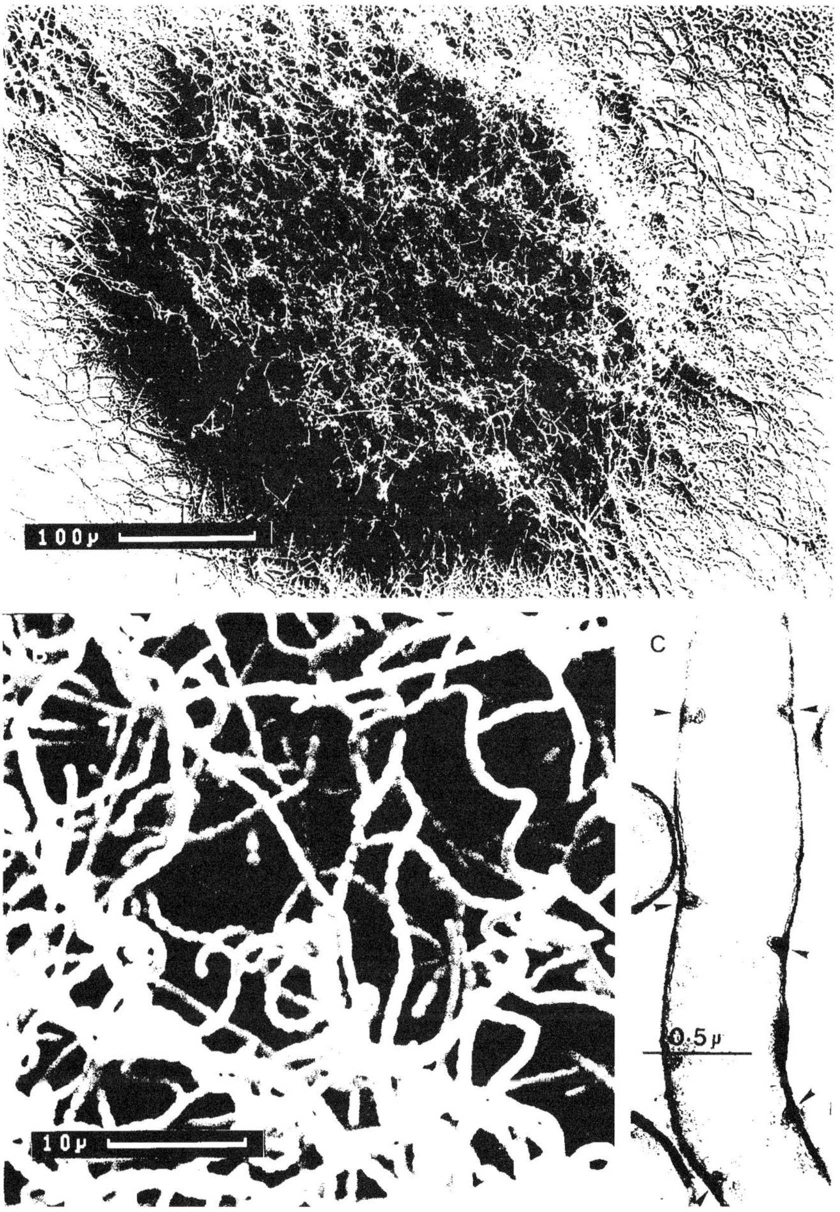

Fig. 6 Multicellular differentiation leads to spore formation in *Streptomyces coelicolor*. A: Scanning electron micrograph of a colony, showing the tangle of aerial mycelium on the older central part of the colony, and the advancing periphery of the vegetative mycelium. B: Long chains of spores form at the tips of aerial hyphae. C: Regularly spaced initiation of septation (arrows) during sporulation of an aerial hypha.

Aerial mycelium development requires several physiological operations: the perception by the colony of the 'need' for aerial growth; the initiation of an aerial branch, and its growth into the air away from nutrients; signalling to halt aerial extension growth and DNA replication, and to initiate sporulation septation; correct spatial control of sporulation septation; the provision and proper control of the septum components; and the morphogenesis of prespore compartments into spores. These operations pose specialized metabolic problems, with alternating phases of rapid vegetative growth followed by metabolic shiftdown in the substrate mycelium, rapid aerial growth, metabolic shiftdown again in the aerial hyphae as sporulation is initiated, and, finally, an almost total metabolic shutdown as spores mature. Although evidence is accumulating that, just as in *B. subtilis*, a fall in GTP concentration may be a key early signal for morphological differentiation leading to aerial growth (36), analysis of the interplay of metabolic changes with the regulation of the morphogenetic processes is only just beginning.

The morphological and physiological differences between spore formation in *Streptomyces* and *B. subtilis* are reflected by marked differences in the underlying regulation.

3.1 Mutations affecting sporulation in *S. coelicolor*

Just as in *B. subtilis*, genetics has provided the key to investigating *Streptomyces* sporulation. Most studies have involved *S. coelicolor* A3(2), because it has well-developed genetics. A valuable resource has been a collection of mutants defective in aerial mycelium formation or sporulation, analogous to the *spo* mutants of *B. subtilis*. These have provided the raw material to identify and clone genes and to begin to investigate their interdependence (37) (Fig. 7).

Mutants that lack aerial mycelium have a 'bald' appearance—hence the term *bld* genes. Perhaps surprisingly, very few *bld* mutants are unconditionally bald. Most of them can form a normal aerial mycelium when grown on defined agar media in which the normally supplied carbon source, glucose, is replaced by certain alternatives, such as mannitol. An interplay of most *bld* genes with carbon metabolism is also revealed by their apparently constitutive, glucose non-repressible expression of some genes for catabolism of carbon sources (38). Many *bld* mutants are also defective in the production of secondary metabolites, such as antibiotics, so we also return to them in a later section dealing with the regulation of antibiotic production.

Other mutants produce an aerial mycelium that, lacking pigmented spores, stays white on prolonged incubation, defining the *whi* genes. The *whi* genes can broadly be subdivided into 'early' and 'late' classes. The early *whi* genes are needed either for sporulation septa to form or for their correct spacing; these genes appear generally to be regulatory in character. The late *whi* genes are needed for spore maturation; for example, some of their products carry out enzymatic functions in spore pigment biosynthesis.

Other genes with important roles in sporulation have been discovered by 'reverse' genetics, in which oligonucleotides encoding conserved regions of selected proteins

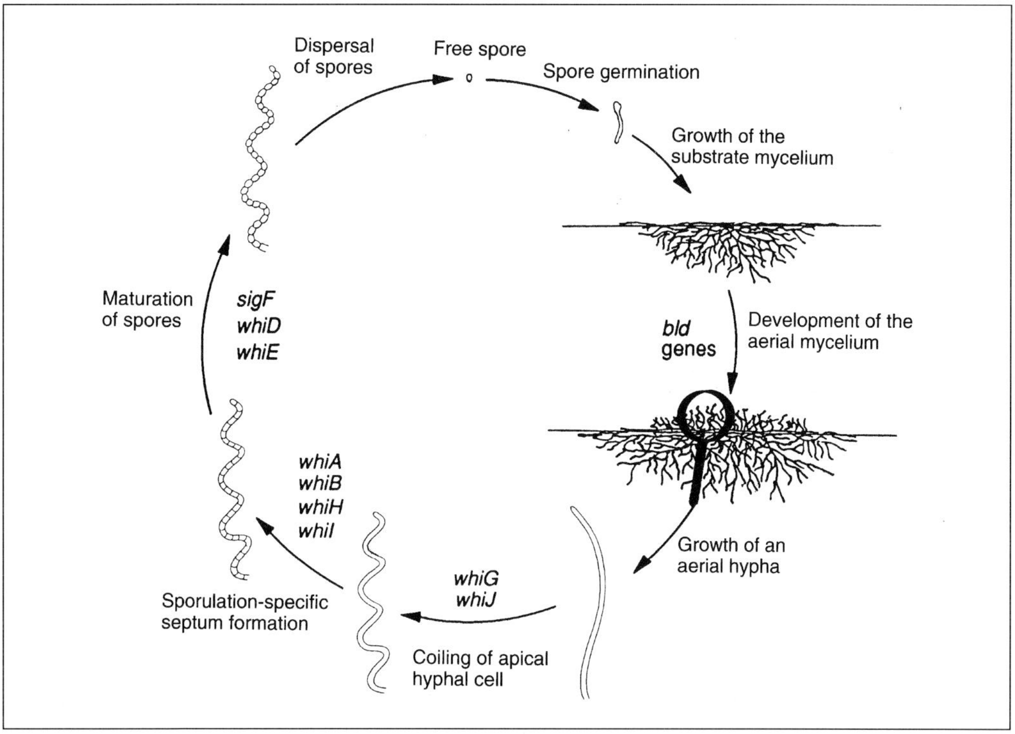

Fig. 7 Diagrammatic life cycle of *Streptomyces coelicolor*. Genes involved at each step are indicated (drawn by N. J. Ryding).

of interest are used to detect the relevant genes, or by the morphological effects of cloning certain genes on high copy number plasmids. These cloned genes are subsequently used for the deliberate construction of mutants, whose phenotype may reveal the role of the protein in question.

3.2 Initiation of aerial growth: evidence for the involvement of extracellular signals and specialized codon usage

So far, *bld* mutants have told us more about cell–cell signalling and signal transduction than about the direct regulation of expression of genes for aerial growth.

The most well-defined extracellular signalling system for aerial growth was discovered not in *S. coelicolor*, but in *Streptomyces griseus*, which produces an extracellular diffusible lipid-soluble γ-butyrolactone, called A-factor, late in growth (39). A-factor is an autoregulator that must accumulate to nanomolar concentrations in order for aerial growth or secondary metabolism of *S. griseus* to begin. Because of the ease with which A-factor diffuses through membranes, cells can effectively 'estimate' its external concentration by a cytoplasmic binding protein. The external A-factor concentration is often thought of as an indicator of cell population density, like the homoserine lactones produced by many Gram-negative bacteria or the oligopeptide sporulation factors of *B. subtilis* (see above). Genetic evidence suggests that the A-factor binding protein is a repressor of a gene that activates regulatory cascades leading to aerial mycelium formation and antibiotic production. Summarizing this evidence, we can say that mutations that eliminate production of A-factor cause a bald, antibiotic-negative phenotype which can be phenotypically corrected either by adding A-factor as a pure compound or by growth near an A-factor-producing colony. Wild-type morphology and antibiotic production are also restored by mutations that eliminate the binding protein.

Suppression just of the sporulation deficiency of an A-factor-less mutant can be achieved by introducing certain DNA fragments on a high copy number plasmid (40). Some of the genes identified in this way (the members of the *amf* gene cluster) appear to be involved in signal transduction, encoding an ATP-dependent membrane-bound transporter complex and a response regulator of the type phosphorylated by histidine protein kinases in response to extracellular signals. These observations indicate that extracellular signals other than A-factor are also important, and that their transduction inside the cell involves the phosphorylation of regulatory proteins. However, the signals themselves, the phosphotransfer pathway, and the target genes, all remain to be identified.

A-factor itself is not produced by *S. coelicolor* A3(2), but this organism does make related compounds, which do not seem to control morphological differentiation under the conditions tested so far (although evidence has now been obtained for a role in antibiotic production: E. Takano and M. J. Bibb, personal communication). However, other extracellular factors are important for aerial growth of *S. coelicolor* in some growth conditions, since *bld* mutants of different genetic classes can often restore sporulation to each other when grown close together (41-43). From such experiments, evidence has emerged of a cascade of five extracellular signal exchanges leading to the secretion of SapB, a modified peptide of only 17 amino acids. SapB appears to be made non-ribosomally, and is thought to coat the surface of the colony and permit aerial growth. This complex signalling system is observed only on certain media, again suggesting that some aspects of aerial growth can be achieved in more than one way. Only a few of the *bld* loci have been extensively characterized. Two of these (*bldK* and *ram*) encode ATP-dependent transporters, consistent with roles in production or uptake of extracellular small molecules (shown to be oligopeptides in the case of *bldK*) (41, 44). The *ram* locus is a homologue of the *amf* locus of *S. griseus* (see above), and, like *amf*, also encodes a response regulator (44). The place of *ram* mutants (which have a bald phenotype) in the extracellular complementation hierarchy has not been described, but *bldK* mutants are blocked very early in the signal cascade: the first signal is probably an oligopeptide that is recognized by the *bldK* system (41).

It is less obvious why *bldA* mutants fit into the signal cascade, since the *bldA* product is the only tRNA in *S. coelicolor* capable of translating the UUA (leucine) codon (45). This codon is very rare in the mRNA of *Streptomyces* spp. because of their very G+C-rich DNA: leucine codons starting with CU are nearly always used. The *bldA* gene can be mutated or deleted without affecting vegetative growth, implying that no genes essential for growth contain in-frame TTA triplets. However, *bldA* mutants lack aerial mycelium and make no antibiotics, so in-frame TTA triplets are expected to occur in a gene or genes specifically needed for these processes. No clearcut *bldA* target has yet been found among *S. coelicolor* genes involved in aerial mycelium formation.

3.3 The early and later stages of sporulation in aerial hyphae of *S. coelicolor* appear to be directed by different sigma factors

Once aerial hyphae have formed, the first step in committing them to sporulation is the accumulation of RNA polymerase containing a sporulation-specific sigma factor, σ^{WhiG}, to a level sufficient to direct the transcription of certain, as yet largely unidentified, sporulation genes. If the *whiG* gene, which encodes σ^{WhiG}, is inactivated, aerial hyphae grow to a considerable length and remain morphologically featureless: conversely multiple copies of *whiG* (or artificial *whiG* overexpression from a regulatable strong promoter) cause sporulation to take place when aerial hyphae are still very short, and—more strikingly—to occur also in the older parts of the substrate mycelium ('ectopic sporulation') (37).

What regulates the appearance of active σ^{WhiG}? Surprisingly, there is a rather constant level of *whiG* mRNA at different stages of colony development (46). Post-transcriptional regulation is therefore probably important. It is particularly appealing to consider the possibility of control at the level of σ^{WhiG} activity, via an anti-σ^{WhiG} protein. As we have already described, the B. subtilis σ^{F} is controlled by the SpoIIAB anti-sigma protein. Regulation by a different kind of anti-sigma protein has been observed for σ^{FliA} of *Salmonella typhimurium* (47), a sigma factor particularly closely related to σ^{WhiG} by molecular phylogeny. Eσ^{FliA} is responsible for the transcription of genes for late stages of the biogenesis of flagella, and even though σ^{FliA} protein is present at earlier stages in the construction of flagella, it does not form part of an active RNA polymerase holoenzyme because it is bound instead to an anti-sigma protein. Since homologues of σ^{FliA} and its antagonist also control flagellum biosynthesis in *B. subtilis* (48), this anti-sigma mechanism seems to be widespread. σ^{WhiG} contains appropriately placed regions of homology to the part of σ^{FliA} known to interact with the anti-sigma protein (49), so an anti-σ^{WhiG} protein with at least a localized resemblance to anti-σ^{FliA} is expected (though not yet discovered). Moreover, σ^{WhiG} appears to be regulatable by anti-σ^{FliA} protein *in vivo* when the *whiG* gene is expressed in a suitable strain of *S. typhimurium* (J. Nodwell, personal communication).

There are at least five other early *whi* genes, all of which are likely to be involved in a complex regulatory network (Fig. 8). One of them, *whiH*, is a direct target for transcription by σ^{WhiG} RNA polymerase holoenzyme (50) while another, *whiB*, appears to be transcriptionally independent of the other *whi* genes (51); so entry into

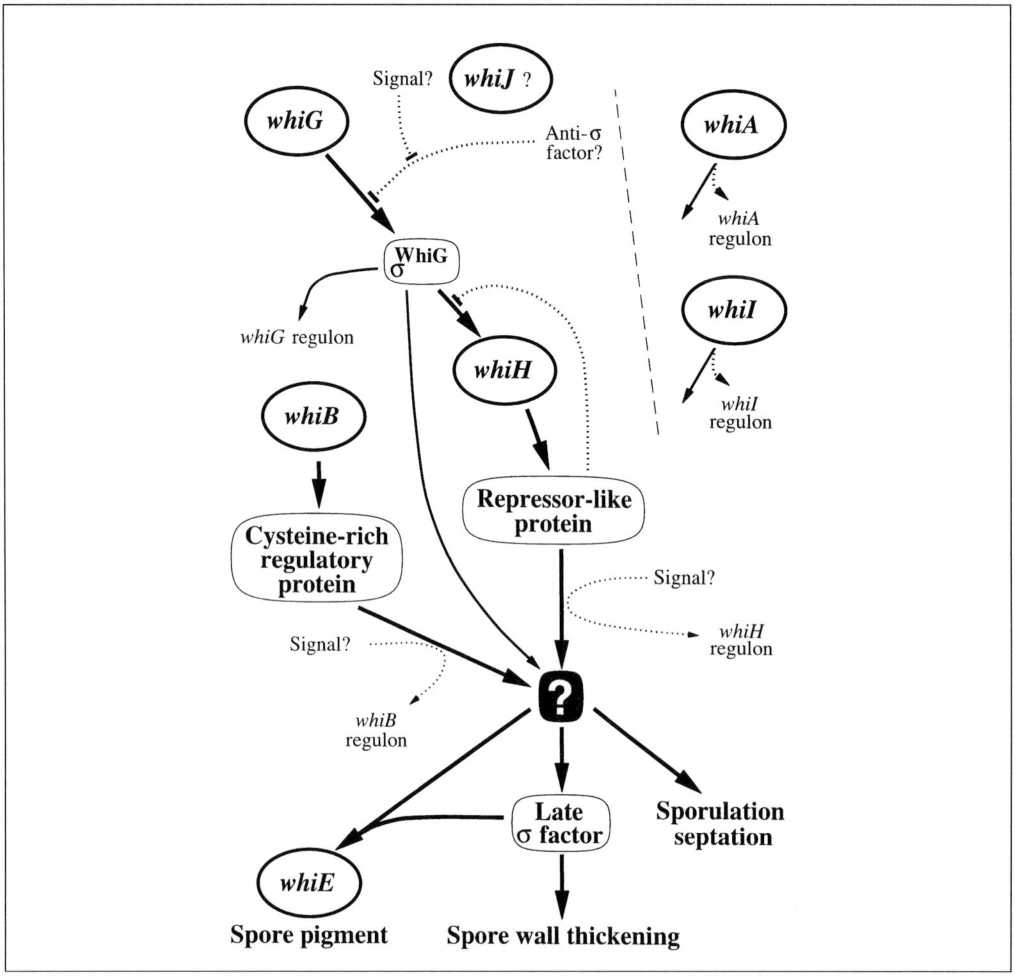

Fig. 8 The interplay of several different kinds of regulatory proteins controls sporulation in *Streptomyces coelicolor*. The arrows indicate dependence relationships, broken lines and question marks being used to indicate unproven relationships. Information is from refs. 37, 46, 50 and 57.

sporulation seems to involve the convergence of at least two transcriptional regulatory pathways. The functions of the products of three early *whi* genes, *whiA*, *whiB*, and *whiJ*, cannot be guessed from comparisons with the protein databases (37; N. J. Ryding and N. M. Hartley, personal communication), but the product of *whiH* belongs to a family of DNA-binding regulatory proteins known from diverse other bacteria, many of whose members bind various primary metabolites (50). *whiI* mutants are complemented by DNA encoding a member of a family (which includes *Spo0A*) consisting of response regulators whose activity is modulated by phosphorylation, often in response to extracellular signals (J. A. Aínsa, personal communi-

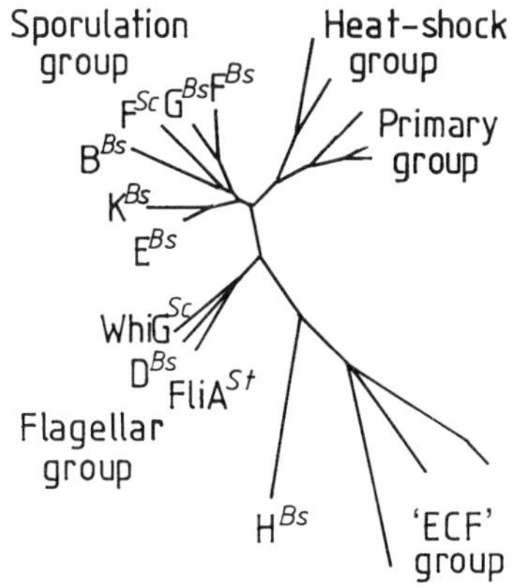

Fig. 9 Phylogenetic relationships of sigma factors involved in sporulation in *Bacillus subtilis* (Bs) and *Streptomyces coelicolor* (*Sc*). Most of the sigma factors involved in sporulation, together with the late sporulation sigma factor σF of *S. coelicolor*, form a well-defined subgroup, but there are two exceptions: σ^{WhiG} of *S. coelicolor* is clustered with σ^{D} of *B. subtilis* and σ^{FliA} of *Salmonella typhimuruim* (*St*); and σ^{H} of *B. subtilis* forms a 'group' of its own. For several of the sigma factor groups (*Primary, Sporulation, Flagellar, ECF*), an ancestral form was apparently already present in the last common ancestor of *B. subtilis* and *S. coelicolor* (data from ref. 52)

cation). Probably, some or all of the early Whi proteins provide a means of executing appropriate gene regulation in response to different physiological signals. In this sense, their role may be analogous to that of the *spo0* and ancillary genes that take part in the sporulation phosphorelay of *B. subtilis* (Fig. 3). As a result of the combined action of the early Whi proteins, the *sigF* gene encoding a late sigma factor (52) is switched on (46). Since a *sigF* mutant produces thin-walled, unpigmented spores, it can be anticipated that some of the genes for spore wall thickening and pigmentation may prove to be direct targets for the σ^{F} form of RNA polymerase.

Although σ^{WhiG} does not closely resemble any of the sigma factors involved in *B. subtilis* sporulation (Fig. 9), σ^{F} turns out to be phylogenetically related to the *B. subtilis* forespore-specific sigma factors σ^{F} and σ^{G}, and to σ^{B}, the *B. subtilis* stress sigma factor (52). The late stages of sporulation in these diverse organisms may therefore have originated in a process present in their last common ancestor some 2 billion years ago. Although σ^{B}, σ^{F}, and possibly σ^{G} of *B. subtilis* are regulated by the interplay of anti-sigma and anti-anti-sigma proteins (see above), no such regulation has yet been found for σ^{F} of *S. coelicolor*.

3.4 The genetic determination of physiological processes that occur in two distinct spatial locations

In addition to the regulatory role of the early *whi* genes in switching on *sigF*, their action also controls metabolic and morphological processes. Several of these processes have counterparts at earlier stages of growth, raising interesting questions about the nature of their temporal and spatial regulation. For example, in *Streptomyces* colonies, septa form in two situations; they are deposited, one at a time, some tens of microns behind growing tips during growth, and synchronously in large numbers, right up to the aerial hyphal tip, during sporulation. Studies of FtsZ, a protein that plays a direct and important part in septation, have revealed that a single *ftsZ* gene is probably involved in both vegetative and sporulation septation, because its inactivation eliminates both classes of septa, and no other *ftsZ*-like sequences can be detected by Southern blotting (53). Presumably, therefore, the early *whi* gene network has a spatially specific effect on the regulation of FtsZ synthesis or activity that is superimposed on, or replaces, the regulation characteristic of growth.

An alternative strategy could also be imagined: there might be two separately regulated *ftsZ* genes. In fact, examples of genetically distinct isoforms of proteins in these different locations have been found. One of them involves storage metabolism. Just as in most bacteria, streptomycetes limited for one class of nutrient exhibit the ability to convert excess nutrients of other classes into storage compounds. The most studied of these is glycogen, which accumulates inside substrate hyphae at the surface of colonies after 2–3 days, just as aerial hyphae begin to emerge. The rapidly growing aerial hyphae do not show such storage reservoirs, and probably some of the glycogen reserves from the substrate hyphae are mobilized to support aerial growth. When aerial growth of a hypha ceases, and sporulation septation begins, a second phase of glycogen synthesis results in deposits forming specifically in prespore compartments, only to be degraded again during spore maturation. At least one step in glycogen synthesis, the formation of α-1,6 branch points in the α-1,4-linked polyglucan chains, has turned out to involve spatially specific enzyme isoforms controlled by two similar *glgB* genes located far apart on the chromosome (54).

A similar situation arises in the biosynthesis of fatty acids and of polycyclic aromatic polyketide compounds such as the grey spore pigment and the blue antibiotic actinorhodin. These pathways share a common biochemistry, in which C2 units derived from malonyl CoA are sequentially condensed to give chains of moderate length. It has emerged that all the steps in condensation of C2 units during carbon chain growth are carried out by pathway-specific enzymes, with complex gene clusters for each pathway. Normally, mutations in actinorhodin pathway genes cannot be complemented by the genes for spore pigment biosynthesis or vice versa; however, some of the proteins involved in the two pathways have been shown to be interchangeable if their genes are artificially expressed together by being placed under suitable transcriptional regulation (55, 56). It therefore seems that there is temporal and/or spatial separation of the different pathways, and that this is regulated at the transcription level. Indeed, transcription of actinorhodin biosynthesis genes has a complex control that is largely independent of the sporulation-specific regulatory *whi* genes (see below), whereas the spore pigment biosynthetic genes are expressed only when the early *whi* genes are intact (57). In contrast, in its last step in delivery from intermediary metabolism to the condensation machinery for fatty acid, actinorhodin polyketide, or spore pigment polyketide biosynthesis, acetate is transferred from malonyl CoA to a pathway-specific acyl carrier protein-isoform by a single transferase enzyme that seems to be used in all three situations (58).

4. Genetic regulation of antibiotic production in sporulating bacteria

Antibiotic production is found more frequently in sporulating than in non-sporulating microbes, and especially in soil organisms: examples include mycelial fungi and Gram-negative myxobacteria (which produce fruiting bodies containing millions of spores), as well as the bacilli and streptomycetes that are the subject of this chapter. It is a curious fact that—at least in laboratory conditions—most antibiotics, like spores, are produced during the stationary phase, and the genes for production are switched off during rapid growth. This seems to rule out the idea that antibiotics have evolved to give the producing organism a competitive advantage when conditions for growth in soil are good. Nevertheless, we suppose that the biochemically specific and powerful effects of antibiotics on sensitive organisms are the results of natural selection. Probably, their adaptive benefits vary, for example, some may be autoregulators perhaps serving as hormone-like signals within the producing population, while others might be produced to defend colonies at a vulnerable stage in their life cycle, as when the ageing vegetative mycelium of a streptomycete is lysing to provide nutrients for the developing aerial mycelium.

In both streptomycetes and *B. subtilis*, the existence of mutant classes simultaneously deficient in sporulation and antibiotic production provides evidence of common elements in the regulation of the two processes. Here we discuss the basis of the connection in examples from each genus.

4.1 Regulation of surfactin biosynthesis in *B. subtilis*

B. subtilis produces the lipopeptide surfactin, a detergent and antibiotic used in the food industry. The peptide bonds linking the seven amino acids of surfactin are formed non-ribosomally by peptide synthetases encoded by the *srfA* operon (59). Surfactin production, like sporulation and the development of competence for genetic transformation by free DNA, is induced at the end of the vegetative growth phase and only in dense cultures, in response to two ribosomally synthesized and post-transcriptionally modified peptide pheromones ComX and CSF (competence-stimulating factor) (13, 60) (Fig. 10). ComX is secreted constitutively, and CSF is produced at the onset of sporulation when AbrB repression ceases (see Section 11.2.4). ComX binds to a transmembrane sensor ComP, a histidine protein kinase, which responds by phosphorylating the cytoplasmically located regulator ComA (61). CSF also contributes to ComA phosphorylation by a signal pathway that has not yet been elucidated. ComA-phosphate induces transcription of the *srfA* operon resulting in surfactin biosynthesis (61, 62). Surprisingly *srfA* transcription depends

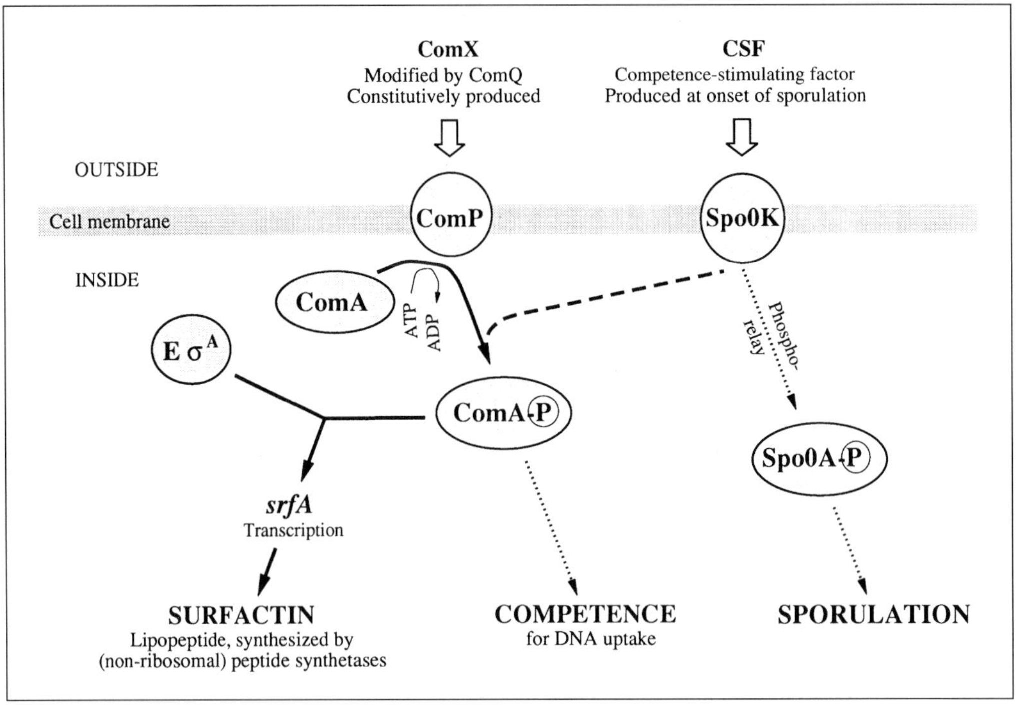

Fig. 10 Extracellular signals and signal transduction regulate surfactin synthesis in *Bacillus subtilis*.

on σ^A, the main vegetative sigma factor (63). Both pheromones must be present at high concentration to achieve sufficient ComA phosphorylation for significant surfactin production.

ComA-phosphate is also required for induction of competence (for genetic transformation); and CSF induction additionally contributes to *Spo0A* phosphorylation and therefore to sporulation (see Section 11.2.4). The *comX* gene, encoding a precursor of the ComX pheromone, is part of an operon (*comXQPA*), that also specifies the ComX-dependent ComA–ComP signal transduction system. The role of the *comQ* gene product is to process the ComX precursor.

4.2 The expression of antibiotic production genes in *Streptomyces* spp.

During the second half of this century (post-penicillin), streptomycetes have provided most of the antibiotics in industrial production. Many thousands of different antibiotics have been discovered by screening independent *Streptomyces* isolates. Most streptomycetes can make several unrelated antibiotics, production of each being by a set of genes organized into a cluster. The clusters are generally in the size range 10–100 kb, usually comprising several adjacent, mostly polycistronic transcription units. For antibiotics with anti-bacterial action (as opposed, for example, to anti-fungals), one or more resistance genes are usually present in the cluster; otherwise, production would be suicidal! Most clusters also contain at least one pathway-specific positively acting regulatory gene, which is needed for the transcription of some or all of the biosynthetic genes in that cluster. No sigma factors are known to be specific for antibiotic biosynthesis, so the idea has grown up that the natural control of antibiotic biosynthesis levels may be exerted entirely through the pathway-specific activators (64). Often, overproduction of the activators leads to significant increases in antibiotic yield (65).

The pathway-specific activators discovered so far in various streptomycetes fall into several major classes. Here we focus on the product of the *act*II-ORF4 gene, which is needed for expression of the *act* genes encoding production of the blue antibiotic, actinorhodin. ActII-ORF4 and similar proteins have been termed '*Streptomyces* antibiotic regulatory proteins', or SARPs (66). They regulate about half of the sufficiently characterized antibiotic pathways in various streptomycetes. Structural comparisons have indicated the presence of a 'winged helix' DNA-binding fold in SARPs, which resembles that of the well-characterized *E. coli* regulatory protein OmpR (66). SARPs are thought to bind to 7-basepair repeat sequences in target promoters, resulting in transcriptional activation (66).

The amount of ActII-ORF4 is at least partially regulated at the transcription level; *act*II-ORF4 mRNA increases sharply in concentration as cultures enter stationary phase, and engineered premature transcription of *act*II-ORF4 results in premature actinorhodin synthesis (67). Production or activation of ActII-ORF4 protein also depends on several pathway-non-specific genes, whose action influences most or all of the (at least three) other antibiotic pathways present in *S. coelicolor* (64, 68). Since two sensor protein histidine kinase/response regulator gene pairs (69, 70), and a protein serine-threonine kinase and its target (71), have been shown to influence antibiotic (including actinorhodin) production by *S. coelicolor*, some kind of phosphorylation cascade, perhaps analogous to the sporulation phosphorelay (see above), may respond to, and integrate, physiological information relevant to the decision to activate antibiotic production.

These are not the only gene products that exert pleiotropic effects on the expression of antibiotic production genes in *S. coelicolor*. Of the various *bld* genes needed for aerial mycelium to grow (see above), at least seven are needed also for antibiotic production, presumably providing the means by which morphological and physiological differentiation can be integrated (68). In most cases, the connection between the *bld* genes and antibiotic production is still obscure. However, as already mentioned, *bldA* encodes the only tRNA capable of efficient translation of UUA codons, and in this case the connection is often straightforward, since the TTA codon is present in about half of the known pathway-specific activator genes for antibiotic production (45, 68). For example *actII*-ORF4 contains such a triplet, and site-directed mutagenesis reveals that this is the sole reason why *bldA* mutants make no actinorhodin (72). It therefore seems likely that, at least in some conditions, the availability of the *bldA* gene product as a mature, charged tRNA may govern antibiotic synthesis through its involvement in translation of the mRNA for pathway-specific activators (67, 73).

Another cluster of genes (the *red* genes) directs *S. coelicolor* to produce a red antibiotic (a series of related molecules collectively called undecylprodigiosin). Production requires an ActII-ORF4-like pathway-specific regulator, encoded by *redD* (74). Unlike actII-ORF4, *redD* does not contain a TTA codon; yet expression of *red* genes is *bldA*-dependent (75). This is because redD transcription (unlike that of *act*II-ORF4) is itself subject to pathway-specific regulation (76), by another regulatory gene (*redZ*) that does contain a TTA codon (77). The RedZ protein is an unusual member of the response regulator protein family, which lacks the aspartate residue that is the diagnostic target for phosphorylation of these proteins. It is interesting that a similar two-step regulation is also found in the daunorubicin pathway of *Streptomyces peucetius*, again involving an atypical response regulator-like protein (DnrN: 78, 79) and an ActII-ORF4 homologue (DnrI: 80, 81). Probably such regulatory mini-cascades provide extended opportunities for signal input into the decision to produce the relevant antibiotics.

In summary, available evidence supports a general model in which both transcriptional and translational controls operate on pathway-specific regulatory genes such as *act*II-ORF4 and *redZ*. These genes thereby integrate a variety of physiological signals that have been transmitted through a number of proteins and a tRNA, in some cases by the specific phosphorylation of transmitter proteins.

What kind of signals are perceived by this sensory network? Concentrations of ppGpp increase in certain conditions of nutrient limitation, and may contribute information to the network (64, 82), but proteins binding either ppGpp or other likely intracellular signalling molecules have yet to be described. However, there is more clearcut evidence from various streptomycetes that extracellular signals are also important. The best-known case is that of A-factor in *S. griseus*, the producer of streptomycin. The importance of A-factor for sporulation was discussed in Section 11.3.2.

Streptomycin production is switched on when the pathway-specific activator StrR is produced in sufficient amounts to bind to its specific target sites upstream of transcription units encoding enzymes of streptomycin biosynthesis (83). (The *strR* gene, incidentally, also contains a TTA codon; 84.) Transcription of *strR* is apparently activated by the binding of a transcription factor to the region upstream of its promoter (39). In turn, this transcription factor is present (or active) only when sufficient A-factor has accumulated to release repression of a master regulatory gene by the A-factor-binding protein (Section 11.3.2).

Probably, A-factor-like signals are widespread among streptomycetes, but they show some degree of strain specificity. For example, the similar, but not identical, lactone esters that stimulate virginiamycin biosynthesis by *S. virginiae* can neither substitute for, nor be replaced by, A-factor (85).

5. Perspectives

We have seen in this chapter that the genetic control of differentiation of a vegetative cell into a form adapted for survival during periods of little or no growth is genetically and physiologically more complex than the regulatory circuits that govern primary metabolism. Many of the classes of regulatory elements familiar from simple regulatory systems are involved: sigma factors, repressors, activators, and response regulators combine to bring about the orderly expression of genes for the enzymes and structural proteins that manifest the observable phenotype. A significant part of this complexity is involved in initial commitment to the process. It is as if sporulation is not undertaken lightly, and organisms have evolved a variety of sensors to check that no other response will more appropriately meet the physiological and environmental challenge to which the cell finds itself exposed. Only when most or all of these sensors receive bad news are the final responses put into operation. Integration of such information may often involve the additive phosphorylation of a central regulatory protein by converging protein kinase cascades. However, the resulting gradual increase in the inflow of phosphate groups may not provide a very sharply defined switch point. Perhaps to avoid this problem, there is evidence that in *B. subtilis*, constitutively produced counteracting phosphatases may keep pace with the rate of phosphorylation of *Spo0A*, up to a threshold rate of phosphate input. Beyond this, sufficiently sharp increases in the overall level of *Spo0A*-phosphate might take place to permit activation of target genes (16).

Sporulation of *Streptomyces* involves two separate stages of commitment—the first to aerial growth, and the second to sporulation septation. In the first of these, extracellular signals are important, and the decision to develop aerial branches seems closely related to the decision to produce antibiotics. The second is achieved by the combined activity of at least six regulatory genes, whose co-evolution appears to have been separate from that of early sporulation genes of *B. subtilis*. Perhaps this reflects the multicellular spatial element of regulation that is important in *Streptomyces* but not relevant in *Bacillus*. On the other hand, the later stages of sporulation in both systems may turn out to have a common evolutionary origin, to judge from the resemblance of the *S. coelicolor* σ^F to σ^B, σ^F, and σ^G of *B. subtilis*.

Acknowledgements

We thank our colleagues (named in the text) for allowing us to cite their unpublished results, and Tobias Kieser and David Hopwood for comments on the manuscript.

References

1. Errington, J. 1993. *Bacillus subtilis* sporulation: regulation of gene expression and control of morphogenesis. *Microbiol. Rev.* **57**, 1–33.
2. Mandelstam, J. 1976. The Leeuwenhoek lecture, 1975. Bacterial sporulation: a problem in the biochemistry and genetics of a primitive developmental system. *Proc. Roy. Soc. Lond.* **B193**, 89–106.
3. Ryter, A. 1965. Etude morphologique de la sporulation de *Bacillus subtilis*. *Ann. Inst. Pasteur* **108**, 40–60.

4. Piggot, P. J., and J. G. Coote. 1976. Genetic aspects of bacterial endospore formation. *Bact. Rev.* **40**, 908–962.
5. Moran, C. P., Jr. 1989. Sigma factors and the regulation of transcription, p. 167–184. In I. Smith, R. A. Slepecky, and P. Setlow, Regulation of prokaryotic development: a structural and functional analysis of sporulation and germination. American Society for Microbiology, Washington, D.C.
6. Weir, J., M. Predich, E. Dubnau, G. Nair, and I. Smith. 1991. Regulation of *spo0H*, a gene coding for the *Bacillus subtilis* σ^H factor. *J. Bacteriol.* **173**, 521–529.
7. Perego, M., G. B. Spiegelman, and J. A. Hoch. 1988. Structure of the gene for the transition state regulator, *abrB*: regulator synthesis is controlled by the *spo0A* sporulation gene in *Bacillus subtilis. Mol. Microbiol.* **2**, 689–699.
8. Mandic–Mulec, I., N. Gaur, U. Bai, and I. Smith. 1992. Sin, a stage-specific repressor of cellular differentiation. *J. Bacteriol.* **174**, 3561–3569.
9. Burbulys, D., K. A. Trach, and J. A. Hoch. 1991. Initiation of sporulation in *B. subtilis* is controlled by a multicomponent phosphorelay. *Cell*, **64**, 545–552.
10. Satola, S. W., J. M. Baldus, and C. P. Moran, Jr. 1992. Binding of Spo0A stimulates *spoIIG* promoter activity in *Bacillus subtilis*. *J. Bacteriol.* **174**, 1448–1453.
11. Dunn, G., P. Jeffs, N. H. Mann, D. M. Torgersen, and M. Young. 1978. The relationship between DNA replication and the induction of sporulation in *Bacillus subtilis. J. Gen. Microbiol.*, **108**, 189–195.
12. Grossman, A. D., and R. Losick. 1988. Extracellular control of spore formation in *Bacillus subtilis. Proc. Natl. Acad. Sci. USA* **85**, 4369–4373.
13. Ireton, K., D. Z. Rudner, K. J. Siranosian, and A. D. Grossman. 1993. Integration of multiple developmental signals in *Bacillus subtilis* through the Spo0A transcription factor. *Genes Dev.* **7**, 283–294.
14. Lopez, J. M., C. L. Marks, and E. Freese. 1979. The decrease of guanine nucleotides initiates sporulation of *B. subtilis. Biochim. Biophys. Acta*, **587**, 238–252.
15. Mitani, T., J. E. Heinze, and E. Freese. 1977. Induction of sporulation in *B. subtilis* by decoyinine or hadacidin. *Biochem. Biophys. Res. Commun.* **77**, 1118–1125.
16. Perego, M., P. Glaser, and J. A. Hoch. (1996). Aspartyl-phosphate phosphatases deactivate the response regulator components of the sporulation signal transduction system in *Bacillus subtilis. Mol. Microbiol.* **19**, 1151–1157.
17. Strauch, M. A., M. Perego, D. Burbulys, and J. A. Hoch. 1989. The transition state transcription regulator AbrB of *Bacillus subtilis* is autoregulated during vegetative growth. *Mol. Microbiol.* **3**, 1203–1209.
18. Losick, R., and P. Stragier. 1992. Crisscross regulation of cell-type-specific gene expression during development in *B. subtilis. Nature* **355**, 601–604.
19. Duncan, L., and R. Losick. 1993. SpoIIAB is an anti-σ factor that binds to and inhibits transcription by regulatory protein σ^F from *Bacillus subtilis. Proc. Natl. Acad. Sci. USA* **90**, 2325–2329.
20. Alper, S., L. Duncan, and R. Losick. 1994. An adenosine nucleotide switch controlling the activity of a cell-type-specific transcription factor in *B. subtilis. Cell* **77**, 195–205.
21. Diederich, B., J. F. Wilkinson, T. Magnin, S. M. A. Najafi, J. Errington, and M. D. Yudkin. 1994. Role of interactions between SpoIIAA and SpoIIAB in regulating cell-specific transcription factor σ^F of *Bacillus subtilis. Genes Dev.* **8**, 2653–2663.
22. Min, K.-T., C. M. Hilditch, B. Diederich, J. Errington, and M. D. Yudkin. 1993. σ^F, the first compartment-specific transcription factor of *B. subtilis*, is regulated by an anti-σ factor that is also a protein kinase. *Cell* **74**, 735–742.

23. Duncan, L., S. Alper, F. Arigoni, R. Losick, and P. Stragier. 1995. Activation of cell-specific transcription by a serine phosphatase at the site of asymmetric septation. *Science*, **270**, 641–644.
24. Magnin, T., M. Lord, and M. D. Yudkin. 1997. Contribution of partner switching and SpoIIAA cycling to regulation of σ^F activity in sporulating *Bacillus subtilis*. *J. Bacteriol.* **179**, 3922–3927.
25. Arigoni, F., K. Pogliano, C. D. Webb, P. Stragier, and R. Losick. 1995. Localization of protein implicated in establishment of cell type to sites of asymmetric division. *Science* **270**, 637–640.
26. Lewis, P. J., T. Magnin, and J. Errington. 1996. Compartmentalized distribution of the proteins controlling the prespore-specific transcription factor σ^F of *Bacillus subtilis*. *Genes to Cells* **1**, 881–894.
27. Najafi, S. M. A., D.A. Harris, and M. D. Yudkin. 1997. Properties of the phosphorylation reaction catalysed by SpoIIAB that help to regulate sporulation of *Bacillus subtilis*. *J. Bacteriol.* **179**, 5628–5631.
28. LaBell, T. L., J. E. Trempy, and W. G. Haldenwang. 1987. Sporulation-specific σ factor σ^{29} of *Bacillus subtilis* is synthesized from a precursor protein, P^{31}. *Proc. Natl. Acad. Sci. USA* **84**, 1784–1788.
29. Karow, M. L., P. Glaser, and P. J. Piggot. 1995. Identification of a gene, *spoIIR*, that links the activation of σ^E to the transcriptional activity of σ^F during sporulation in *Bacillus subtilis*. *Proc. Natl. Acad. Sci. USA*, **92**, 2012–2016.
30. Londoño-Vallejo, J.-A., and P. Stragier. 1995. Cell-cell signalling pathway activating a developmental transcription factor in *Bacillus subtilis*. *Genes Dev.* **9**, 503–508.
31. Sun, D., R. M. Cabrera-Martinez, and P. Setlow. 1991. Control of transcription of the *Bacillus subtilis spoIIIG* gene, which codes for the forespore-specific transcription factor σ^G. *J. Bacteriol.* **173**, 2977–2984.
32. Setlow, P. 1988. Small, acid-soluble spore proteins of *Bacillus* species: structure, synthesis, genetics, function, and degradation. *Annu. Rev. Microbiol.* **42**, 319–338.
33. Kroos, L., B. Kunkel, and R. Losick. 1989. Switch protein alters specificity of RNA polymerase containing a compartment-specific sigma factor. *Science* **243**, 526–529.
34. Kunkel, B., K. Sandman, S. Panzer, P. Youngman, and R. Losick. 1988. The promoter for a sporulation gene in the *spoIVC* locus of *Bacillus subtilis* and its use in studies of temporal and spatial control of gene expression. *J. Bacteriol.* **170**, 3513–3522.
35. Zheng, L., R. Halberg, S. Roels, H. Ichikawa, L. Kroos, and R. Losick. 1992. Sporulation regulatory protein GerE from *Bacillus subtilis* binds to and can activate or repress transcription from promoters for mother-cell-specific genes. *J. Mol. Biol.* **226**, 1037–1050.
36. Okamoto, S., M. Itoh, and K. Ochi. 1997. Molecular cloning and characterization of the *obg* gene of *Streptomyces griseus* in relation to the onset of morphological differentiation. *J. Bacteriol.* **179**, 170–179.
37. Chater, K. F. 1993. Genetics of differentiation in *Streptomyces*. *Ann. Rev. Microbiol.* **47**, 685–713.
38. Pope, M. K., B. D. Green, and J. Westpheling. 1996. The *bld* mutants of *Streptomyces coelicolor* are defective in the regulation of carbon utilization, morphogenesis and cell–cell signalling. *Mol. Microbiol.* **19**, 1151–1157.
39. Horinouchi, S. and T. Beppu. 1994. A-factor as a microbial hormone that controls cellular differentiation and secondary metabolism in *Streptomyces griseus*. *Mol. Microbiol.* **12**, 859–864.
40. Ueda, K., K. Miyake, S. Horinouchi, and T. Beppu. 1993. A gene cluster involved in aerial

mycelium formation in *Streptomyces griseus* encodes proteins similar to the response regulators of two-component regulatory systems and membrane translocation. *J. Bacteriol.* **175**, 2006–2016.

41. Nodwell, J. R., K. McGovern, and R. Losick. 1996. An oligopeptide permease responsible for the import of an extracellular signal governing aerial mycelium formation in *Streptomyces coelicolor*. *Mol. Microbiol.* **22**, 881–893.
42. Willey, J., R. Santamaria, J. Guijarro, M. Geistlich, and R. Losick. 1991. Extracellular complementation of a developmental mutation implicates a small sporulation protein in aerial mycelium formation by *S. coelicolor*. *Cell*, 65, 641–650.
43. Willey, J., J. Schwedock, and R. Losick. 1993. Multiple extracellular signals govern the production of a morphogenetic protein involved in aerial mycelium formation by *Streptomyces coelicolor*. *Genes Dev.* **7**, 895–903.
44. Ma, H. and K. Kendall. 1994. Cloning and analysis of a gene cluster from *Streptomyces coelicolor* that causes accelerated aerial mycelium formation in *Streptomyces lividans*. *J. Bacteriol.* **176**, 3800–3811.
45. Leskiw, B. K., M. J. Bibb, and K. F. Chater. 1991. The use of a rare codon specifically during development? *Mol. Microbiol.* **5**, 2861–2867.
46. Kelemen, G. H., G. L. Brown., J. Kormanec, L. Potúcková, K. F. Chater, and M. J. Buttner. 1996. The positions of the sigma factor genes, *whiG* and *sigG*, in the hierarchy controlling the development of spore chains in the aerial hyphae of *Streptomyces coelicolor* A3(2). *Mol. Microbiol.*, **21**, 593–603.
47. Brown, K. L., and K. T. Hughes. 1995. The role of anti-sigma factors in gene regulation. *Mol. Microbiol.* **16**, 397–404.
48. Mirel, D. B., P. Lauer, and M. J. Chamberlin. 1994. Identification of flagellar synthesis regulatory and structural genes in a σ^D-dependent operon of *Bacillus subtilis*. *J. Bacteriol.* **176**, 4492–4500.
49. Kutsukake, K., S. Iyoda, K. Ohnishi, and T. Iino. 1994. Genetic and molecular analyses of the interaction between the flagellum-specific sigma and anti-sigma factors in *Salmonella typhimurium*. *EMBO J.* **13**, 4568–4576.
50. Ryding, N. J., G. H. Kelemen, C. A., Whatling, K. Flärdh, M. J. Buttner, and K. F. Chater. 1998. A developmentally regulated gene encoding a repressor-like protein is essential for sporulation in *Streptomyces coelicolor* A3(2). *Mol. Microbiol.* **29**, 343–357.
51. Soliveri, J., K. L. Brown, M. J. Buttner, and K. F. Chater. 1992. Two promoters for the *whiB* sporulation gene of *Streptomyces coelicolor* A3(2), and their activities in relation to development. *J. Bact.* **174**, 6215–6220.
52. Potuckova, L., G. H. Kelemen, K. C. Findlay, M. A. Lonetto, M. J. Buttner, and J. Kormanec. 1995. A new RNA polymerase sigma factor, σ^F, is required for the late stages of morphological differentiation in *Streptomyces* spp. *Mol. Microbiol.* **17**, 37–48.
53. McCormick, J., E. P. Su, A. Driks, and R. Losick. 1994. Growth and viability of *Streptomyces coelicolor* mutant for the cell division gene *ftsZ*. *Mol. Microbiol.* **14**, 243–254.
54. Bruton, C. J., K. A. Plaskitt, and K. F. Chater. 1995. Tissue–specific glycogen branching isoenzymes in a multicellular prokaryote, *Streptomyces coelicolor* A3(2). *Mol. Microbiol.* **18**, 89–99.
55. Kim, E. S., D. A. Hopwood, and D. H. Sherman. 1994. Analysis of type II polyketide β-ketoacyl synthase specificity in *Streptomyces coelicolor* A3(2) by *trans* complementation of actinorhodin synthase mutants. *J. Bacteriol.* **176**, 1801–1804.
56. Yu, T. W., and D. A. Hopwood. 1995. Ectopic expression of the *Streptomyces coelicolor whiE* genes for polyketide spore pigment synthesis and their interaction with the *act* genes for actinorhodin biosynthesis. *Microbiology*, **141**, 2779–2791.

57. Kelemen, G. H., P. Brian, K. Flärdh, L. Chamberlin, K. F. Chater, and M. J. Buttner. 1998. Developmental regulation of transcription of *whiE*, a locus specifying the polyketide spore pigment in *Streptomyces coelicolor* A3(2). *Journal of Bacteriology* **180**, 2515–2521.
58. Revill, W. P., M. J. Bibb and D. A. Hopwood. 1995. Purification of a malonyltransferase from *Streptomyces coelicolor* A3(2) and analysis of its genetic development. *J. Bacteriol.* **177**, 3946–3952.
59. Cosmina, P., F. Rodriguez, F. de Ferra, G. Grandi, M. Perego, G. Venema, and D. van Sinderen. 1993. Sequence and analysis of the genetic locus responsible for surfactin synthesis in *Bacillus subtilis*. *Mol. Microbiol.* **8**, 821–831.
60. Magnuson, R., J. Solomon, and A. D. Grossman. 1994. Biochemical and genetic characterization of competence pheromone from *B. subtilis*. *Cell.* **77**, 207–216.
61. Roggiano, M., and D. Dubnau. 1993. ComA, a phosphorylated response regulator protein of *Bacillus subtilis*, binds to the promoter region of *srfA*. *J. Bacteriol.* **175**, 3182–3187.
62. Nakano, M. M., and P. Zuber. 1993. Mutational analysis of the regulatory region of the *srfA* operon in *Bacillus subtilis*. *J. Bacteriol.* **175**, 3188–3191.
63. Nakano, M. M., L. Xia, and P. Zuber. 1991. Transcription initiation region of the *srfA* operon, which is controlled by the *comP–comA* signal transduction system in *Bacillus subtilis*. *J. Bacteriol.* **173**, 5487–5493.
64. Bibb, M. 1996. The regulation of antibiotic production in *Streptomyces coelicolor* A3(2). *Microbiol.* **142**, 1335–1344.
65. Chater, K. F. 1990. The improving prospects for yield increase by genetic engineering in antibiotic-producing streptomycetes. *Bio/Technology* **8**, 115–121.
66. Wietzorrek, A., and M. J. Bibb. 1997. A novel family of proteins that regulates antibiotic production in streptomycetes appears to contain an OmpR-like DNA-binding fold. *Mol. Microbiol.* **25**, 1181–1184.
67. Gramajo, H. C., E. Takano, and M. J. Bibb. 1993. Stationary-phase production of the antibiotic actinorhodin in *Streptomyces coelicolor* A3(2) is transcriptionally regulated. *Mol. Microbiol.* **13**, 837–845.
68. Champness, W. C., and K. F. Chater. 1994. Regulation and integration of antibiotic production and morphological differentiation in *Streptomyces* spp. In: Regulation of Bacterial Differentiation (Piggot, P., Moran, C. P., Youngman, P., Eds), pp 61–93. Washington DC: *American Society for Microbiology*.
69. Brian, P., P. J. Riggle, R. A. Santos, and W. C. Champness. 1996. Global negative regulation of *Streptomyces coelicolor* antibiotic synthesis mediated by an *absA*-encoded putative signal transduction system. *J. Bacteriol.* **178**, 3221–3231.
70. Ishizuka, H., S. Horinouchi, H. M. Kieser, D. A. Hopwood, and T. Beppu. 1992. A putative two-component regulatory system involved in secondary metabolism in *Streptomyces* spp. *J. Bacteriol.* **174**, 7585–7594.
71. Matsumoto, A., S.-K. Hong, H. Ishizuka, S. Horinouchi, and T. Beppu. 1994. Phosphorylation of the AfsR protein involved in secondary metabolism in *Streptomyces* species by a eukaryotic-type protein kinase. *Gene* **146**, 47–56.
72. Fernández-Moreno, M. A., J. L. Caballero, D. A. Hopwood, and F. Malpartida. 1991. The *act* cluster contains regulatory and antibiotic export genes, direct targets for translational control by the *bldA* transfer RNA gene of *Streptomyces*. *Cell* **66**, 769–780.
73. Leskiw, B. K., R. Mah, E. J. Lawlor, and K. F. Chater. 1993. Accumulation of *bldA*-specified transfer RNA is temporally regulated in *Streptomyces coelicolor* A3(2). *J. Bacteriol.* **175**, 1995–2005.
74. Narva, K. E., and J. S. Feitelson. 1990. Nucleotide sequence and transcriptional analysis of the *redD* locus of *Streptomyces coelicolor* A3(2). *J. Bacteriol.* **172**, 326–333.

75. Guthrie, E. P., and K. F. Chater. 1990. The level of a transcript required for production of a *Streptomcyes coelicolor* antibiotic is conditionally dependent on a tRNA gene. *J. Bacteriol.* **172**, 6189–6193.
76. White, J., and M. Bibb. 1997. The *bldA*-dependence of undecylprodigiosin production in *Streptomyces coelicolor* A3(2) involves a pathway-specific regulatory cascade. *J. Bacteriol.* **79**, 627–633.
77. Guthrie, E. P., C. S. Flaxman, J. White, D. A. Hodgson, M. J. Bibb, and K. F. Chater. 1998. A response regulator-like activator of antibiotic synthesis with an amino-terminal domain that lacks a phosphorylation pocket. *Microbiol.* **144**, 727–738.
78. Otten, S. L., J. Ferguson, and C. R. Hutchinson. 1995. Regulation of daunorubicin production in *Streptomyces peucetius* by the $dnrR_2$ locus. *J. Bacteriol.* **177**, 1216–1224.
79. Furuya, K., and C. R. Hutchinson. 1996. The DnrN protein of *Streptomyces peucetius*, a pseudo-response regulator, is a DNA-binding protein involved in the regulation of daunorubicin biosynthesis. *J. Bacteriol.* **178**, 6310–6318.
80. Stutzman-Engwall, K. J., S. Otten, and C. R. Hutchinson. 1992. Regulation of secondary metabolism in *Streptomyces* spp. and overproduction of daunorubicin in *Streptomyces peucetius*. *J. Bacteriol.* **174**, 144–154.
81. Tang, L., A. Grimm, Y. X. Zhang, and C. R. Hutchinson. 1996. Purification and characterization of the DNA-binding protein DnrI, a transcriptional factor of daunorubicin biosynthesis in *Streptomyces peucetius*. *Mol. Microbiol.* **22**, 801–813.
82. Takano, E., and M. J. Bibb. 1994. The stringent response, ppGpp and antibiotic production in *Streptomyces coelicolor* A3(2). *Actinomycetologica*. **8**, 1–10.
83. Retzlaff, L., and Distler, J. 1995. The regulator of streptomycin gene expression, StrR, of *Streptomyces* griseus is a DNA binding activator protein with multiple recognition sites. *Mol. Microbiol.* **18**, 151–162.
84. Distler, J., K. Mansouri, G. Mayer, M. Stockmann, and W. Piepersberg. 1992. Streptomycin biosynthesis and its regulation in prokaryotes. *Gene* **115**, 105–111.
85. Miyake, K., S. Horinouchi, M. Yoshida, N. Chiba, K. Mori, N. Nogawa, N. Morikawa, and T. Beppu. 1989. Detection and properties of A-factor-binding protein from *Streptomyces griseus*. *J. Bacteriol.* **171**, 4928–4302.

12 Evolution of prokaryotic regulatory systems

SIMON BAUMBERG

1. Introduction

The preceding chapters provide a wealth of detail about the often complex and intricate mechanisms found to control gene expression in present-day prokaryotes. This brief concluding chapter, as a contrasting afterword, consists principally of unanswerable questions, of the type: 'How did things get the way they are?'. The questions can be grouped as follows:

1. How did the interacting molecules and sequences—proteins, short lengths of DNA or RNA—evolve?
2. What were the selective pressures that led to the evolution of regulatory systems?
3. How can we account for the differences in mechanism for control of the same regulated system in different organisms or groups?

Much the same questions were asked in an earlier review on this topic (1).

2. Origins of regulatory molecules and sequences

Some of the mechanisms discussed in earlier chapters need no explanation under this heading. Transcriptional attenuation mechanisms of the type exemplified by several enterobacterial amino acid biosynthetic systems (Chapter 1) involve no specific components at all. Similarly, translational repression of *rps* and *rpl* operon expression (Chapter 5) makes use of pre-existing binding sites of ribosomal proteins within mRNA. Also, it does not seem necessary to explain the evolution of the nucleic acid sequences, whether relatively small recognition sites for protein binding or more extensive segments such as those that give rise to extensive secondary structure in RNA, exemplified by attenuation leader regions with their complex alternative secondary structures, and protein-binding stem-loop structures as with the *E. coli bgl* and the *Bacillus subtilis trp* clusters (Chapters 1, 5). However, most regulatory systems involve protein components with domains that bind to specific nucleic acid sequences, and it is on these that the discussion will focus.

A typical 'classical' regulatory polypeptide—a repressor, activator, or the simpler kind of anti-terminator such as BglG—binds to its nucleic acid site, to a low molecular weight effector (generally though not invariably), and to other polypeptides in order to form homo- or hetero-oligomers. DNA-binding domains usually belong to one of a small number of motifs, of which the helix–turn–helix appears to be the commonest (see Chapter 2). It may be supposed that solutions to the problem of how a polypeptide can specifically recognize a short DNA sequence have arisen only a few times in evolution, and have then reassorted with appropriate effector binding domains. How these DNA-binding domains arose in the first place is a question of the same kind as for substrate-binding domains in enzymes, for example. A somewhat more concrete question can be asked about effector ligand-binding domains, namely whether these derive from binding domains for the same or similar molecules that had arisen in enzymes or transport proteins. (This question implies that regulatory proteins evolved after the pathways which they regulate: a reasonable but again unproven assumption.) An example is the LacI repressor (Chapters 2 and 4). Each subunit (four of which combine to give the active tetramer) is made up of three domains: the N-terminal headpiece (amino acids 1–59), which binds to the operator; the core (amino acids 60–330), which binds inducers and permits dimerization; and the C-terminal (amino acids 331–360), which permits the aggregation of two dimers to give the active tetramer. The sequence of the core shows weak homology to three sugar-binding periplasmic proteins, for arabinose (ABP or AraF), glucose/galactose (GGBP or MglB), and ribose (RBP or RbsB) (2). These proteins function both in sugar transport and (GGBP and RBP only) as chemoreceptors in chemotaxis. Since their crystal structures were known, that of the LacI core could be modelled on them: the results were consistent with genetic and chemical data, and accorded with the idea that the sugar-binding domains were of similar structure (3). This does not however prove that sugar-binding domains from such proteins were recruited for regulatory purposes; both might have evolved from a common precursor of undetermined function.

3. Selective pressures leading to the evolution of regulatory systems

The answers to the implied question may seem self-evident. Metabolic regulatory systems are virtually without exception predicted to lead to economy in the use of cellular materials and/or energy; many other systems, as in the phage lytic and lysogenic cycles or in sporulation in *B. subtilis* or *Streptomyces coelicolor* (Chapter 11), provide canalized sequential patterns of gene expression. The universality, diversity, and complexity of the regulatory systems described in the preceding chapters strongly suggests significant advantage in their acquisition and retention. There are questions, however, particularly in regard to metabolic systems. End-product control in biosynthetic systems, whether via repression or attenuation—exemplified, for instance, in many amino acid biosynthesis pathways in *E. coli* and *Salmonella*—is less than universal. Whereas in virtually all strains of these organisms, arginine biosynthetic enzymes are repressed by arginine, in the B strain of *E. coli* they are constitutive at mid-level, this phenotype resulting, however, from only a slight change in the genetic system found in the repressible *E. coli* K-12 (4). Also, if we move to the more distant enterobacterium *Proteus mirabilis*, lack of repressibility of the enzymes of arginine biosynthesis becomes the norm (5). It is not known in this case whether any of the regulatory components found in *E. coli* and *Salmonella* are present.

If repressibility of the arginine biosynthetic enzymes is sufficiently advantageous to most *E. coli* strains as to lead to the preservation of the repression apparatus, why does this not apply to *E. coli* B? And why should repression in this system be less valuable to *Proteus mirabilis* than to other enterobacteria? It could be argued that the particular environmental niche occupied by these organisms makes repression of these enzymes relatively unimportant, but in the absence of any explanation for this, or any evidence that it may be true, the explanation is unconvincing. A further, though anecdotal, aspect of this argument is that regulatory mutants generally show no tendency to overgrowth by wild-type revertants (6). Very few quantitative studies have been done on competition between such mutants and the corresponding wild types. One instance in which a selective advantage of the wild type was indeed shown was between a tryptophan non-repressible mutant of a normally repressible strain of *B. subtilis* (7). As the mutant in this case was uncharacterized, it is possible that other perturbing factors were at work. Nevertheless, if the counter-intuitive finding were ever to be made in such competition experiments that deregulated mutants were no less fit than the regulated wild type, it could always be argued that the experimental environment was too artificial: an argument that could be extended almost indefinitely, since laboratory tests with *E. coli* in mice would no doubt use inbred strains rather than genetically far more diverse natural populations, and so on.

Where the same regulatory mechanism is found in widely differing organisms, for instance repression of arginine biosynthetic enzymes by aporepressors of the ArgR family in both enterobacteria and bacilli (8, 9), it is likely that the regulatory system evolved initially in an ancestral organism whose environment differed from both present-day groups. A way out of the possible problem that constitutive mutants do not seem at a selective disadvantage is to propose that the systems evolved at a time and in an environment where small savings in cell material and energy were profoundly significant—yet another 'just-so' story?

A coherent attempt to explain why one form of regulatory mechanism has been selected in preference to another was made by Savageau (10, 11). He suggested that a repressor-controlled mechanism would be selected for a given operon in an environment where there was a low demand for its expression, and an activator-controlled mechanism where there was a high demand for its expression. Although this theory is a valiant attempt at synthesis of a variety of observations, it is not clear that it can do justice to the wide variety of mechanisms that have now been shown to exist.

4. Why are there different control mechanisms for the same system in different organisms?

To look at variation in control mechanisms used for a single system across the range of bacteria, it makes sense to take a housekeeping system: a good example is tryptophan biosynthesis (12). In the enterobacteria, two systems apply independently: the tryptophan aporepressor with bound tryptophan binds to the operator overlapping the promoter (Chapters 2 and 4), while when tryptophan is plentiful, transcripts may prematurely terminate within the leader region by a classical attenuation mechanism (Chapter 1). In the pseudomonads, both repression and attenuation are apparently absent; most of the genes show little control, but the *trpBA* cluster encoding the final enzyme, tryptophan synthase, is induced by that enzyme's substrate, indoleglycerol phosphate (13). *B. subtilis* shows a straightforward anti-termination system; the TRAP protein in the presence of tryptophan binds to an mRNA leader region and prevents it from forming a transcription-anti-terminating stem-loop (14; Chapter 1). In *Corynebacterium glutamicum*, control may again involve both repression and attenuation (15). In *Streptomyces coelicolor*, the sequence upstream of the *trpE* gene has features consistent with an attenuation mechanism, but the unlinked *trpCBA* cluster shows no control by tryptophan (16; D. A. Hodgson, personal communication). One can make similar points about systems other than metabolic ones. Why, for example, do some phages achieve sequential gene expression through formation of successive anti-termination proteins, as does coliphage λ (Chapter 1), while others, such as *B. subtilis* SPO1, deploy alternative σ factors? Why, in *Neisseria gonorrhoeae* (Chapter 9), is *pil* variation achieved by gene conversion, but *opa* variation by slippage resulting in frameshifting?

How should we interpret such differences? The two main possibilities are: (i) that each regulatory mechanism permits greater fitness of the organism than could be achieved by an alternative mechanism (the 'just-so' story again), or (ii) that each mechanism started off by a random event in an ancestral organism and was steadily elaborated and improved by selection. There is in no case any experimental evidence to decide between these alternatives, although it would in principle be possible to engineer an attenuation system of the *E. coli* type into *B. subtilis*, or the TRAP system into *E. coli* and study the results. It is however of interest that a particular type of mechanism is found repeatedly in a given organism or group. Taking amino acid biosynthesis systems as an example again—and bearing in mind that not all groups of organisms have that many well-examined instances—the enterobacteria largely use attenuation and occasionally repression (17), the pseudomonads seem never to use attenuation and in most cases show no end-product control of gene expression (18), the bacilli use protein-mediated anti-termination (19), the corynebacteria appear to use both attenuation and repression (see e.g. 20), and the streptomycetes (the least well-studied group) probably show a mixture of attenuation and repression mechanisms not unlike the enterobacteria (D. A. Hodgson, personal communication; C. A. Potter, H. F. Craster and S. Baumberg, unpublished results).

5. Final considerations

The first chapter ended with a suggested deficiency—of integration as opposed to reduction—among the wealth of fascinating detail that has accumulated on the great diversity of mechanisms for regulation of gene expression in the prokaryotic world. As a counter to complacency, it is perhaps as well to end with another such suggestion, this time to the effect that we do not really understand how this diversity came about or how it may be maintained.

References

1. Baumberg, S. (1981) The evolution of metabolic regulation. In *Molecular and cellular aspects of microbial evolution. Symposium 32 of the Society for General Microbiology* (ed. M. J. Carlile, J. F. Collins, and B. E. B. Moseley), pp. 229–272. Cambridge University Press, Cambridge, UK.
2. Müller-Hill, B. (1983) Sequence homology between Lac and Gal repressor and three sugar-binding periplasmic proteins. *Nature*, **302**, 163–164.
3. Nichols, J. C., Vyas, N. K., Quiocho, F. A., and Matthews, K. S. (1993) Model of lactose repressor core based on alignment with sugar-binding proteins is concordant with genetic and chemical data. *J. Biol. Chem.*, **268**, 17602–17612.
4. Tian, G. L., Lim, D. B., and Maas, W. K. (1994) Explanation for different types of regulation of arginine biosynthesis in *Escherichia coli* B and *Escherichia coli* K12 caused by a difference between their arginine repressors. *J. Mol. Biol.*, **235**, 221–230.
5. Prozesky, O. W. (1969) Regulation of the arginine pathway in *Proteus mirabilis*. *J. Gen. Microbiol.*, **55**, 89–102.
6. Clarke, P. H. (1979) Regulation of enzyme synthesis in the bacteria: a comparative and evolutionary study. In *Biological regulation and development: Volume 1, Gene expression* (ed. R. F. Goldberger), pp. 109–170. Plenum Press, New York.
7. Zamenhof, S. and Eichhorn, H. H. (1967) Study of microbial evolution through loss of biosynthetic function: establishment of 'defective' mutants. *Nature*, **216**, 456–458.
8. Maas, W. K. (1994) The arginine repressor of *Escherichia coli*. *Microbiol. Rev.*, **58**, 631–640.
9. Miller, C. M., Baumberg S., and Stockley, P. G. (1997) Operator interactions by the *Bacillus subtilis* arginine repressor/activator: novel positioning and DNA-mediated assembly of a transcriptional activator at catabolic sites. *Mol. Microbiol.* **26**, 37–48.
10. Savageau, M. A. (1976) *Biochemical systems analysis: A study of function and design in molecular biology*. Addison-Wesley, Reading, Massachusetts.
11. Savageau, M. A. (1979) Autogenous and classical regulation of gene expression: a general theory and experimental evidence. In *Biological regulation and development: Volume 1, Gene expression* (ed. R. F. Goldberger), pp. 57–108. Plenum Press, New York.
12. Crawford, I. P. and Stauffer, G. V. (1980) Regulation of tryptophan biosynthesis. *Ann. Rev. Biochem.*, **49**, 163–195.
13. Chang, M., Essar, D. W., and Crawford, I. P. (1990) Diverse regulation of the tryptophan genes in fluorescent pseudomonads. In Pseudomonas*: Biotransformations, pathogenesis, and evolving biotechnology* (ed. S. Silver, A. M. Chakrabarty, B. Iglewski, and S. Kaplan), pp. 292–302. American Society for Microbiology, Washington, DC.
14. Babitzke, P. (1997) Regulation of tryptophan biosynthesis: Trp-ing the TRAP or how *Bacillus subtilis* reinvented the wheel. *Mol. Microbiol.*, **26**, 1–9.

15. O'Gara, J. P. and Dunican, L. K. (1994) Direct evidence for a constitutive internal promoter in the tryptophan operon of *Corynebacterium glutamicum*. *Biochem. Biophys. Res. Commun.*, **203**, 820–827.
16. Hood, D. W., Heidstra, R., Swoboda, U. K., and Hodgson, D. A. (1992) Molecular genetic analysis of proline and tryptophan biosynthesis in *Streptomyces coelicolor* A3(2): interaction between primary and secondary metabolism—a review. *Gene*, **115**, 5–12.
17. Neidhardt, F. C. (editor-in-chief) (1996) Escherichia coli *and* Salmonella typhimurium*: Cellular and molecular biology* (2nd. edn), Section B1. American Society for Microbiology, Washington, DC.
18. Clarke, P. H. and Ornston, L. N. (1975) Metabolic pathways and regulation. In *Genetics and biochemistry of* Pseudomonas (ed. P. H. Clarke and M. H. Richmond), pp. 191–340. Wiley, London.
19. Sonenshein, A. L. (editor-in-chief) (1993) Bacillus subtilis*: the model Gram-positive organism*, Section II. American Society for Microbiology, Washington, DC.
20. Sahm, H., Eggeling, L., Eikmanns, B., and Kramer, R. (1995) Metabolic design in amino acid-producing bacterium *Corynebacterium glutamicum*. *FEMS Microbiol. Rev.*, **16**, 243–252.

Index

Note: page numbers in *italics* refer to figures and tables